D0251545

WHY POPCORN COSTS SO MUCH AT THE MOVIES

And Other Pricing Puzzles

Richard B. McKenzie

WHY POPCORN COSTS SO MUCH AT THE MOVIES

And Other Pricing Puzzles

Copernicus Books
An Imprint of Springer Science + Business Media

Richard B. McKenzie
The Paul Merage School of Business
University of California
Irvine, CA, 92697-3125
USA
mckenzie@uci.edu

ISBN 978-0-387-76999-8 e-ISBN 978-0-387-77001-7
DOI 10.1007/978-0-387-77001-7

Library of Congress Control Number: 2008923102

Typesetting and production: le-tex publishing services oHG, Leipzig, Germany
Jacket design by James Sarfati

Printed on acid-free paper

9 8 7 6 5 4 3 2 1

springer.com

Preface

~

HOW PRICES MATTER

Prices are ubiquitous, so much so that their importance to the smooth operation of a market economy (even one constrained by extensive political controls as is the case in China) can go unnoticed and unheralded. Prices are what all trades, whether at the local mall or across the globe, are built around. They facilitate trades among buyers and sellers who don't know each other, meaning they make less costly, or more socially beneficial, the allocation and redistribution of the planet's scarce resources. Indeed, as the late Friedrich Hayek is renowned for having observed, prices summarize a vast amount of information on the relative scarcity and, hence, the relative cost of resources (with much of the information subjective in nature) that can be known only by individuals scattered across markets and cannot be collected in centralized locations, except through market-determined prices.[1]

Because they summarize, and largely hide from view of buyers, so much information spread among people throughout the world, prices can be puzzling. Why prices are what they are, and change for reasons that are obscured by a multitude of economic events that can extend backward in time and forward into the future, can be mysterious. Explaining many puzzling prices can be detective work that the modern-day Sherlock Holmes would surely find challenging.

But the national economic planners of the past failed to appreciate the mystery of prices. Instead, they saw prices as nothing more than tags on goods and services—$1.99 or $599—that could be dictated or declared with the stroke of administrative pens. All they thought they had to do was write out a few numbers. Voila! A Price. Professor Hayek received a Nobel Prize in economics in part for pointing out the fundamental error in national economic planning, that knowledge of what people want and are capable of producing in all of its various forms is nowhere known to anyone or any small group of planners. Once more, the myriad knowledge needed by planners to do their jobs is so enormous that it cannot all be absorbed by the planners themselves, even with the help of the most powerful computers (which economic planners in the former Soviet Union did not have).

So much relevant knowledge to the efficient operation of an economy is highly detailed, is local in nature, and is subjective, which means so much production and consumption-relevant knowledge cannot be known to outsiders, no matter how hard they try. To Hayek,

> The economic problem is thus not merely a problem of how to allocate "given" resources—if "given" is taken to mean given to a single mind which deliberately solves the problem set by these "data." It is rather a problem of how to secure the best use of resources to any of the members of society, for ends whose relative importance only these individuals know. Or, to put it differently, it is a problem of the utilization of knowledge not given to anyone in its totality.[2]
>
> The only way this vast knowledge can be revealed is to give the people who possess knowledge the right incentives to make use of what they know and to communicate what they know to all relevant others through the pricing system.
>
> Fundamentally, in a system where the knowledge of the relevant facts is dispersed among many people, prices can act to coordinate the separate actions of different people in the same way as subjective values help the individual to coordinate the parts of his plan...The mere fact that there is one price for any commodity—or rather that local prices are connected in a manner determined by the cost of transport, etc.—brings about the solution which...might have been arrived at by one single mind possessing all the information which is in fact dispersed among all the people involved in the process.[3]

Buyers need not know the relative scarcities of myriad resources or the considerable intricacies of producing goods as simple as a pencil or as complex as a computer. The late journalist and market advocate Leonard Read penned a wonderfully brief but insightful essay, "I, A Pencil," on how unexpectedly complicated the production of pencils is.[4] As a consequence, Read argued that no one in the world knows how to make a pencil, at least not totally from scratch. Yet, tens of millions of pencils are produced each year for world consumption. The miracle of pencil production is guided by the forces of market competition—and market-determined prices.

To determine what they want, all buyers have to do is compare prices, along with the features, of alternative goods. Prices, in other words, economize on the knowledge buyers need to have to make tolerably informed purchases. Again, in Hayek's words,

> The most significant fact about this [pricing] system is the economy of knowledge with which it operates, or how little the individual participants need to know in order to be able to take the right action. In abbreviated form, by a kind of symbol, only the most essential information is passed on, and passed on only to those concerned. It is more than a metaphor to describe the price system as a kind of machinery for registering change, or a system of telecommunications which enables individual producers to watch merely the movement of a few pointers, as an engineer might watch the hands of a few dials, in order to adjust their activities to changes of which they may never know more than is reflected in the price movement.[5]

In facilitating trades, prices can extend the scope of markets. In doing that, prices allow people to move away from self-sufficiency and narrow the scope of things they do, thus enabling people everywhere to reap the benefits of specialization. And an expansion of markets can result in greater competitive pressures for producers to become ever more cost-effective in production.

Most people intuitively grasp that product innovations, largely unfettered by government controls, can improve human welfare. Apart from the products to which they are associated, prices, too, can be innovative (as shown throughout this book) and can contribute to the growth in human welfare—until someone takes a page from the training manuals of economic planners of the past or gets the not-so-bright idea that they know better than markets what prices should be and that prices should be controlled by governments.

Back in August, 1973 President Richard Milhouse Nixon realized that the federal government could no longer control the price of gold at $35 an ounce. So he freed gold, leaving its price to be determined by unfettered market forces. Then what did he do? Something inexplicable, given his admission that the government could no longer control the price of a single commodity. He froze the prices of everything else—gazillions of goods and services—in the economy. Why? Because the inflation rate had reached a staggering (for the times) 3.76% for the previous seven-plus months of 1973. The result was an economic mess, and a recession—caused partially by people wasting time sitting in their cars in notoriously long lines at gas stations and by people having to adhere to silly rules only bureaucrats could love when people could fill their gas tanks. Several years later, President Gerald Ford thought he could beat back the upward price spiral of the 1970s by passing out (what else?) "WIN" buttons (for "whip inflation now"). Readers who lived through the WIN program understand that the buttons constituted a waste of valuable resources. The button's only effect on prices was to drive up the price of the metal used in them. Sloganeering will never cure inflation, or the high price of anything. The market forces behind prices are simply too powerful.

Perhaps the inflationary spiral and the price-control debacles of the 1970s brought home lessons that were grudgingly learned by the public, Congress, and succeeding presidents. Inflation is mainly a monetary phenomenon, meaning that it can only be contained in the long run by controlling the growth of money. If the flow of new dollars is curbed, then the upward pressure of prices will be abated. Price controls can only mask, for a time, upward pressure on prices that growth in the number of dollars in circulation can bring. Broadly applied (or even narrowly focused) price controls can do only economic damage in the long run.

Perhaps because in part of lessons learned from the inflationary spiral of the 1960s and 1970s that gave rise to price controls and revealed their follies and because a growing array of studies that showed how misguided government regulatory efforts had been, prices in a variety of industries—most notably airlines, trucking, natural gas, and electricity—were deregulated in the 1970s and 1980s. However, as will be seen in this volume, the lessons from price control debacles in the 1970s have not always been remembered by contemporary policy makers. They continue to employ price controls that have, often in unrecognized ways, perverse consequences. Will we ever learn? Maybe this volume will help drive home the lesson again.

For decades, I have taught my students the basics of microeconomic theory, mainly revolving around how prices in competitive markets are determined by the forces of supply and demand and how monopolies can, by restricting market supply, charge higher-than-competitive prices and reap higher-than-competitive profits. The lessons learned from those lines of analysis are important, and should always be taught and never forgotten. But those lines of argument elevate in largely unrecognized ways and leave unaddressed a host of interesting pricing puzzles, a number of which are addressed in this volume. The world is literally abuzz with interesting, but deceptively unsophisticated, pricing issues that standard "price theory" within economics never comes close to addressing—unfortunately. This book seeks to remedy that deficiency.

On passing through theater turnstiles, moviegoers are often astounded at the price of a large tub of popcorn, which can, in some parts of the country, rival the prices of whole meals at casual restaurants. No doubt, many moviegoers mutter under their breath a seemingly innocuous question, "Why does movie popcorn cost so much?" Most are convinced they have an explanation: Theaters are greedy monopolists that unabashedly turn the price screw as much as they can on trapped theater patrons. Nonetheless, their presumed answer to the popcorn-pricing puzzle has an ounce of truth, but only an ounce (since almost all firms in the USA, and for that matter, world economy) have some control over the prices they charge. But as we will see, that pat answer is, for the most part, as wrong as it is appealing and widely believed.

Popcorn is hardly the only pricing puzzle associated with the movie business. Have you not noticed that all movies—whether an expected mega-blockbuster film like *Spider-Man* or *Harry Potter* or a recognized niche film like *Miss Potter*—carry the same ticket prices? Astounding, to say the least. Don't movie studios and theaters know to charge more when the demand for a movie is high than when the demand is low or when the production costs run into hundreds of millions than when production costs are tens of millions? Venues for rock concerts know to do that. They vary their ticket prices radically, depending on the popularity of the stars on stage. Tickets for concerts by Paul McCartney carry much higher prices than tickets for concerts by Lorena McKennitt. What's so different about the movie market?

Why Popcorn Costs So Much at the Movies, and Other Pricing Puzzles seeks to unravel an array of pricing puzzles from the one captured in the book's title to why so many prices end with "9" (as in $2.99 or $179) to why ink cartridges can cost as much as printers to why stores use sales, coupons, and rebates. Along the way, I explain how the 9/11 terrorists have killed—through the effects of their heinous acts on the relative prices of various modes of travel—more Americans since 9/11 than they killed that fateful day, and the terrorists have been dead since 9/11.

Moreover, I detail how the Transportation Security Administration can cause, via the pricing effects of its policy decisions, the deaths of Americans simply by elevating the security alert status at the nations' airports. I also explain how well-meaning efforts to spur the use of alternative, supposedly environmentally friendly fuels—ethanol and biofuels, in particular—have caused, through the effects on grain prices, malnourishment and starvation among millions of desperately poor people around the world—and have given rise to the deforestation of rainforests in Malaysia and Indonesia. How can this be? If you think you already have an answer, read on. The solutions to this and other such puzzles are more sophisticated and surprising than you likely now think.

We end with unraveling a conundrum that has bedeviled societies for a long time, why men earn more on average than women everywhere—around globe, across industries and cultures. Can the male/female wage gap be summarily dismissed, as many are inclined to do, by chalking up the differential to rank male chauvinistic discrimination, all organized to hold the economic lot of women *everywhere* down? No doubt, rank discrimination does explain *some* of the wage gap, but, as we will see, far from *all* of the gap. As we will also see, some of the wage gap can be attributed to evolutionary forces in our distant past that are not likely to subside completely anytime and anywhere in the near term, or, for that matter, long term. And in case you are concerned, this is not a line of argument I relish. Indeed, I wish it were possible to expect the wage gap to evaporate, and the sooner, the better, but I have to follow the logic and evidence on this issue. That's the only way to understand why things—from pay gaps to queues—are the way they are.

Our inquiries will be mainly economic in the sense that the economic way of thinking about prices, and all other related matters, will be front and center in the discussions of all puzzles. At the same time, I insist that satisfactory explanations for various pricing strategies necessarily requires a multidisciplinary approach, and so I draw freely on the findings from the disciplines of psychology, sociology, demography, evolutionary biology, and evolutionary psychology, as well as behavioral economics (which stands astride economics and psychology) and neuroeconomics (which stands astride neurobiology and economics).

The respected nineteenth-century economic journalist and satirist Frédéric Bastiat (1801–1850) observed with his customary poignancy, "There is only one difference between a bad economist and a good one: the bad economist confines himself to the visible effect; the good economist takes into account both the effect that can be seen and those effects that must be foreseen." [6] In no small way, this volume is dedicated to uncovering the unheralded explanations for why prices are what they are and the unseen effects of prices, as well as explaining how firm and government policies affect prices and, therefore, people's behavior often in unrecognized and unanticipated ways. The "law of unintended consequences" stalks the pages of every chapter in the book. I seek to pique your interest in the various pricing puzzles considered by confronting you with twists and turns in arguments that are novel and unsuspected. Indeed, the puzzles covered were selected for inclusion in this book because their solutions are counterintuitive and go against conventional wisdom. While I cite a mountain of evidence for the many logical deductions drawn, I must confess to being partial to the economic *logic* embedded in the arguments, as distinct from the economic and other data used to test claims in the arguments. Both logic and references to real world happenings are needed for a proper, complete analysis, but I also suspect that it will be the economic logic, and the many demonstrations of how it can be used to unravel and solve puzzles, that will most likely impress you (and other readers), and stay with you after this book has long been closed.

To some (especially young readers and reviewers), this book might appear to emerge only *because* of the success of other economists who have sought to apply economic reasoning broadly, as Steven Levitt, an economist, with wordsmithing help of journalist Stephen Dubner, has done in the wildly successful book, *Freakonomics.*[7] I salute Levitt and Dubner and others for reaching a broad audience for economics as a way of thinking. I have recommended their book to my classes.[8]

However, readers should understand that *this* book emerges from a career of applying economic reasoning to an unchecked range of topics outside the proverbial disciplinary box (whatever the "box" is conceived to be). If this book has antecedents, it is in the work of George Stigler, James Buchanan, Gordon Tullock, and Gary Becker (especially Gary Becker) whose work, one or two generations removed, inspired, albeit unknowingly and indirectly, the work of Levitt and his followers. My own first effort in treating economics as a discipline unbounded by the topics considered, undertaken with Gordon Tullock, a distinguished economist, was in a book that was widely adopted and translated precisely because it broke ranks with the then stodgy view of what the discipline of economics could be. The book was *The New World of Economics,* first published in 1975.[9] In *The New World,* Tullock and I applied economic reasoning to an array of topics considered at the time "unusual," and for some critics, beyond the pale: riots and panics, presidential elections, dying, marriage and divorce, exploitation, education, lying and cheating, *and* sex (not prostitution but the normal kind). Over the five editions that book went through in its thirty-year run, Tullock and I, along with the economists I mentioned above, probably helped to convince any number of budding economists that economics is not so much defined by *the* core problem—scarcity—that economists had long held dear as by the methods of analysis used to think through issues. The only limit we imposed on ourselves was whether or not the economic methods yielded insights that might have gone unnoticed if other analytical methods were used.

I would like to think this book is a natural and improved extension of *The New World,* informed by advancements in economic reasoning since that book was last published in the early 1990s. I am indebted to my mentors, both those whose classes I took and those whom I knew by their written works, for the motivating mantra they left with me, that economics can be very interesting, and at times exciting and energizing, if not fun. Perhaps this book will have similar effects on others.

I am more immediately indebted to several key people who read and commented on the book when it was in manuscript form: Dwight Lee, George Selgin, Otto Reyer, Robert Daley, and Kathryn McKenzie. Their criticisms and suggestions for improvement helped me improve the substance and organization of the book. My wife Karen did her usual excellent job of editing preliminary drafts of the book.

Irvine, California
September 2007

Richard McKenzie

NOTES

1. Hayek (1945).
2. Hayek (1945, pp. 519–520).
3. Hayek (1945, p. 526).
4. Read (1958).
5. Hayek (1945, pp. 526–527).
6. Bastiat (1845).
7. Levitt and Dubner (2005).
8. See books that take an expansive view of the domain of economics by Landsburg (2006), Cowen (2007), and Frank (2007).
9. McKenzie and Tullock (1975 with the latest edition published in 1994).

Contents

Chapter 1

PRICE AND THE "LAW OF UNINTENDED CONSEQUENCES"

Economics is as much a communicable disease as it is a discipline. Economics is a way of thinking about everything and coming to a sense of understanding life better. When you catch it, the way of thinking (by way of learning a few basic but powerful economic principles), it is hard not to see most of life's large and small events as economic puzzles worthy of reflection and solution.

I admit it, I am an economist with this affliction: I am constantly puzzling over everything I read in the newspapers, watch on television, and hear others say, especially when the comments are about why prices are what they are (and not something else). But then I puzzle over observed prices when many others seem to miss their importance. I understand all too well that prices are the products of so-called market forces, but leaving the explanation at that superficial level of analysis is hardly satisfying, especially since my affliction is terminal. I feel a compulsion to understand exactly what market forces are at work on the prices I see. And when I see prices that don't make sense, my compulsion goes into overdrive. I must understand why prices are what they are.

Chalking supposedly ill-conceived prices up to people's stupidity (or to their unthinking or irrational behavior) is hardly satisfying, not that I don't recognize that people—both buyers and sellers—do a lot of stupid things as they go about their daily business. Most ill-conceived prices are quickly corrected, mainly because ill-conceived prices imply that someone can make them better—*and profit by doing so*. The ill-conceived prices we often notice are ones that are systemic and have staying power, or else we would not have time to pay much attention to them, or need to explain them. I can't help but search for explanations for persistent "ill-conceived prices"—that, to me, by their very persistence suggests that they are not nearly so ill-conceived as thought. Indeed, "ill-conceived prices" often do have rational, albeit counterintuitive, explanations, as will be shown throughout this volume (with "rational" explanations being grounded with due consideration given to costs and benefits facing market participants). Finding explanations for observed prices is a form of economic detective work, which can be fun, especially when the sources of observed prices and their consequences are as unintended as they are unexpected.

Prices have been at the heart of economic inquiries for a very long time, but prices can still be mysterious. Satisfying explanations for the many prices we see all around us can be as surprising as they are elementary. Pricing strategies can also have consequences that are ...well, perverse—again, as will be shown time and again throughout this book. For a start, try to understand my professional affliction by considering a puzzle embedded in Apple's price for the iPhone on its release in mid-2007 (and its one-third reduction in the price of the top model two months later), Audible's announced clearance sale, and the proposed price control for brothel prostitution in post-war Japan.

Early in 2007, Steve Jobs, founder and CEO of Apple, announced that his company would enter the mobile phone business with the introduction of the iPhone by mid-2007. The iPhone would be a multipurpose device, one that could be used to make calls, to listen to music, to store pictures and videos, and to surf the web, all with the typically sleek Apple design touch.

In making his announcement, Jobs set off a worldwide media feeding frenzy about the iPhone that reached a crescendo in late June 2007. And sure enough, as the June 29 released date approached, Apple devotees around the world began forming lines outside of Apple stores.[1] To hold their places in line, many slept for several nights on the concrete sidewalks and put up bravely with the discomfort from rain.

Just before midnight on June 28, the queues outside of many Apple stores wound around several blocks—in spite of some technology reviewers' warnings that the iPhone had problems (a not-so-user-friendly virtual keyboard and connection incompatibilities, for example) and in spite of iPhone's high initial prices, $499 for the model with 4 gigabytes of memory and $599 with 8 gigabytes. The early less-than-stellar reviews of the iPhone notwithstanding, people in the long queues were convinced that the iPhone would be as cool as the phenomenally successful iPod, and would set the standard for the next generation of cell phones just as the iPod had set the standard for MP3 players a half-dozen years earlier.

When the doors of the Apple (and AT&T) stores swung open one minute after midnight on June 29, the throngs of "Appleholics" poured in to snatch up their iPhones. During the first weekend, Apple reportedly sold at least a half of a million, and maybe three quarters of a million, iPhones, several times Apple's and everyone else's aggressive sales projections made earlier in the year from market research, but the company could have sold more.[2] Any number of Apple (and AT&T) stores quickly ran out of both iPhones models before 1:00 a.m., and surely before the sun came up.[3]

The iPhone's introduction, and its immediate market mega-success, is surely puzzling to many economists, if not everyone else, for several reasons. Aren't markets supposed to clear? If they are, then the long queues at the Apple stores for the iPhone's release must have been an unintended consequence, or was it? When Jobs saw the media feeding frenzy build early in 2007, why didn't he order an even higher price in anticipation of long queues on the release date to ensure that many people wouldn't waste time camping out for days—and, not immaterially, Apple's profits would rise? Immediately after that last weekend of June, reports surfaced that the 8-gig model, which was in especially short supply, began showing up on eBay at prices a third higher than the posted retail price at Apple stores. eBay reported that the highest bid for an iPhone that first weekend was a remarkable $12,500.[4] Why did Jobs leave money literally on the sidewalks for "technoscalpers" to pick up, or did he? Did Jobs know something that is not apparent to microeconomic textbook authors (who write glowingly about how price hikes can, and will, relieve market shortages)?

Then, I can't help but wonder why Apple charged only 20% (or $100) more for the iPhone with 8 gigabytes of memory than the 4-gigabyte model? Why not more, especially since the excess demand of the 8-gig model was greater? Does anyone really think that the price difference is really attributable to the cost difference in memory? If cost doesn't explain the price difference, then what was behind Apple's pricing strategy?

During the first week of September after the iPhone's release, Jobs did what he had never done before: he lowered the price of the 8-gig iPhone by $200, causing the price of Apple stock to fall immediately by 5%—because, according to media reports, the price reduction indicated that the iPhone was not selling as well as anticipated, as reported by the *Wall Street Journal*.[5] Might it not be the case that the market got it wrong? Perhaps Apple hiked the price of the iPhone on its release in anticipation of the initial surge in demand—and in anticipation of the price reduction two months later and in anticipation of encouraging a "tipping" of the media player market even more in Apple's favor.[6]

Even more perplexing, why did the prices for all iPhones end with "9"? For that matter, why have the prices of almost all Apple products, from iPods to iTune songs, ended with "9"? Do Jobs and the obviously very smart marketing people at Apple really think that their buyers are so dumb that they can't see that prices of $499, $599, or $399 are just a dollar short of $500, $600, and $400, especially since they were obviously smart enough to earn enough to pay the considerable purchase prices of their iPhones? If the $1-off prices were intended to fool people, then it is hard to see how, since so many print and online news reports of the iPhone's release dispensed with the 9s, giving the prices of the two models at $500 and $600.

Shortly after the iPhone was released in the summer of 2007, I went to Audible.com to download additional audio books to my iPod, which I listen to while riding my bike (a modern form of multitasking that has increased both the books I have "read" and the amount of exercise I get, a true win-win). I was struck by the banner announcement on Audible's web page: "SUMMER CLEARANCE SALE ... 25% Extra Off ... Selections from Thousands of Titles." I couldn't help but wonder, "Audible is clearing out its inventory? How can that be? It *doesn't* have an inventory, other than the master copies of audio books from which it duplicates the copies its subscribers download (at a close to zero cost to Audible, I might add, since its "inventories" are non-material, or are nothing more than electrons in a server's hard drives). Surely Audible is not giving up its masters. There would be no need." If my mental muttering has merit, then why would Audible announce a "summer clearance sale"?

Only a marketing gimmick, you might be thinking? Maybe so, but maybe Audible's clearance sale suggests that similar sales conducted by brick-and-mortar retail stores may be motivated by some economic motive that is independent of the stores' interest in clearing out inventories that are, supposedly, unwanted because they represent mistakes in ordering. If inventory clearance doesn't explain many seasonal (winter, summer, or after-Christmas) inventory clearance sales, then what does? Might not after-Christmas sales be as planned as carefully as the before-Christmas non-sales, which suggests that "sales" may have a hidden logic beyond the obvious, that stores use them to move unwanted goods?

If you find such questions uninteresting, you probably bought the wrong book. If you find them intriguing and enticing, then read on, because addressing those kinds of questions is what this book is about—but also much more, as another puzzle dealing with ... (oh no!) *sex* reveals. By the time you finish this book, you should have a far deeper understanding of why Jobs and Apple chose the pricing strategy they did, without my ever providing an explanation—not directly, at least.

Rendigs Fels, an economist at Vanderbilt University, recalls in a puzzle he repeatedly gave his introductory economics classes during his long and heralded teaching career, how when he was stationed in Yokohama, Japan after World War II, he was put in charge of imposing and enforcing price controls throughout the Japanese economy. "One day the medical officer of our company came to see me,"

Professor Fels writes. "He was worried about the health of the American troops. They were picking up girls on the street instead of patronizing the brothels, where the girls were given a medical inspection once a week. The medical officer thought the soldiers were picking up girls on the street because the brothels' prices were too high. Since I was in charge of price control, he wanted me to take action."[7]

Professor Fels initially thought that it would be a good idea to require Yokohama brothel prostitutes to charge no more than their counterparts in the streets. He figured that if brothel prostitutes were "cleaner" than streetwalkers and brothel prices were lowered, more troops would substitute the services of brothel prostitutes for the services of streetwalkers. Accordingly, venereal disease among the troops would decline.

Professor Fels set aside his plan, but only because he worried that newspapers back in the States would report unfavorably that "a United States Army officer was reducing prices in brothels for the benefit of American troops." He muses, "Years later, when I finally saw the light, I became shocked at the deficiency of my economics training" (in spite of having earned a Ph.D. in economics from Harvard before going to war). He concluded that medical officer's proposal to control the prices of brothel prostitutes "would have had the exact opposite effect of the one he intended."[8]

Talk about an unintended consequence... surely the professor would not have intended his price control to cause more American troops to come down with various venereal diseases.

How is it that the good professor could have possibly reasoned that lower brothel prices would have had a truly perverse and deadly effect, increasing the spread of VD among American troops?

If you don't understand how that can be true, or find the good professor's delayed insight as mysterious, know that this book (and especially this and the following chapters) is founded on the proposition that a little elementary economic reasoning can go a long way in unraveling such mysteries, and can help us understand how prices, especially ones intended to override market forces, can have unintended—but still fascinating, if not amusing—consequences. Again, read on. Unraveling the Fels puzzle should be a snap by the time you complete this book—with no (direct) help from me.

HYBRIDNOMICS: HOV-LANE ECONOMICS, CALIFORNIA STYLE

In order to encourage sales of fuel-efficient, environmentally friendly hybrid cars, Congress authorized a tax credit for hybrid automobiles (which use a combination of gas and electric powered motors) of up to $3,150, with the credit varying with the hybrid's EPA fuel efficiency and the year of production.[9] The California legislature upped the ante for owning hybrids, authorizing the state's Department of Motor Vehicles to distribute 85,000 stickers to hybrid owners, but only to owners of cars that had an EPA fuel efficiency rating (given the rating methods in place at the time) of at least 45 miles per gallon. Hybrid owners with the stickers can drive alone in all of the state's High Occupancy Vehicle (HOV) lanes formerly restricted to cars with two or more passengers.

The tax credit and HOV-lane sticker privilege did what they were supposed to do. They drove up the demand for the Toyota Prius and Honda Civic hybrids (the only cars that qualified for stickers at the time), but the sticker privilege surely had market consequences that were unexpected and unintended. For example, because of the stickers, the small Prius in 2006 was selling for over $30,000, and had waiting lists until early 2007. The Civic hybrid carried a dealer "added premium" to the manufacturer's suggested list price of as much as $4,000 (with the hybrid Civic total price more than $7,500 higher than the quoted price of a non-hybrid Civic).

No doubt, there were many hybrid buyers who did not have warm and fuzzy feelings for the environment. They saw in the tax credit and HOV-lane privilege reductions in the *effective price* (dealer price minus tax and commute savings) of the hybrid. The tax credit that accompanied the hybrid purchase lowered the after-tax purchase price of the hybrid. The reduction in buyers' time cost of their commutes to and from work also lowered the *effective* price commuters had to pay for their cars. Commuters' demand for hybrids, inflated by the tax credit and the lower commute times drove up the dealer prices for hybrids and drove out of the hybrid market many dedicated environmentalists (but not sufficiently dedicated or wealthy to pay the hybrid premiums commuters were willing to pay).

At the end of January 2007, the DMV ran out of stickers, leaving more than 800 new Prius and Civic hybrid owners, who had bought their hybrids at premium prices and who had applied for the stickers, with the tax credit but without the right to drive alone in the state's HOV lanes.[10] They gambled and lost on the stickers, and we can feel their pain.

Now with no more stickers to distribute, what can be expected to happen in the California market for hybrids? No doubt some of the effects we can list were unanticipated and unintended.

First, we should expect a drop in the demand for new hybrids at dealers, along with a drop in their negotiated sale prices. Buying a new hybrid Civic instead of a non-hybrid Civic has been difficult for even warm-hearted environmentalists to justify, since the hybrid would very likely have to be driven over 500,000 miles (or driving the car for more than 42 years at 12,000 miles a year!) before the savings in gas could offset the added purchase price plus the cost of replacing the hybrid battery (most likely every 10 years) and the added interest and sales taxes on the added purchase price.[11] However, those added car costs can be easily justified by a commuter who earns $40 an hour and who, with the stickers, can save an hour a day commuting to and from work. Such drivers can cover the added hybrid costs through lower commute costs within a year.

Since the HOV-lane stickers stay with the hybrids, the demand for used hybrids with stickers can be expected to rise, along with their prices, perhaps dramatically. Used hybrids with stickers can be expected to sell for more than hybrids comparably equipped with approximately the same miles on them but without the HOV-lane stickers. Hardly surprisingly, by spring 2007, *USA Today* reported that Kelly Blue Book had found a $4,000 difference in used Priuses with and without stickers.[12] No doubt the hybrid/non-hybrid price differential will rise with the growth in California's population and the count of cars on the state's freeways and will fall as the expiration date for the HOV-lane stickers draws closer (now set for 2011)—and, of course, will rise with any extension in the expiration date for the stickers.

The growing number of drivers with long commutes and high opportunity costs, meaning high hourly earnings, can be expected to be lead bidders for used hybrids. They can be expected to buy hybrids from owners who bought their hybrids for environmental reasons and from owners who have lower cost savings from using the HOV lanes, because they have lower wage rates and/or shorter commutes.

As a consequence of the used hybrid sales, we should expect the HOV lanes to become more crowded since the lanes will be dominated to a greater extent by people with longer commutes (while all other lanes will become marginally less crowded), which will, of course, undercut (albeit marginally) the value of the stickers and the price of used hybrids. Given the market value of stickers (equal at least to the $4,000 price differential between hybrids with and without stickers) and the fact that the DMV appears to have distributed stickers that are far from counterfeit proof (even though the stickers are designed, supposedly, to crumble if tampered with), no one should be surprised if a healthy black market for stickers emerges, with the counterfeit stickers dampening the rise in the prices of used hybrids. No one should be shocked if the theft rate for hybrids with stickers exceeds by a healthy margin the theft rate for hybrids without stickers. Indeed, by mid-2007, reports had surfaced that two to three dozen sets of California HOV-lane stickers were being stolen from hybrids each month.[13]

The impact of used hybrid sales on automobile pollution is more difficult to assess. On the one hand, the people who buy used hybrids to speed up their commutes will reduce pollution, since they will be driving the less-polluting hybrids and will spend less time on their commutes with their engines running. On the other hand, the more crowded HOV lanes will mean that other non-hybrid HOV-lane users will, because of the greater crowding, have longer commutes with their non-hybrid engines running all the while. The slowing of traffic in the HOV lanes can also lead to less carpooling (again, albeit marginally).

Should hybrid owners with stickers be allowed to sell their stickers as separate items, that is, without selling their cars? Of course so, *if* the goal of government is to make sure that the scarce HOV-lane slots are used by drivers with the most urgent need to travel faster, but pollution control might be the more important government goal.

On first thought, it might seem that pollution would remain unchanged, since the stock of stickers and hybrids will remain at 85,000. However, you can bet current hybrid owners with stickers would love to be able to sell their stickers separate from their cars, since they would not then have the hassle of buying another car and since the demand for and price of their HOV-lane sticker advantage would be heightened by the added value commuters with Hummers (and all other large and small cars) would put on the stickers. Hummer dealers could also see an advantage in independent sticker sales since people could buy Hummers with the intent of going into the "used sticker" market to reduce their commute times.

If stickers could be sold independently of the hybrids, we might see another marginal increase in the crowding of the HOV lanes because of the likelihood that some of the used sticker buyers would have cars larger than the relatively small Prius and Civic that would be replaced in the HOV lanes.

The impact of shifting to independent HOV-lane sticker sales on pollution is, again, problematic. If current Hummer owners move into the HOV lanes, they might pollute less, since they would have lower commute times; but, again, the added crowding could add to the pollution coming from all the non-hybrid cars using the HOV lanes for daily commutes However, independent sticker sales could spur sales of cars and trucks larger than the current crop of hybrids. Such sticker sales could also cause large car buyers to move farther from work.

However, hybrid owners need to be aware that their cars resale prices will wane with time, given that the stickers are (according to current law) scheduled to expire in 2011. Hence, the stickers' value to both commuters and environmentalists will decrease as the expiration date is approached. Of course, the California legislature can simply terminate the stickers at any time between now and

2011 or it can do away with HOV-lane privileges for everyone at any time, as one think tank has proposed.[14] If such a proposal gains media attention and political support, you can bet that the price gap between hybrids with and without stickers will narrow. If the legislature extends the expiration date for the stickers, hybrid owners with stickers should expect an immediate increase in the resale prices of their cars.

Regardless, as I write this section, this is clearly a pretty good time for me to sell the new hybrid Civic I bought (a year before I wrote this section), mainly because, when I bought it, I was interested in experiencing the hybrid technology and because I expected to see the used price of my Civic jump once all authorized stickers were distributed. I only subsequently realized that the car was not a good deal for me (even though I bought it in North Carolina, which doesn't have HOV-lane stickers for hybrids, at $4,200 below the best quoted price at California Honda dealers). It has the stickers, and I rarely use the HOV lanes, since I live less than a mile from my university office. Thank goodness for the restricted supply of HOV-lane stickers. I got a windfall from the restricted supply. Moreover, if I sell now (mid-2007), I can capture in the resale price of my hybrid almost 4 years of value of the stickers to car buyers who commute long distances to work. I also sense that the media and state legislators have begun to take seriously arguments that California's 1,200-plus miles of HOV lanes have done little to increase carpooling, and the HOV lanes would be better used to reduce highway congestion if they were opened to all drivers (at least during non-rush hours of the day). If HOV-rights are dissolved (or to the extent that their dissolution is seriously threatened by legislative action), you can bet that the premium I can get for my hybrid with stickers will dissipate.

AIR TRAVEL SAFETY FOR INFANTS AND TODDLERS

Historically, parents have been able to buy airline tickets for themselves and hold their infants and toddlers under 2 years of age on their laps during flights. Under the banner of saving children's lives, back in the late 1980s, the National Transportation Safety Board and Los Angeles Area Child Passenger Safety Association petitioned the Federal Aviation Administration to end the free ride for infants by requiring the use of child-restraint systems in paid seats for infants.[15] James Kolstad, chairman of the NTSB, maintained that "the economic cost of the extra passenger seat ... [is] a very small price for preventing injuries and saving lives."[16]

In case the FAA resisted changing its child-seating rules, then-Representative Jim Lightfoot (R-Iowa) and Senator Kit Bond (R-Missouri) introduced legislation to mandate the use of safety seats by infants and toddlers on airplanes.[17] Congressmen Lightfoot was spurred to introduce his bill by the death of two infants in the crash of United Airlines flight 232 in Sioux City, Iowa, in July 1989 (a fiery runway crash, with the plane somersaulting down the runway, that has been aired repeatedly around the world in the years since it happened because of how fiery it was). Lightfoot spoke for his supporters within policy circles and the general public when he reasoned that rules requiring the use of safety seats in automobiles should be extended to airlines because "the potential for injury in an aircraft flying at 550 miles per hour is much greater than the potential for injury in an automobile traveling at 50 miles per hour."[18]

The FAA, the 50 or so members of Congress, the National Transportation Safety Board and everyone else who at the time supported the rule change were rightfully concerned with the safety of traveling children. However, what proponents of child-seat rules, both back then and since, have not considered is the prospect that the obvious effects from the rule change might not be all of the effects, and some effects might be unanticipated, unintended, and even perverse.

The more notable unanticipated and unintended effect was that the infant-seat requirement would increase the total price for families of travel by air, encouraging families to travel by automobile instead. The basic problem with that effect is that auto travel is far more dangerous than flying. At the time Lightfoot and Bond introduced their bill to regulate infant safety in the air, automobile transportation was at least thirty to forty times as hazardous in terms of the death-rate per mile traveled.[19] In a study prepared for the FAA at the time the Lightfoot/Bond legislation was considered, Department of Transportation researchers concluded that mandatory infant safety seats could have prevented at most only one infant death since 1978. All other infant fatalities in airline crashes occurred in sections of planes where no one survived.[20] On the other hand, nearly 1,200 children under 5 years of age were killed in automobile accidents in 1988.[21] That means that there were approximately one-quarter more automobile deaths of very young children in 1988 alone than there were total deaths of children and adults on scheduled airlines during the entire 1980–1988 period.[22]

According to the FAA's own (admittedly rough) calculations at the time of the congressional debate, mandated safety seats for infants could increase the average air travel cost of a family of four (two parents with one child over three and one infant) by at least 21%—assuming that airlines charged half fares for infants and do not raise their fares across the board because of rule-induced increased demand.[23] That cost increase could reduce the boardings of infants by about 18%, or 700,000, again according to FAA estimates. Nevertheless, the FAA figures that airlines would be able to sell 3.3 million additional seats each year to infants' parents at a cost of $205 million (equal to about $325 million in 2007 dollars), a handsome sum that explains the airlines' interest in the proposed rule.[24]

The precise effect on air travel safety of requiring seats for infants and toddlers has been debated ever since Congressman Lightfoot and Senator Bond introduced their legislation in 1990, and will probably be debated again. My own econometric research (undertaken with colleagues at the University of Mississippi and Clemson University) on the impact of airline deregulation documents a point that the FAA and Congress must keep in mind: air and highway travel are interchangeable modes of transportation for many families. Changes in airline fares significantly alter the amount of highway traffic, and highway accidents, injuries, and deaths are highly correlated with the amount of highway travel and congestion.[25] Our research suggests that there is every reason to believe that increases in air travel costs for families, as a result of the proposed safety seat requirement, should have the opposite effect of the one intended: The infant safety-seat proposal would have, on balance, increased infant travel deaths.[26]

The FAA subsequently drew the same general conclusion—that an infant-seat requirement would cause more infant travel deaths than it would save, although its estimates of the infant lives lost was much more conservative than the estimates my colleagues and I developed.[27] In essence, the infant-seat proposal to save infant lives is probably a proposal to sacrifice lives of relatively less wealthy people who make their trips by car to save fewer lives of relatively more wealthy people who continue to fly, in spite of the added expense.

From time to time, a Lightfoot/Bond-type proposal has been tendered in the media (which has caused the FAA to make additional pronouncements against requiring infant seats as late as 2005[28]). If such a proposal is ever adopted, an unknown number of the travel victims would surely be infants who would have traveled quite safely on their parents' laps in airplanes. Many of the automobile victims will also be the infants' parents, brothers, and sisters, but many will also be road travelers who may have never contemplated air travel as an alternative means of transportation. They just happened to be in the wrong place at the wrong time on the nation's roads, made marginally more congested by an airline infant safety-seat requirement.

There is one good rule that comes out of this analysis that Congress and all government agencies should heed: do not create a travel-injury problem that is bigger than the one being addressed. The lesson learned is very straightforward: changes in policies that make for changes in prices, whether explicit or hidden, can prove deadly, which is a point fortified in the following discussion of antiterrorism measures.

CHAPTER 1

9/11 TERRORISTS AND AMERICAN DEATHS SINCE 9/11

The overarching lesson of the last section is straightforward: A change in the price of air travel can impact car travel and highway deaths. That lesson should never be forgotten when assessing the consequences of one of the most appalling acts of terrorism in human history committed on September 11, 2001. The nineteen 9/11 terrorists killed more than 2,700 Americans when they commandeered four planes and flew them into buildings and the ground on that surreal day. Such a loss of innocent lives is tragic enough. However, those terrorists have very likely killed (albeit indirectly) more Americans since that fateful day than they killed on that day. [29]

How can that be? The terrorists have been dead since 9/11. The explanation is remarkably straightforward. On 9/11, the terrorists immediately increased the overall price of flying by increasing many potential air passengers' perceived risk of flying. After all, before 9/11, few Americans considered the prospects that a bunch of religious zealots would harbor so much hatred for Americans that they would be willing and able to take over planes only to use them as guided missiles. Since 9/11, most air travelers have understandably feared that copycat terrorists would strike again.

The terrorists, of course, forced the U.S. government to dramatically beef up security checks at airports, the result of which has been an increase in travel time for all passengers. The time spent in security lines at airports has translated into a greater overall cost—and effective price—of air travel relative to ground travel.

Hence, since 9/11, more Americans than otherwise have been more inclined to make their trips by car, leading to more miles driven and greater highway congestion. Since travel by car is far more deadly per mile than air travel, it should surprise no one that automobile accidents, injuries, and deaths have increased as a consequence of the greater cost of air travel imposed by the 9/11 terrorists (independent of other changes—for example, road conditions—that can be expected to affect car-travel deaths).

Garrick Blalock, Vrinda Kadiyali, and Daniel Simon, Cornell University economists, have reported in two working papers the econometric findings of the price tie-in between the 9/11 terrorists' actions and car-travel deaths.[30] They found that the 9/11 events and resulting security measures reduced air travel volume, independent of other forces, by about 5% across all of the nation's airports (and 8% from the nation's major airports). The resulting increase in car travel following 9/11 led to approximately 242 more automobile deaths per month than would otherwise have been predicted for the last quarter of 2001.

As Americans adjusted their travel behavior in subsequent months to accommodate the greater cost of air travel, the increase in the number of car deaths per month attributable to the 9/11 attacks began to taper off. Still, the Cornell researchers were able to surmise that at least 1,200 more Americans lost their lives on the nation's roadways in the twelve months following 9/11 than would have otherwise been predicted.[31] It is no stretch to think that the greater count of American road deaths over the past six-plus years attributable to greater flying risks and 9/11 security measures have surpassed the 9/11 deaths.

The economic tie between air and car travel means that the Transportation Security Administration (TSA) should be ever mindful of the prospects of unintended consequences, the most notable of which is that raising the security alert from, say, yellow to orange can spell greater road deaths, because the security measures can lengthen check-in lines and thus increase the total cost of flying and drive many would-be air travelers to the much deadlier highways. Indeed, the Cornell economists cited above have found that the tighter airport security measures instituted by the TSA after 9/11 also decreased air travel, increased road travel, and led to about a hundred more American road deaths in the twelve months following 9/11 than would have been projected.[32]

The price tie between tighter airport security measures and road deaths means that the TSA has a life-and-death management issue on its hands that has no easy solution. Suppose the TSA has heard of a *potential* terrorist plot to take over a plane. The TSA considers the source reliable, but not perfectly reliable. Should it raise the alert status from, say, yellow to orange? Without the potential for its security measures affecting road deaths, the TSA's decision is perhaps clear—raise the alert status because the only effect will be to inconvenience travelers who will have to stand in longer lines and to suffer more frequent searches. With the price tie of its alert pronouncements to road deaths, the TSA's decision is far more serious, because its decision can lead to more highway deaths, perhaps more deaths than would be suffered if the alert status were not raised and the terrorist plot became a terrorist act, with deaths in the air.

Needless to say, the TSA might at times refuse to raise its alert status because by not doing so, it can save more American lives on the nation's highways than might be lost from terrorists in the nation's airways. But then, the TSA must also be ever-mindful that not raising the alert status can result in additional deadly terrorists' acts on planes, which, again, can drive hordes of Americans to the nation's roadways. Indeed, without an occasional elevation of the alert status, many Americans might drive with greater frequency to their destinations because they fear that the TSA is not doing its job, which is catching wind of terrorists' plots to use planes as missiles.

Clearly, the line of argument developed here speaks to one policy issue: Any waste of scarce TSA manpower on screening aging grandmothers and infants, because of a prohibition on profiling, can be deadly. This is because the tighter security measures and waste of security resources can increase the time cost of air travel. The result can be more car travel—and more road accidents, injuries, and deaths.

Of course, terrorists may figure that they can effectively cause greater deaths of Americans even when they get caught trying to breach airport security defenses. Their failed efforts can keep the terrorist threat alive, and can cause more Americans than otherwise to take to the roads.

By the same token, efficiency improvements in screening passengers, which reduce the time spent in security lines, can save American lives. The price effect of shorter lines can lead to a reverse substitution of air travel for car travel—and fewer accidents, injuries, and deaths on American roads.

In short, the interplay between the full cost of air and road travel cannot—and should not—be overlooked, by homeland security agents or terrorists as they develop their respective defensive and offensive strategies. Regrettably, TSA officials understand all too well that they will catch hell from the media and policymakers if they allow terrorists to slip through and pull off another massacre on board a plane. Those same officials will not likely ever be held responsible for how their airport policies affect highway accidents and deaths. Accordingly, we should not be surprised if TSA officials will want to err on the side of being too cautious, which can translate into more deaths on the nation's roads than will likely be saved in the air.

WATER CRISES IN SOUTHERN CALIFORNIA

In my fully-employed and executive MBA classes in microeconomics at the University of California, Irvine (50 miles south of Los Angeles), I will usually ask at some point in the first lecture, "Why are there water crises in Southern California?" Students seem to draw back, somewhat puzzled, because on the surface the question seems silly. But then, why would I ask it if it were silly? Of course, in spite of their puzzlement, they think they *know* the answer, and more than one student will offer the "obvious" answer, "It doesn't rain much in Southern California!" If I ask how many agree, I usually get a sea of raised hands.

Granted, the prompt answer contains an element of truth. Rainfall in Southern California averages 13 (or fewer) inches a year, making the area close to desert conditions.[33] I usually tell the students that their answer might be an especially good one—in a course in atmospheric physics. But then I remind them that they are in an economics class, and I expect them to offer an explanation that has some tie to the discipline they are studying, a retort that often leaves more of them stumped (as I want them to be).

After dealing with a variety of student efforts to amplify the point that Southern California's intermittent water crises are caused by low rainfall, and at times require an elaborate system for water rationing, I stress, "True, it doesn't rain *water* in Southern California, but it also doesn't rain *Mercedes Benzes* in the area either, and neither does it rain Snickers candy bars, or any other good of value! Have we ever had a *Mercedes Benz* crisis in Southern California?"

The question answers itself and directs student attention (eventually) to a good-old fashioned reason why Southern California sometimes has water shortages (that, in the media, easily get elevated to dire "crises") but never Mercedes Benz shortages. The streets are full of Mercedes Benzes, as are the lots of dealerships—all for a very good reason: The price of Mercedes Benzes is left to move with the forces of supply and demand. If the demand for Mercedes rises or their supply contracts, the price of the cars rises, cutting out any would-be shortage by curbing the number of Mercedes bought and averting anything approaching a shortage, much less a "crisis."

On the other hand, the price of water is stuck at some subsidized level, determined by government officials who are reluctant to change the price of water to accommodate transient changes in the demand for and/or availability of water. If rainfall drops way below average, as it is bound to do from time to time (rainfall for the year when this section was being written was one-fifth the annual average), and the price is not hiked, people can be expected to continue using water as if nothing has happened. After all, the low price of water tells many consumers (especially a large percentage of the population that never pays attention to the news) that water is as abundant as ever. The continuing flow of water out of home faucets can convince uninformed and informed consumers that any shortfall in rainfall in Southern California could be offset by a greater snow pack in the mountains of Northern California where Southern California gets a third of its water.

Southern California water consumers can also reason (if they are aware of the drought) that if they alone curb their consumption, the water tables in the area's reservoirs will not be noticeably affected. Even if a sizable bunch of consumers curb their water use, consumption would not likely be materially affected because other consumers can expand their use of water. And do understand that Southern Californians use water with little thought of how scarce water really is, mainly because its low price—.25 cents per gallon for residential use,[34] which is one-third the price of water in Mississippi where the rainfall is over 50 inches a year—makes it seem abundant (which is the case, given the considerable federal, state, and local government subsidies to draw water from other parts of the state through aqueducts and from other parts of the country through tapping in aquifers that extend up into the upper Midwest). Accordingly, like so many other Southern Californians, my backyard looks for all the world as though I live in the tropics (without the heat and humidity). The water subsidies have actually increased the price of my house (because they have made living in the SoCal desert more affordable than it otherwise would be).

So, when rainfall falls off and people continue to use water without restraint, a "crisis" eventually raises its ugly head in public discussions, with public officials first appealing for voluntary cutbacks in water consumption, which typically have meager impacts.

Indeed, during the water crisis underway as this section was being finalized, the Orange County, California water authorities told everyone that the situation was "dire" (given the combination of little rainfall and the reconstruction of a major water main), pleading with everyone to conserve. What happened? Water consumption rose markedly, as many people washed their cars and watered their lawns, fearing that their faucets would soon run dry or they might soon be told that washing cars and watering lawns is prohibited.[35] All the while, the waterlines around the area's reservoirs were sinking deeper and deeper. Understandably, appeals for voluntary curbs are usually followed by threats of "water police" prowling neighborhoods looking to give tickets to violators of water-use ordinances.

Of course, some state institutions pay lip service to water conservation, with some effect. In the midst of the growing water crisis as this section was being finalized, my university announced reductions in its sprinkling of the campus lawns. At the same time, it continued landscaping newly opened areas of the campus with thousands of water-thirsty shrubs, trees, and flowers.

The more general lesson to be learned from the water-crisis puzzle I pose to my classes is as simple as it is unheralded: Where shortages are evident, it is a good bet that prices have been held in check someway, somehow. The coming water crisis at the time of this writing would all go away if the water authorities had the fortitude to do what businesses—Chevron, as well as Mercedes—do naturally: raise the price! And make no mistake about it, at the same time that a water crisis in Southern California was emerging, the price of gasoline was well above $3 per gallon and rising rapidly (because of ongoing political/military problems in the Middle East and because refineries were being taken offline for repairs). But the price increase (even though it might be temporary) did its job. Even though the number of licensed drivers and the number of vehicles on California roads had both risen by more than 10% over the 2000–2006 period, gasoline consumption had risen far less and showed signs of falling, at least at this writing, according to reports in the *Los Angeles Times*.[36]

ETHANOL SUBSIDIES AND WORLD HUNGER

Following the OPEC oil embargo of 1973, which led to a spike in gasoline prices, price controls on gasoline, and long lines at service stations, Congress legislated the use of ethanol, which is produced from corn, as a gasoline substitute. In 1977, then-President Jimmy Carter made energy independence the "moral equivalent of war," a position that over the intervening decades led to the passage of a variety of federal and state subsidies for the production of corn and ethanol.[37] In 2005, corn farmers in the USA received nearly $9 billion in subsidies from the U.S. Department of Agriculture intended to stimulate corn production, a growing portion of which has been used in ethanol production. Ethanol producers receive slightly more than a half dollar in subsidies (in the form of tax credits) for every gallon produced. The wars and political instability in oil-producing countries of the Middle East and the rapidly modernizing and expanding economies of India and China caused a run-up in the price of oil on world markets in the early 2000s that further increased the demand for oil-substitutes, with ethanol being one of them.

Not surprisingly, by the end of 2006, 110 ethanol refineries were in operation in the USA, many of which were expanding their production capacities. Seventy-three more refineries were being built.[38] In 2006, U.S. biofuels firms produced 5 billion barrels of ethanol. In 2007, production was expected to rise 40% to 7 billion barrels.[39] Also not surprisingly, the growing demand for ethanol has hiked the demand for corn, which has driven up the price of corn by a third in less than a year, from $3 a bushel in the summer of 2006 to $4 a bushel in the spring of 2007, a price level not seen in a decade.[40] Moreover, the prices of other food crops—for example, wheat, peas, sweet corn, and rice—have jumped upward as farmers have moved land into the production of corn, contracting the supplies of other food crops and causing their prices to rise. The growing prices for grains have (literally) fed into upward pressures on chicken and beef prices—and to price increases on (among other products made from grains crucially important in the diets of many poor and rich people alike) tortillas![41]

What is the basic problem with the corn and ethanol subsidies? To fill up an SUV with ethanol, it takes 450 pounds of corn, which contains enough calories to feed a poor person for a year.[42] There are at least a half billion and maybe a billion people in the world who are chronically hungry, which means that they do not get enough calories on a daily basis to remain healthy, many of whom continually face starvation. According to the World Bank, the world poor's consumption of calories declines by .5% for every 1% increase in the prices of basic foods.[43] Moreover, the world's count of "food insecure" people rises by 16 million for every 1% increase in the prices of staple foods.[44] And the various policies designed to encourage use of ethanol could have increased the world price of corn and other grain crops by several percentage points.

No doubt, Jimmy Carter and other political leaders who have pressed for the development of an ethanol industry may have had their hearts in the right place, but they may have overlooked the power of the law of unintended consequences, which in this case can be bleak for many poor people around the world. As applied economists C. Ford Runge and Benjamin Senauer have observed, "The world's poorest people already spend 50% to 80% of their total household income on food. For many among them who are landless laborers and rural subsistence farmers, large increases in the prices of staple will mean malnutrition and hunger. Some of them will tumble over the edge of subsistence into outright starvation, and many more will die from a multitude of hunger-related diseases."[45]

Perhaps the bad things the world's poor will suffer because of the indirect effects of corn and ethanol subsidies could be offset by a couple of potentially positive effects. The rise in the world price of corn, along with the drop in the price of blue agave, a cactus-like plant used in Mexico and elsewhere to make tequila, has caused Mexican farmers to contract their planting of agave in order to make room for corn. The reduction in the supply of agave (from what it would otherwise have been) can be expected to lead to a rise in the price of tequila, and a reduction in its consumption.[46] That price change can be expected to lead to less drunk driving and, very likely, fewer road accidents, injuries, and deaths among Mexicans. Through a change in the world price of tequila, such a positive effect of the hike in the price of corn can be expected to spread across the globe (although the effect might be hard to detect).

The corn and ethanol subsidies harbor the potential for positive environmental—or "green"—effects from ethanol use. A cleaner environment could mean a healthier world population and, hence, more income, and a better life, on balance, for the world's poor. However, Runge and Senauer report that "using gasoline blends with 10% corn-based ethanol instead of pure gasoline lowers emissions by 2%."[47] Then, the crops used to make ethanol require the use of fertilizers and pesticides, and farm machinery that consumes oil-based products as they are used on farms. In short, the environmental effects could be meager, and difficult for the poor of the world to detect.

Then again, the green effects could be significant—and negative. According to reports by the Friends of the Earth (which surfaced just as this section was being drawn to a close), Europe's encouragement of use of biodiesel fuels has led to the destruction of rainforests in Indonesia and Malaysia because of the creation and expansion in those countries of oil-palm farms to satisfy the increased demand for oils that come mainly from palms and rapeseeds and that are used in the production of biodiesel fuels.[48]

Granted, biodiesel fuel can be made, and is being made, by firms such as Metro Fuel Oil Corporation, which in 2007 was awaiting approval to open its plant that would produce 110 million gallons of biodiesel fuel from recycled raw vegetable oils collected from restaurants in the New York City area.[49] The use of such oils could have beneficial green effects since some of the used oils would have been thrown away, but some of the oils could have been recycled for use again in restaurants' deep fryers. That means that the production of biodiesel fuels from used vegetables oil would require the production of more new vegetable oils used in restaurants that, again, could cause some food prices to rise and impose problems for the world's poor.

In short, subsidizing the use of renewable plants to satisfy a portion of the world's energy needs sounds like a nice idea on all fronts, until you consider the price implications and how the world's resources will be shifted, often in unanticipated and unintended ways, in response to price shifts. Those who would like to think biofuels provide the proverbial "free lunch" either for the economy or the environment will be sadly disappointed.[50]

If (or to the extent that) carbon dioxide is a significant culprit in global warming (or any other environmental problem), a more promising solution is one economists have been touting for decades: Tax the carbon dioxide that is emitted from cars (or any other plant and equipment). The greater the carbon dioxide emitted, the greater the tax. The expectation is that the tax will feed into the price of the offending products, and fewer of those products will be bought and used. Greenhouses gases will be reduced. Global warming will be setback into the future, if not eliminated altogether. Okay, the higher prices will affect the poor, and no one wants to hurt the poor. There is an easy solution on that front: Return the carbon taxes paid by everyone to the taxpayers who paid the carbon taxes in the form of tax refunds. People will have more or less the same spendable money incomes, plus a cleaner environment. But because of the carbon tax and the higher prices on the taxed products, people will move their consumption from less environmentally damaging products.[51]

But then, relief for the poor and the environment can come through price adjustments. By late 2007, it was becoming apparent that an overcapacity in ethanol production had emerged since early 2007, with the price of ethanol falling 30% between March and September.[52] That price reduction can dampen the demand for corn and other crops, which can reduce upward pressures on food prices paid by the poor. Nevertheless, the subsidies for ethanol should still leave corn and other grain prices higher than what they would have been.

THE CALIFORNIA ELECTRICITY CRISIS

In 2001, the wholesale price of electricity in California, then newly deregulated, jumped from the convergence of several supply-and-demand forces:

- There was at the time an absence of new generating plants coming on line;
- There was a spike over the previous year in the price of natural gas (which is widely used in the state to fire generators);
- There was also an ongoing drought in the Northwest, which caused the water flow in the Columbia River basin, a major source of hydropower generation in the region, to fall by half;
- The booming California economy caused a doubling of the growth rate in electricity demand from projections of 3 or 4 years before;
- And the now-defunct Enron Corporation, as well as other energy traders, began to drive up the wholesale price of electricity by, in effect, cornering the market (according to critics of California's electricity deregulation record).[53]

All of these market forces threatened the vitality of the world's fifth largest economy—California—because of the then-pending shortages of a critical resource, electricity.

During the early stages of the crisis, the vice president for administration at my university emailed the faculty and staff regularly about pending "rolling blackouts," suggesting in one email that university employees and students should drive carefully because traffic lights might go out without notice. And they did one day early in the emerging crisis, causing the death of a driver in San Diego.

Nevertheless, judging from people's behavior in my immediate area, you would not have believed that there was an electricity "crisis" at all, unless you read the morning papers. In my university building, one out of every three hallway (florescent) lights were turned off late in the afternoon, but only for the last hour or so of the workday. The modest hallway "dim-out" suggested the turned-off lights didn't appear to be needed anyway.

Otherwise, it was hard to detect changes in behavior. Few people seemed to be truly concerned enough to make real sacrifices. But then why should they? Most people seemed to take the view, "Anything I might do to conserve would be of no consequence." The "free rider problem," which economists have spent careers talking about in their classes, was on full display.

When my wife and I went out to dinner in Newport Beach at Christmas time (about the time the electricity shortfall was, we were warned, peaking), the largely empty office buildings surrounding the upscale shopping center where we ate were aglow on practically every floor—as if nothing was wrong. Dozens of palm trees at the entrances of businesses remained wrapped in Christmas lights. I could detect no change in the lights in the windows of people's homes as we drove by. Then, my university lit up a new one-hundred-yard-long grand entrance to the university with a few thousand watts of lighting, probably offsetting the savings from the dimmed hallways of our office buildings.

But why should things have changed? Electricity waste has been a way of life in California. As the crisis began to fester in December 2000, the nearby international headquarters of the Trinity Broadcasting Network, whose religious television sets drip with ornate gold leaf props, had its multi-acre campus ablaze with what appeared to be several million Christmas lights. Throughout the Christmas season, Fashion Island, the shopping center where we ate, turned on nightly the "World's Largest Decorated [and Lighted] Christmas Tree." Because the tree was so massive, 110 feet tall, they had to hang Christmas lights the size of soccer balls, and you can bet there were lots of them.

It's transparently clear that electricity is relatively cheap in the state, given the widespread use, a fact that stands in contrast to what you hear from the talking heads on the tube in local studios, who, by the way, made their dire points about the crisis in front of a few thousand watts of television lights. What I found remarkable is that our electric bill during the crisis for my four-bedroom home in California averaged, at the time, less than $75 a month, maybe two-thirds, if not one-half, what it was when we were back in South Carolina a decade before. Everyone cites our relatively "high" electricity *rates*, but few note how little electricity is needed in such a moderate climate.

I've spent many hours with fellow economists talking about the "tragedy of the commons" that emerges when prices are not allowed to seek their market-clearing level. Typically, the talk is about how, say, cattlemen will invariably overgraze pastures when the property is held in common, meaning no one owns the property and no charge is exacted for access. The "tragedy," underfed cattle because of the overgrazed pastures, is an outcome none of the cattlemen wanted.

If there ever was a tragedy of the commons, I stood witness to its making. The electricity tragedy was man-made by those who were least suspected. Few consumers (or policymakers) seemed to understand that every time they turned on a light, they "overgrazed" the power grid and increased the junk debt of the local power distributors, and the "overgrazing" continued because the retail price of electricity remained regulated, capped throughout the crisis, while the deregulated wholesale prices of electricity rose. Who cares? Indeed, it just struck me that by writing about the crisis in the midst of it on my power-gulping computer, I was adding to it, and the electric power companies' indebtedness (and the threat of their bankruptcy), but by so little that I need not have changed my writing plan. Therein lies the source of a real-life commons tragedy. Economists in other parts of the country only have to appreciate the argument intellectually. I had to live with the consequence of the tragedy that was unfolding around me.

The state rapidly ran through billions of tax dollars to subsidize all the energy waste I saw around me. They only belatedly came to realize how their actions to hold the retail price of electricity down, in the face of the mounting shortage, eliminated any incentives to conserve electricity use all the more.

Never mind, those palm trees couldn't have looked more regal at night. The hot tubs in our development remained heated, at their toasty legal limits, 104 degrees. Yes, the hot tubs are heated with natural gas, but few seem to realize that the high demand for natural gas was a source of the state's electricity crisis, because electricity is produced with furnaces heated with natural gas. Southern Californians—hot tub bathers and all—could have been made to realize the social consequences of their use of electricity and natural gas through a simple change in policy—a substantial hike in the prices of electricity *and* natural gas.

CONCLUDING COMMENTS

The discussions of various topics in this chapter have helped to spotlight an important economic lesson: unless business people and policy makers understand how prices are affected by market and non-market forces, the "law of unintended consequences" will bedevil people's best intentions when setting prices—and especially when they try to subvert market forces. The root problem with water and electricity crises in Southern California was the underlying price controls that have encouraged people to consume more of those scarce resources, which at the time of the controls were more scarce than usual, than they should have been, given limitations in the supplies of those critical resources.

The discussion of infant and toddler seat requirements on airlines explained why policymakers need more than good intentions to save lives; they need to understand the interplay between the prices of various modes of travel. Similarly, the discussion of the 9/11 tragedy exposed how the TSA's changes in the security alert status at airports should be taken with deadly seriousness, because the consequences can indeed be a matter of life and death in ways not widely recognized. Security alerts can change the relative price of air travel vis-à-vis car travel, all without anyone noticing the change or its consequence. The "law of unintended consequences" rules, often with deadly silence. That theme will continue to form the foundation of the discussions of additional pricing puzzles considered in following chapters, especially the next one.

An important purpose of this chapter has been to reassert a point too easily overlooked: A well-functioning market system depends crucially on prices. Prices do far more than alert people to how much they must pay for the things they buy. They are themselves productive by providing incentives for people to choose and buy wisely, by containing a great deal of information that permits people to economize on the amount of information they must gather and absorb, and by helping coordinate close-at-hand exchanges and also complex economic activity of people spread throughout the world. Without prices to "grease the skids" of the economy, we all would be less productive than we are and worse off.

Another, equally important purpose of the discussions in this chapter has been to convince you that a study of prices can help us understand better (not perfectly) why people behave the way they do. An understanding of how prices are determined and changed can help us unravel a host of seemingly obtuse economic puzzles.

What is remarkable about the discussion in this chapter is how much of it has been founded on one economic principle, the "law of demand," that price and quantity are inversely related. If the price of a good is raised, people will consume less of it. If the price is lowered, people will consume more of it. That principle will remain in heavy use throughout this book and will play a key role in our unraveling many pricing puzzles.

NOTES

1 Apple fans also formed long lines around AT&T cellular phone outlets, given that AT&T had acquired exclusive rights to sell the iPhones to its subscribers.

[2] Guglielmo (2007).

[3] Friedman and White (2007). Bloomsburg News reported that a third of the 164 Apple stores ran out by Sunday night after the iPhone was released on the previous Friday, and most of AT&T's 1,800 cell phone stores were out of stock within twenty-four hours (Guglielmo 2007).

[4] Guglielmo (2007). The *Orange County (Calif.) Register's* technology columnist found that on July 2, 2007 Craig's List (craigslist.com) the average price of the 8-gig iPhone model was $781.57 in Orange County alone (Stewart 2007).

[5] Wingfield (2007).

[6] The day after Apple announced its $200-price reduction on the iPhone, Steve Jobs posted an "open letter" on its web site, announcing that early iPhone buyers would receive a $100-store credit, seeking to assuage the complaints of early iPhone buyers (Hafner and Stone 2007). Jobs suggested that the price reduction "benefits both Apple and every iPhone user to get as many new customers as possible in the iPhone "tent". We strongly believe the $399 price will help us do just that this holiday season" (http://www.apple.com/hotnews/openiphoneletter/, accessed September 6, 2007). See also

[7] Fels (1975, p. 32).

[8] Fels (1975, p. 33).

[9] The tax credit for various models and rules for receiving the tax credit can be found on an IRS web page, http://www.irs.gov/newsroom/article/0,,id=157557,00.html.

[10] Radcliff (2007).

[11] To determine how many miles the Civic hybrid would have to be driven in order for the gas savings to cover the added hybrid costs, including the price differential of $7,500 (plus interest and taxes) and the $3,000 cost of a new hybrid battery, I used *Consumer Reports'* estimate of the gas mileage for the Civic hybrid and non-hybrid, 37 MPG and 28 MPG, respectively. I also used the *average* price of regular gas in Southern California in for 2004 through early 2007, or $2.60 a gallon. The actual miles that would have to be driven for the savings in gas cost to more than cover the hybrid costs is 523,000 miles.

[12] Woodyard (2007).

[13] Murph (2007).

[14] Poole (2007).

[15] U.S. Department of Transportation, Federal Aviation Administration 1990.

[16] Kolstad (1989).

[17] See U.S. Congress, House 1990.

[18] Lightfoot (1990). Representative Lightfoot maintained that his proposal does not mandate the purchase of additional seats, only the use of safety seats by infants. He suggested that parents can use their auto safety seats and utilize free empty airline

seats, as they now do. However, it seems unlikely that parents would buy their own tickets—especially cheaper nonrefundable tickets—and take the risk of not being able to board at the last minute because adjoining seats are unavailable for their infants.

[19] For the period 1995 through 2003, for every per hundred million miles traveled there were .03 fatalities by air throughout the world (International Civil Aviation Organization 2003) and .93 fatalities traveled by car on U.S. highways (Bureau of Transportation Statistics 2003), as reported by Sanders, Weisman, and Li (2005).

[20] U.S. Department of Transportation 1990. In a Harvard Medical School study prepared in the late 1980s, Richard Snyder (1988) estimated that the infant seat requirement would save an average of only 0.6 infant lives a year, or three lives in five years. Richard G. Snyder (1988), as reported in Coleman 1990. The FAA found that an additional six infants and toddlers were injured in airline flights between the late 1970s and 1990, as reported in USA Today (editorial, 1990).

[21] National Safety Council (1989). In addition, it should be noted that the annual infant death rate for the era covered, which can be inferred from FAA findings, averaged slightly above one per 100,000 daily boardings of infants each year between 1978 and 1990 (the mean of the estimated range of daily boardings, 5,000 to 10,000, or 7,500, times 12). The death rate per 100,000 infants under one in 1986 was higher in other activities than in air travel: 5 due to mechanical suffocation, 4.2 due to ingestion of food and objects, 3.2 due to fires and burns, and 2.5 due to drownings (National Safety Council 1989, p. 8).

[22] Total airline fatalities on scheduled airlines (U.S. air carriers operating under 14 CFR 121) between 1980 and 1988 came to 975, including the 259 passengers killed in the explosion of the Pan American flight over Lockerbie, Scotland (U.S. Department of Transportation, Federal Aviation Administration 1989, p. 154).

[23] Windle and Dresner (1991) estimated that the cost of air travel per "family travel unit" would be $296 without the child-safety-seat requirement and $358 with the requirement (assuming half-price fares for infants), a 21% increase.

[24] Because some families would not fly, the net increase in revenues to the airlines was estimated to have been $119 million (U.S. Department of Transportation (1990, p. v.)

[25] See McKenzie and Shughart 1988; McKenzie and Warner 1987; and Rose 1987. Indeed, Warner and I (1987) found that airline deregulation, which led to lower airfares and an expansion of flights, increased air travel by an annual average of 11% and reduced passenger car travel by an annual average of just under 4% between 1978 and 1985. Accordingly, airline deregulation significantly reduced highway accidents, injuries, and deaths. McKenzie and Warner estimated that in 1979–1985, after adjusting for a number of factors that affect highway travel safety, airline deregulation reduced automobile accidents by an annual average of over 600,000, lowered

automobile injuries by an average of approximately 66,000, and reduced automobile fatalities by an annual average of almost 1,700.

[26] If the number of automobile trips by families goes up by a third of the estimated reductions in infant boardings and if the average length of the trips is 400 miles one way (800 miles round trip), automobile travel will increase by more than 185 million miles each year. That represents a very small percentage increase in automobile travel. Nevertheless, at the time of the congressional debate, according to our econometric model, the increase could translate into more than 1,600 additional automobile accidents each year, and the increase in accidents could result in more than 175 additional disabling injuries and just under 5 additional deaths each year. If the fare increase had been much greater than the FAA conservatively assumes, the increase in highway injuries and deaths would have, of course, been greater.

[27] In a report prepared for the FAA, Apogee Research (1993) estimated that had the infant-seat requirement been imposed, there would have been a net increase in travel deaths of 8.2 (additional car deaths minus reduction in air deaths). There also would have been an additional 52 serious injuries and 2,300 minor injuries from travel.

[28] Federal Aviation Administration. (2005). Federal Aviation Announces decisions on child safety seats, August 25.

[29] Although I have been making this argument about American deaths after 9/11 from the terrorists' acts on 9/11 for years in my class, I should acknowledge that Nassim Nicolas Taleb has made it in his best selling nonfiction book *The Black Swan* (2007, pp. 111–112).

[30] Blalock, Kadiyali, and Simon, forthcoming (2005a).

[31] Blalock, Kadiyali, and Simon, forthcoming (2005a).

[32] Blalock, Kadiyali, and Simon, forthcoming (2005b).

[33] A desert is generally defined as any area that receives less than 10 inches of rainfall a year.

[34] According to researchers at the Wessex Institute (Hammer 2007).

[35] County officials used lighted overhead freeway signs to alert people to the crisis, with the admonition: "Orange County Water Emergency. Conserve Water," as reported by the Orange County Register (Townsend, Carpenter, and Jolly 2007).

[36] See Douglass (2007).

[37] This section relies heavily on an insightful essay by applied economists C. Ford Runge and Benjamin Senauer (2007).

[38] As reported by Runge and Senauer (2007, p. 5).

[39] As reported by Associated Press (2007).

[40] As reported by Associated Press (2007).

[41] As reported by Runge and Senauer (2007, p. 5).

[42] As reported by Runge and Senauer (2007, p. 2).

[43] As reported by Runge and Senauer (2007, p. 5).

[44] Runge and Senauer (2007, p. 5).

[45] Runge and Senauer (2007, p. 5).

[46] Reuters (2007b).

[47] Runge and Senauer (2007, p. 6).

[48] Pontoniere (2006, p. 1)

[49] Rivera (2007, p. A14).

[50] Janet Larsen, research director for the Earth Policy Institute, an environment-friendly organization, told the *New York Times*, "Turning food crops into fuel crops does not make sense, economically or for the environment" (Rivera 2007, p. A10).

[51] Admittedly, there are a couple of problems with the proposed taxation/distribution scheme skipped over in the body of the chapter to avoid choking the discussion there with details. One obvious problem is that the price increase on the taxed product will not likely equal the tax of any given size. The solution there is simple: Continue to raise the tax until the price is raised enough to achieve the desired environmental effect. The second problem is that the return of the taxes paid to consumers will mean that consumers will be able to buy more of the taxed products than they would have been able to buy had the taxes been kept by the government. No problem, all this refinement in argument means is that the tax needs to be raised further until the combination of the higher price and return taxes achieves the desired environmental effect. I understand also that there will be costs in collecting the taxes and then distributing the revenues back to buyers. No one ever said that greater environmental cleanliness would ever be costless on all fronts, and of course, my proposed solution is no solution at all if the tax collection/distribution costs are greater than the gains from greater environmental cleanliness.

[52] Etter and Brat (2007).

[53] For a review of a variety of forces at work in the California electricity market leading up to the crisis of 2001, see Weare (2003). The importance of Enron's trading practices on California's electricity crisis remains in serious dispute. See Weare (2003) and Taylor and VanDoran (2002).

Chapter 2

PRICING LEMONS, VIEWS, AND UNIVERSITY HOUSING

Prices capture a whale of a lot of information on the scarcity of the resources that go into the production of products and on how much people value various goods. Prices enable buyers to economize on their time. By not having to know much, if anything, about production conditions in various parts of the world or about consumer tastes other than their own, buyers can focus their time and energy on comparing prices and attributes of goods they want to buy that, with as much income as many buyers have these days, is not always an easy problem.

Buyers can be forgiven if they are lulled into not understanding why many prices are a mystery in that that they don't seem to reflect production costs and consumer values, as reflected in the precipitous drop in the resale price of new cars as they exit the dealer lots. They might also be forgiven if they accept, without reflection, many comments on prices that, because they are heard so frequently, seem indisputable, such as in the comment real estate agents often parrot, "Houses with views sell quicker than houses without views."

In this chapter, I attempt to explain the wisdom of another quip economists often make, "If everyone believes it and says it, doubt it!" You will find that the "law of unintended consequences" will remain with us as we consider several pricing puzzles and frequently heard glib comments about prices, which are puzzling only because so many people believe the comments in spite of the fact that the comments are often patently misguided.

CHAPTER 2

THE PRICING OF LEMONS

I'm a great believer in how important economic lessons can be learned from unraveling puzzles. For a long time economists were puzzled by the fact that new cars drop precipitously in value once they are driven off dealer lots.

One well-worn explanation is that many car buyers yearn for the "new-car smell" and are willing to pay a premium for new cars over what they are willing to pay for used cars, even cars that may have only recently left dealers' showrooms. Another explanation for the new/used-car price differential is that car dealers are in the business of making markets for their cars with glitzy showrooms and glossy advertisements. Car owners are not in a position to maintain the demand for their cars that the dealers created. As a consequence, car values drop on leaving dealer lots because the demand for the cars drops.

Such explanations cannot be summarily dismissed, but we must wonder if they are the whole story, especially since the resale price of a car just driven off a dealer's lot can be 20% (or upwards of $10,000 for some luxury cars) below its purchase price. Economist George Akerlof has offered perhaps a far more telling explanation for the price gap between comparably-equipped new and used cars.[1] To keep the analysis simple (as does Akerlof), suppose there are two types of used cars, good ones (which have low maintenance costs) and bad ones (which have high maintenance costs),—with the bad ones commonly known as "lemons." Buyers will discover which cars they have from using their cars. Hence, they will have information, drawn from their experience, about their cars' quality that potential buyers of used cars will not have. Information on car quality will be decidedly one-sided—or "asymmetric"—meaning buyers and sellers do not go into potential deals with the same level of information.

Buyers in the used-car market can be expected to reason that new-car buyers who learn they have good cars will keep their cars. On the other hand, buyers who learn they have lemons will want to lower their car maintenance costs by putting their cars up for resale. Hence, the available used cars can be disproportionately dominated by lemons. That is to say, used-car buyers will have to worry that they will likely buy problem cars, or cars with nontrivial repair costs. The price of used cars must drop if buyers are to be enticed into buying used cars. Of course, as the price of used cars drops, car owners with problem cars, which are not total lemons, can be expected to pull their cars off the resale market, because they can be better off incurring their modest repair costs than suffering the lost resale value. This means that the available stock of used cars for sale will become even more heavily dominated with (serious) lemons, again, given that the better-used (problem) cars will be retained by the owners. A drop in the price of used cars can, in other words, lead to a further drop.

This line of argument draws into question a frequently heard claim that "used cars are better deals than new cars" because of the dramatic price difference between them. If that were the case, and everyone knew that were the case, then the demand for used cars would rise while the demand for new cars would fall, causing the prices of used and new cars to converge, until used cars were not the "better deal" they are claimed to be.

Sure, used-car buyers can pay a much lower price than they would have to pay for new cars, but they must also suffer the normal wear and tear attributable to the miles put on the used cars. More importantly, used-car buyers have to suffer the risk cost associated with buying in a market potentially dominated by lemons that can translate into high repair costs (especially when the warranties on the used cars have expired).

Granted, the new/used-car price differential might be expected to exceed the expected repair cost, but that still doesn't make used cars "better deals." The problem of asymmetric information can't be denied; it is a real problem that used-car buyers have to consider as best they can. The prospects that used-car buyers just might buy cars with repair costs far higher than "average" can weigh down the price they are willing to pay for used cars.

In the so-called "lemon problem" (as with all "problems"), there is money to be made by entrepreneurs who can solve the problem. Individual used-car sellers might have a credibility problem with potential buyers the sellers do not know, but sellers can elevate the price they can charge by, for example, allowing potential buyers to have the cars they are considering inspected by mechanics. Used-car sellers might only try to sell their cars to relatives and friends where their word on the quality of their cars would carry more weight, because of the potential ostracism sellers might suffer if they are not true to their word. And sellers can also pay for extended warranties, which is a means sellers can use to ease the risk facing the buyers. Presumably, the added price used-car sellers charge for their cars because of the warranties will at least cover the price of the warranty.

Alternately, used-car sellers can sell their cars to reputable dealers who can pay premium prices for used cars because they can get even greater premium prices from the resale of their used cars. Dealers can charge premium prices to the extent that they have established reputations for honest dealing, a line of reasoning that explains why so many new-car buyers trade-in their used cars when they buy new ones. New-car buyers can get better deals on their trade-ins from the dealers than they can get from individuals, and the dealers can make money on the trade-ins because they solve, to a degree, the lemon problem, or rather the underlying asymmetric information problem in the used-car market.

Akerlof points out that the problem of selling health insurance to the elderly has features of the lemon problem. As people age, those who see themselves as being most in need of expensive and frequent healthcare are the ones who are most likely to buy health insurance. Healthy people will be less inclined to buy health insurance. This is especially true because health insurance providers will have to charge premiums that reflect the relatively high costs of healthcare provided to policyholders that, as a group, will tend to need lots of healthcare, which makes them, for all intents and purposes, "human lemons." As the price of health insurance is raised to accommodate the so-called problem of "adverse selection" (or the tendency of people to buy insurance when they expect to be beneficiaries), healthier people will drop out of the insurance market, leaving policyholders even more dominated by people who expect to need lots of healthcare. The price of insurance will have to rise again to reflect the growing adverse selection problem.

Akerlof notes in passing that the "lemon problem" in healthcare is an argument for some form of national health insurance for the elderly. That could be the case, but what Akerlof doesn't mention is that public provision of healthcare can give rise to other problems. If people know that they will not have to pay for their health insurance when they become elderly (and will not likely have to pay a premium in line with their state of health when they are elderly), they can have less incentive to take care of themselves before they have access to public provided health insurance. In addition, if healthcare for the elderly is heavily subsidized, then we should expect the elderly to demand more healthcare than they otherwise would, and that increase in demand can push up healthcare prices for everyone, including the young. The result can be an increase in the health insurance prices the young face, with many of them deciding not to buy health insurance because their expected healthcare costs are lower than their insurance premiums.

Insurance companies have found ways of solving the adverse selection problem in health insurance, at least somewhat. First, they provide health insurance policies to workers through their employers. Such a distribution channel has one largely unrecognized advantage: It reduces the pool of policyholders who can't meet a minimal health standard, being able to work and hold a job. In other words, group health insurance policies narrow the adverse selection problem.

Second, health insurance typically gives policyholders a menu of policy options, with a key differentiating feature being the size of the deductible, after which all care costs are covered by insurance. The policyholders who seek a small deductible are self-identifying themselves as people who see themselves as likely needing a great deal of care (including lots of office visits that require only small "co-pays"). The policyholders who select a high deductible are self-identifying themselves as likely needing little care. The insurance company can simply charge the low-deductible group far more than they charge the high-deductible group. This line of argument helps explain why in moving from a deductible of $250 a year to $1,000 a year, the premium drops by substantially more than $750 a year. This is because the policyholders move from a high healthcare-cost group to a low healthcare-cost group.

HOW PRICES ADJUST TO ADVANTAGES AND DISADVANTAGES OF PROPERTY

One of the unheralded advantages of prices is that through market forces, they capture the advantages and disadvantages of property, in the process giving a market value to the advantages or disadvantages. Prices adjust until buyers are more or less indifferent between properties. In this section we consider three real-world cases of how property prices can neutralize the advantages and disadvantages of different properties: 1) property inside and outside floodplains, 2) property with and without views, and 3) property that is owned and rented.

PROPERTY INSIDE AND OUTSIDE FLOODPLAINS

Should we feel sorry for our fellow Americans in the Midwest (or elsewhere) who are, from time to time, flooded out of their homes by nearly forty days and nights of continuous rain and snow? Of course we should. Vivid reports of mounting property losses from floods on television and in newspapers do weigh heavily on just about everyone's emotions. No one wants to see others suffer, and the outpouring of aid for flood victims is understandable—as a raw emotional response. We all are, or should be, our brothers' and sisters' keepers—*to some reasonable extent*, with "reasonable" meaning the consequences of helping victims guiding and constraining our judgments.

We can't dismiss the question—should help be provided?—summarily, as if the only answer is that we should help, because that question leads, inexorably, to the tougher questions of how much help should be rendered and in what form. Those decisions must be grounded in a hard-nosed assessment of the real damage incurred by flood victims—and potentially caused by the relief itself. Such an assessment may cause us to reach a paradoxical conclusion: on balance, many flood victims may not be *victims* to the extent media reports indicate, at least as measured by their *net* losses—in spite of the fact that many have experienced sizable property losses. The paradox can be unraveled with a little reflection on the economics of floods (and other similar natural disasters), and how the consequences of floods and relief for victims can be captured in prices.

By virtue of an area's designation as a "floodplain," people who live in them, or who might contemplate living in them, know that floodplains are prone to floods with varying frequency and duration (but most often with *expected* frequency and duration). The residents (and prospective residents) might not know exactly when the floods will come or how severe they will be when they come, but that should not stop them from considering the prospect of floods and the damage that must be endured when the floods do occur. The prospects of floods, without much question, temper the market's demand for pieces of property in floodplains, causing their market values to be lower than property with similar attributes but without the prospects of floods and the damage that goes with them.

This being the case, when viewing alternative pieces of property, some in and some outside of floodplains, prospective buyers should not be willing to pay as much for floodplain property as for other property that is deemed safer. Indeed, prospective buyers should lower the price they are willing to pay for floodplain property by an amount at least equal to the *expected* losses during floods (with the actual losses, measured in dollars, discounted for risk and time). The greater the frequency and duration of floods, the greater the expected damage, and thus the lower the expected floodplain property prices.

To illustrate, if a house on a "safe" piece of land outside of a floodplain costs $100,000 and if the expected losses from floods on a comparable house and piece of land inside the floodplain is $20,000 over the foreseeable future, the floodplain property should sell for $80,000 (more or less). If the floodplain property had a price of $90,000, the total cost, including the loss from expected floods, would be $110,000, which means the prospective buyer would turn to the property outside the floodplain. Hence, the price differential between the property inside and outside the floodplain can be expected to diverge until it is (roughly) $20,000. With the price gap of $20,000, the floodplain property owners can endure $20,000 of losses without actually being any worse off than they would have been had they chosen to buy outside the floodplain.

Clearly, some floodplain property owners will suffer heavier losses than were expected, mainly because floods cannot be predicted precisely, or may occur more frequently and/or be more severe than expected. By the same token, some property owners, in spite of their losses during floods, can be net gainers, mainly when their losses turn out to be less than expected, that is, lower than the discount they received on the price of their property for buying in a floodplain.

For example, suppose the owners in the above example who bought the floodplain property for $80,000 suffer only $12,000 in flood-related losses. In effect, they realize an economic gain, on balance, in the instance of that flood because their flood-related losses are $8,000 less than the $20,000 premium they would have had to pay for property outside the floodplain. Ironically, those who bought outside the flood-prone area and are not flooded lose, in this example, more than the victims of the flood; the non-victims lose the premium paid on their property, $20,000. (I know some readers may be thinking that flood victims must work to clean up their property. True enough. Such clean-up costs will simply increase the price gap between the property inside and outside the floodplain. The basic point is left undisturbed.)

Flood insurance might seem to be an obvious way for the floodplain property owners to protect themselves against losses. The problem private insurance companies face in making available flood insurance is that the likely flood victims know who they are, and they will be the only ones wanting to buy flood insurance. People outside the floodplain know they are safe. Why should they pay flood insurance premiums? Again, the problem of adverse selection (a form of the lemon problem) rears its head. The floodplain property owners are unwilling to pay more for flood insurance than their expected losses from floods. Hence, the insurance companies can't charge more than their expected payouts that will equal the victims' expected losses, which means the companies can't make a profit, if all they had to cope with was the problem of adverse selection. Insurance companies face the added problem of "moral hazard," or the tendency of policyholders to change their behavior, which in this case would mean putting more property at risk because their prospective flood losses are lowered due to their flood insurance coverage.

Because of the problems of adverse selection and moral hazard, if flood insurance is going to be provided, it generally must be heavily subsidized, which it is in the USA. Premiums of flood insurance policies written under the National Flood Insurance Program of 1968 are 35–40% of what the true risk premiums would be to cover expected damage. Accordingly, it should be no shock that in 2003, payments for flood losses amounted to a half a billion dollars more than the premiums collected.[2] The problem with so many government aid programs is that they force the Americans who paid premiums for their property outside floodplains to cover the losses of people who bought discounted flood-prone property. One must wonder, then, who are actually the victims, those who live inside floodplains or those who live outside them?

The point of following this line of argument is not to say that no aid should be provided. Rather, it is to stress that aid should be provided very judiciously and with great caution and restraint. If the losses of flood-prone property owners are fully covered by aid from, say, federal and state treasuries, the real benefits of the relief effort are likely be short-lived—not because the aid will dry up (pardon the pun) but because property values will adjust to account for the expected aid in the future. Prospective buyers of property inside and outside floodplains can be expected to take into account the expected aid for flood victims in their purchases. The demand for floodplain property will rise, as will its market value, in line with the expected aid. Future prospective owners of floodplain property will no longer get discounts for their expected losses on the floodplain property they buy. The expected (discounted) value of the future aid will be captured, in effect, in the current prices of floodplain property. The gainers from the aid will not necessarily be the owners who incur the losses when the floods actually occur (they've had to pay upfront, before the advent of the flood, a premium for their property because of the aid they receive), but rather the former property owners who receive a price for their property that was inflated by the prospective aid going to current or future owners.

In fact, when aid is routinely offered to victims of floods, it can actually raise the number of victims and the amount of their losses during floods. This is because of the problems of adverse selection and moral hazard. Knowing that all or a significant portion of their losses will be covered, more people will be willing to move to floodplains, to build bigger and more expensive houses there, and to stock them with more expensive furniture. They may even be less inclined to try to save their property in times of floods. They can also be less inclined to self-protect themselves with flood insurance, which means that flood insurance must be even more heavily subsidized to get floodplain property owners to buy the insurance. Why? They can expect some, if not all, of their prospective losses will be covered by disaster relief programs. Only by public policymakers and agency administrators (and charity groups) being extremely cautious and conservative in the allocation of aid can we reduce the perverse incentives inadvertently fostered by aid programs.

Victims of major natural disasters—whether in the form of floods, earthquakes, or hurricanes—receive a great deal of attention in the media and from government agencies because they are easy to identify and their numbers are large. They are natural candidates for government largess. However, many other people in the country are victims of a series of minor natural and man-made disasters, with their total losses often exceeding the losses of victims of major floods. Nevertheless, the victims of a string of minor losses are often ignored by government and the media, though their numbers are large, precisely because they are not so easily identified and their relatively small losses in each isolated minor disaster are not headline makers. We must be cautious in giving aid to the victims of floods because the aid may not be allocated evenhandedly across all victims of all major and minor disasters. Those who suffer unacknowledged minor disasters may actually be double victims, for not only do they lose when they endure their own losses in minor disasters, but they are also called on to aid the victims of major disasters.

Floods have a way of destroying property. Hard-headed thinking has a way of throwing cold water on emotional responses to losses that are suffered and widely reported. There is no clear argument against aid, but there are very good reasons for exercising considerable restraint, especially when many flood victims are fully capable of buying property outside of potential disaster zones, but choose not to do so. Unless carefully crafted, aid programs can create policy disasters that are no less threatening and damaging than the natural disasters themselves. Disaster aid that is routinely given and becomes expected by property buyers can entrap policymakers because, as noted, the future value of the aid can become captured—or to use the jargon of finance, capitalized— in the value of the property. When aid is capitalized in the value of the property, then any withdrawal of aid can undermine the value of the property, which means that the withdrawal of aid can destroy the market value of property as surely as can natural disasters.[3]

Our consideration of aid for flood victims elevates a lesson that has wide applicability: Prices today can capture expected gains and losses going forward. Change the streams of prospective current *and* future gains and losses on properties, and today's prices of those properties can capture the change.

HOUSES WITH AND WITHOUT VIEWS

This lesson lays open the folly in many widely heard and believed claims. Consider the often-repeated claim of real estate agents who glibly announce that "houses with views sell more quickly than houses without views." Perhaps that is sometimes the case (just as the opposite is sometimes the case), for reasons unassociated with the presumed value of the view, but should we expect the claim to be systematically reflective of the housing markets because of the difference in views houses have?

I have no qualms with the equally often-made claim that houses with views sell for higher prices than comparable houses without views. Of course, houses with views will sell for more—precisely because of the (presumed) value of the views of, say, the ocean or a mountain valley. (Similarly, no one would doubt that houses with views of garbage dumps will sell for less than houses without such views.) Indeed, we would expect comparable houses to have price differences that approximate the market value of the view, which will be affected by the relative scarcity of such views. The greater the abundance of (good) views, the lower the market value of views, and the lower the view premium that will be captured in the value of the property with views.

My question is, however, why houses with views should be expected to sell systematically *faster* than houses without views? If houses with views did sell faster, might we not expect their owners to hike their prices even more to slow the pace of their sales to the pace of sales for houses without views? Might not owners of properties without views lower their prices to speed up the sale of their properties?

Granted, there is one possible reason houses with views might sell more quickly, *but not so much because of the views in and of themselves* (without their implication for the value of the property). Because of their relatively higher prices, owners of houses with views might have more equity in their houses than do owners of houses without views. They might want to unload their houses with greater urgency because of the greater cost of delaying their sales, with the greater cost equal to the time-value of their relatively greater equity. But then, buyers of houses with views might be expected to be as reluctant to tie up substantial equity in a house, through a quick purchase, than the sellers are to get their equity out of their houses. Maybe buyers and sellers of houses with views have different discount rates—that is, they place different time values on tied-up home equity. Otherwise, we should expect, as a rule, the prices of houses with and without views to adjust so that their speed of sale is very close.

HOUSES OWNED AND RENTED

Consider another claim. "Buying a home is a better deal than renting an apartment. The interest on a home mortgage is tax deductible, and the value of homes can appreciate." I am sure every reader has heard the argument. If the argument carried the weight of truth that the proponents suggest, we must wonder about the sanity of the hordes of apartment renters around the country. Many renters can afford to buy their own homes but choose not to do so, for good economic reasons apart from the fact that they don't want to put up with the problems of maintaining owned homes. If there were a decidedly large tax advantage to buying homes, then we would expect two consequences that would narrow, if not eliminate, the relative value of owning a home vis-à-vis renting an apartment: First, the demand for owned homes would rise, along with their prices. Home sellers would capture much, if not all, of the tax advantage. Second, the demand for rental apartments should fall, along with their rents. Besides, people who press the argument about the tax deduction of mortgage interest often fail to acknowledge that owners of apartment complexes have mortgages, and they can deduct their interest payments from their rental charges. Apartment owners' tax advantage should show up, through competition for renters, in lower rents.

Granted, homeowners can see their property values appreciate, but they can also see them depreciate. Such downside risk should temper people's enthusiasm for buying the argument, stripped of qualifications, that owning a home is a better deal than renting. Moreover, if homeowners can be confident that their home values will appreciate, then surely the sellers can work from the same expectation, which means sellers can be expected to capture some, if not much, of the expected appreciation in the selling prices. Also, it makes sense to rent for a longer period than otherwise when renters expect housing prices to fall or even when they expect the appreciation of housing at some point in the future to spike upward. Renters, in other words, can be affected by what they *expect* to happen to housing prices in the future.

All of this is not to say that homeownership is never a better deal than renting. It is to say, however, that market-induced adjustments to prices help us understand a would-be puzzle, why so many people continue to rent in full knowledge of the ownership "advantages" they forego.

WHY RETIREMENT DOES NOT CURB THE RETIREES' FOOD CONSUMPTION

Many social scientists have observed what for them has been a puzzle: after retirement, people drastically cut their expenditures on all goods, but especially food. Indeed, two economists, Mark Aguiar and Erik Hurst, found that people's food expenditures rise from the time they are in their early twenties until their early fifties, but their food expenditures fall by 17% at retirement. While high-income people spend more on food and tend to eat healthier both before and after retirement, the food expenditures of all income classes decline markedly at retirement.[4]

Some researchers, finding even larger drop-offs in food expenditures, have concluded that the pre/postretirement drop-offs in food and other expenditures prove that people do not plan for their retirement very well. They've also concluded that people are obviously not as rational in their behavior as economists conventionally assume. If the subjective value of food declines with the amount consumed, the value of the last dollar spent on food postretirement has to be greater than the value of the last dollar spent on food before retirement. People could improve their welfare by consuming less food in their preretirement years and save more to boost their consumption of higher-valued food in retirement. Researchers inclined toward social activism have used the decline in retiree's expenditures on food and other goods to support their political case for forcing (or inducing) people to save more for retirement than they are inclined to save voluntarily.

Economists who have based their theoretical careers on the assumption that people are rational (or more rational than retirees seem to be) see the findings of people's lifetime consumption patterns as a major puzzle. Rational people should tend to even out their consumption of goods over the course of their lives, following what has been dubbed the "permanent-income hypothesis," which is based on the work of the late Milton Friedman, a Noble Prize-winning economist.[5]

The problem with this analysis is that it fails to recognize important points about prices and retirement:

- First, the *effective* prices of so many goods people consume are not captured by what's on price tags alone, mainly because things people buy are really inputs (or resources) into what people produce at home for themselves (a point stressed most prominently by economist Gary Becker, another Noble Laureate[6]). The prices of home-produced goods can rise and fall with the prices of inputs *and* the opportunity costs of people's time.

- Second, on retirement, people who retire knowingly give up some income to gain more time to do what they want. Retirees may have less income to spend on food, but they have more time to search out food bargains and to produce their own meals. This means that retirees' *consumption* of food can differ markedly from their *expenditures* on food.

Once these points are recognized and accommodated in analysis, perhaps people's lifetime consumption patterns are not the mystery (or as out of sync with rational precepts) we have been led to believe. Indeed, Aguiar and Hurst have found that after retirement people devote, on average, 53% more time to shopping for food and to preparing their own meals than they did before retirement.[7]

One explanation for why people increase their food expenditures through their early fifties is that they are substituting prepared foods and meals out for time-intensive and (because of the opportunity value of their time) higher-cost meals at home. Along the way, with less time spent searching for good deals on food purchases, they probably pay higher prices than they would have to pay if they had more time for searching out deals. When people retire, they will understandably become more price sensitive, since they will have more time to check out prices and features of alternative goods they want to buy and will thus have more knowledge of which goods have lower prices (given their qualities and features). One explanation for "senior citizen discounts" is that stores understand that seniors are more price sensitive, with the senior citizen discounts feeding declines in their *expenditures*, not their *consumption*.

Aguiar and Hurst have found, contrary to conventional wisdom, people's *consumption* of food remains more or less flat from their early twenties through their late forties but then trends upward, albeit slightly, through their early seventies (the last age the researchers have the necessary data to make the required consumption calculations). While it is true that retirees spend less on meals out than they did before retirement, the reduction is largely in expenditures at fast-food restaurants, not sit-down restaurants. Moreover, retirees do not tend, as a group, to lower the healthiness of the food they consume.[8]

Clearly, while people face difficult problems in planning for retirement, they seem to be doing much better than many people have surmised by considering misleading *expenditure* figures.

UNIVERSITY MISPRICING

Like so many other state-funded universities, my university—the University of California at Irvine—wants to believe that it can pursue higher academic standards through price controls on student and faculty housing. This on-campus housing will, supposedly, have the effect of indirectly subsidizing student education and faculty salaries. The presumption is that the subsidies can increase the "quality" (however the university wants to define "quality") of its students and faculty who can do great work on campus for the benefit of the rest of the world. Unfortunately, the university's controlled prices for student and faculty have had much the opposite effects of those intended. To be more direct, the implicit housing subsidies embedded in the price probably have in unexpected ways undermined the overall quality of the university's students and faculty.

STUDENT HOUSING SUBSIDIES

The University of California-Irvine provides a limited number of graduate students with on-campus apartments at monthly rental prices that are several hundreds of dollars below the rental prices in Irvine and other surrounding Orange County communities. For example, at the time of this writing in early 2007, a two-bedroom graduate student apartment on campus rented for $600 a month. A similar size nonuniversity apartment across the street from the university rented for $1,990. Two-bedroom apartments a mile down the road from the university rent for more than $2,500 a month, partially because the apartments are nicer, but also because the apartment complexes seek to price (potentially unruly) students out of their apartment complexes, increasing the net value of the apartments to the nonstudent residents who pay the premium rents.

The university argues that by controlling the prices of its on-campus apartments, it can attract better Ph.D. students from the best undergraduate programs in the country and can pay them less than otherwise for their teaching and research assistantships. Moreover, the reputation of the university will be enhanced by the high-quality graduate students who help UC-Irvine faculty do their top-academic-journal research and who after graduation go out into the academic world and develop stellar scholarly records of their own, reflecting academic glory back on the graduates' degree-granting university.

Although the university seems convinced that much of what it does represents a positive contribution to society, it may take more credit than it deserves for the success of its graduate students. After all, high quality graduate students might be able to build substantial scholarly records even if they got their advanced degrees elsewhere, making the marginal contribution of UC-Irvine's programs more debatable than the university might want to concede.[9] Indeed, if the university didn't offer the students the price break on housing, thus lowering the overall costs of their degrees at UC-Irvine, at least some of the graduate students might have chosen to go to more highly rated universities (say, Stanford or Harvard) with fewer benefits but with better graduate educations and, as a consequence, might have been, after getting their degrees, in a position to develop even more stellar scholarly records.

This line of argument suggests that the UC-Irvine rental subsidies could be marginally undercutting the extent of some students' career successes. Put another way, some students might be better off—given that with the rental subsidies they are able to maintain higher living standards while in graduate school—even though they might do less well in their careers were the rental subsidies not available. Alternatively, for those students whose parents are covering the graduate student bills of their children, the graduate student rental subsidies can show up in a higher living standard not for the students, but for the parents, with the parents' higher living standards captured, for example, in bigger and better cars or more frequent and longer vacation trips.

But then, there is a good chance that the university's rental price controls are themselves impeding the university's efforts to achieve the highest academic standards it can *with the available housing resources.* This is because with the rents well below market, graduate students have an incentive to "buy" more apartment space than they need, or at least more space than they would buy were they forced to pay market rents. A married couple with a child might rent from the university a two-bedroom apartment at $600 a month when one bedroom would do— if they had to pay the outside rental rate of $1,990 a month. Because of the subsidy, the available university land and floor space could be, and probably is, allocated among a smaller number of students than would be the case were rental rates set at market.

More importantly, graduate students get the $1,390 monthly subsidy for a two-bedroom apartment *only for as long as they are in school.* With the total housing subsidy tied to the students' length of stay, students are given a financial incentive to extend their graduate careers longer than otherwise, denying in the process the use of the limited number of apartments to other incoming students. Indeed, some married couples lucky enough to get one of the apartments have become "serial graduate students." After one spouse has strung out his or her graduate career for as long as possible, the other spouse applies for graduate admission, thus extending the couple's collection of the implicit monthly subsidies. As a consequence, 20% of the graduate students in the rent-controlled apartments have "squatted" in their apartments for twelve or more years.[10] Their extended stays no doubt have reduced the university's ability to attract good graduate students. The available housing has been taken by graduate student "squatters."

The university could easily remedy the "squatting" problem. The university could restrict the number of years students can stay in the apartments, but such a restriction has an obvious flaw: Some students in some programs need more time to finish their degrees than others. Would the university really want all students to be treated equally in terms of their tenure in student housing? If so, what should the restriction in years be? The number of years required to obtain a Ph.D. in management or the number of years required to get a degree in rocket science?

The university can rationalize the system by simply raising its rents to market levels. Those who valued on-campus apartments at less than the market rental rate, $1,990 a month, would look elsewhere for cheaper, far-removed-from-campus, and lower quality apartments, freeing university housing for use by students for whom location adds more value than the added rent. The squatting problem would go away, since students would not have the built-in subsidy incentive they now have to extend their graduate careers any longer than is really necessary. Apartments would be freed up for use by more and larger generations of graduate students who could be expected to complete their degrees in shorter time frames.

Now, it might be thought that the higher rental rates would scare off good graduate students. They could, and will, *if* there are no offsets to the higher rents set at market rates. Fortunately, the university could relieve the problems created by charging market rents simply by using its higher rental revenues to hike the payments made to students under its fellowships and teaching and research assistantship programs. That is to say, if the monthly rent for on-campus two-bedroom apartments is raised from $600 to $1,990, the university could award students $16,680 a year (12 × $1,390) more in scholarships or hike their pay by that amount under teaching and research assistantships. Granted, students may have to pay taxes on their additional income, but it should be stressed that the $16,680 in cash is worth more to students than the $16,680 embedded in the controlled rental prices, especially since graduate students typically have low incomes and are in low tax brackets. Cash would be preferred by students simply because the students would then have more choice over housing: they could decide to pay market rental rates for on-campus apartments or go off campus to comparable apartments at more or less the same rental rates. Of course, given that students could choose among on-campus and off-campus apartments, we might anticipate that the competition among housing developments on and off-campus would elevate the quality of apartments on campus over what the quality level would be when students have to take their subsidies only through renting on-campus housing. This means that by switching from in-kind/apartment embedded subsidies to cash subsidies, the university should be able to attract higher quality graduate students than with the in-kind rental subsidies.

Indeed, given that the cash is preferable to the embedded rent subsidy, the university can potentially raise the rent by $1,390 a month and then give higher quality students, say, $1,200 a month in cash with the result being that the students are better off than they would have been with the $1,390 a month in the rental subsidy. In this example, the university would then have $190 a month from each student given the cash subsidy to offer additional graduate students fellowships and assistantships. The shift from embedded rent subsidies to cash subsidies is a potential win-win university policy change for everyone.

Why then don't state universities like UC-Irvine change their rent policies? The best answer is that university officials haven't read this book. Better yet, because the price of education (as well as housing) is subsidized, university officials are protected from competitive market pressures to find the most efficient pricing policies, but I am hardly satisfied with these answers. I was in one of my university's many administrators' meetings in which the topic of the shortage of graduate student housing was a prominent item on the agenda. The administrators barked one after the other:

- "We need more graduate student apartments to attract more and better graduate students."

- "We don't like the way the limited supply of apartments is allocated across departments."

- "We have a shortage of teaching assistants because of the university's apartment shortage."

- "Too many students are in their apartments for far too long."

When I interjected how many of the voiced concerns could be attributed to the rent controls and explained how market-based rents combined with more generous fellowships and assistantship payments could partially remedy many, if not all, of the problems mentioned, the administrators paused, but in short order continued their complaining about the shortage of student housing, dismissing totally my proposal as "free-market ideology." My proposal has nothing to do with any ideology, free-market or otherwise. It has everything to do with getting prices right (even in institutions that are as socialistic in basic structure as public universities), and, in the process, advancing the university's declared goals.

But then, the meeting gave me good reason to question if this analysis of the issue was complete, mainly because even the graduate students on the committee summarily dismissed the proposal, which I had assumed they would eagerly support. Why? A potential answer came from one of the executive MBA students when I related the meeting and arguments made at the meeting on graduate student housing. The student asked an insightful question: "What percent of graduate students actually seek on-campus housing?" Just for the sake of following the logic implied in the question, suppose 40% of graduate students don't want on-campus housing, perhaps because they live in the area and have a working spouse with sufficient income to live away from campus (in a location closer to the working spouse's job, for example). Many graduate students might oppose the switch from the in-kind to cash subsidy system because the cash subsidy could be spread over far more graduate students, resulting in a substantial decrease in the subsidy going to students who are in a position to claim the in-kind/on-campus housing subsidy.

If instead of giving out cash subsidies, the university were to pass out "housing vouchers" (which give holders, say, 3 years of on-campus housing), then the vouchers could be sold by the students. Again, the housing rights would very likely be split among a larger number of graduate students, with the students who can claim the on-campus apartments receiving less in subsidies than they would receive under the current system. In short, these graduate students (who can be a majority of graduate students and who can be expected to be disproportionately represented on committees that consider the way the available apartments are allocated) have good reason to want to focus the subsidies on themselves through unlimited in-kind housing subsidies. In short, all of the grumbling about graduate student housing boils down to on-campus politics giving rise to some bad economics in the form of behavior-distorting prices.

FACULTY HOUSING SUBSIDIES

My university provides good analytical fodder for my classes and this book. This is because, like so many public institutions, it does many things that are not thought through, in this case the well-intended goal of providing faculty with reasonably priced housing (in a very high housing cost area of the country).

The university arose rapidly in the late 1960s on 1,500 acres of orange groves and pastures in Orange County, California. The university's land was given to it by the Irvine Company, which owned, in the early 1960, about 180,000 acres of prime Orange County land and which expected a new University of California campus to increase the commercial and residential value of the Irvine Company's remaining acreage. This remaining land would eventually be developed into the City of Irvine, which at this writing has close to 200,000 residents.

By the mid-1980s, having expanded to a student body of more than 10,000, UC-Irvine was facing growing pains, one of which was peculiar to the then (and for decades since) "hot" housing market in Southern California. The price of housing in Irvine and surrounding communities was rising far more rapidly than were the state-controlled salaries of UC-Irvine professors. To continue to attract and retain top-quality faculty (in pursuit of its goal of becoming one of the top 50 research universities in the country, which it has since achieved), the university came up with an idea that many administrators and faculty members at the time considered ingenious: the university could use a few hundred of its then unused acres on the perimeter of its core campus to build faculty housing. The single-family houses and townhouses could be sold to faculty members at the cost of construction (not market prices). If the difference between construction costs and market value of a 2,000 square-foot house was $100,000 in 1990, the embedded subsidy on the house itself then amounted to about $6,000 a year (assuming a mortgage interest rate of 6%).

By the dictates of the land grant and charter, the university could not legally sell its land to existing or prospective faculty, but it could legally lease the land to the faculty member for 99 years at far below market—that is, subsidized—rates. A lot that might cost $250,000 in the Irvine community adjacent to the university property in 1990 might be leased to a faculty member as if the lot cost only $30,000. At 6%, the $220,000 differential between the actual land cost and the university lease value represents a covert annual subsidy of $13,200, an add-on to the faculty salary.

Total house and land subsidy in our example (which was close to reality in 1990): $19,200 a year ($6,000 in house subsidy and $13,200 in land subsidy), the equivalent to about a 50% increase in effective income for a full professor in the humanities and a 20% increase in effective income for a full professor in the business school. Again, the presumption was that the subsidy would enable the university to continue growing with better faculty than could otherwise be hired.

To make the plan work, the university, however, had to incorporate some resale restrictions. Otherwise, the initial new faculty members who bought their houses at cost (and leased the land far below market rates) could be expected to turn around and sell their houses to other incoming faculty or to people in the community at market prices. The faculty could run off with the capital gains that were supposed to go to a series of faculty members over the following decades. There were five major kickers to the housing contracts the university signed with faculty residents in what has become known as "University Hills" (and sometimes referred to as the "Faculty Ghetto"):

- First, the faculty members who bought University Hills homes could only resell their homes for what they paid for them, plus an appraised value of any improvements and an appreciation in the initial value of the homes equal to the increase in the consumer price index between the date of purchase and the date of resale. For example, if a professor bought a house in 1990 at $200,000, never improved the house (beyond regular maintenance), and wanted to move to another university in 2007, that professor could only resell the house for $318,000 (given that the CPI rose by about 59% between 1990 and 2007).

- Second, the professor had to offer the house for sale first to existing or prospective UC-Irvine faculty members. If no faculty member wanted to buy the house, then the house could be offered to staff members. Only when no faculty or staff member wanted to buy the house could the house be offered for sale to people outside of UC-Irvine, and then the "outsiders" would be required to follow the resale restrictions. (Because there has always been an excess demand among UC-Irvine faculty and staff members, no University Hills track house has ever been sold to an outsider.)

- Third, faculty (or staff) members who leave the university without retiring from the university system have to sell their houses, following the above rules. However, retiring faculty members can stay in their houses for as long as they live. Their surviving spouses can also remain in their University Hill houses for as long as they live.

- Fourth, faculty members can rent their houses, but for no more than 2 years in sequence (which means that faculty members could only rent their houses when they go on sabbatical or on leave from the university).

- Fifth, faculty members' University Hills houses must always be their "primary" residence (which effectively requires faculty members to live in their houses more than 50% of any year).

University Hills housing was initially, no doubt, a factor in attracting good faculty members because of the implied housing subsidy, which is, effectively, an expensive fringe benefit. However, the improvement in faculty quality probably has not been as great as the embedded housing subsidy, taken by itself, might imply. This is because the subsidy has likely taken the pressure off the State of California to raise faculty salaries and other fringe benefits. That is, faculty salaries and fringe benefits have risen in real dollar terms over the last decade but, very likely, not by as much as they would have risen had the housing subsidies not increased the supply of qualified faculty members and held faculty salaries and fringe benefits down (below what they would otherwise have been).

However, given points made in our earlier discussion about the relative value of in-kind and cash subsides, it should be noted that to attract and keep any given quality faculty, salaries need not have been raised in 1990 by as much as the housing subsidies, which in the above example was the equivalent of $19,200 a year. This is because the housing is an in-kind benefit that is tied to the consumption of a given good, housing. A salary increase of $19,200 would surely be preferred by most existing or prospective faculty members over the exact same in-kind, housing subsidy. As with the student renters, the faculty member could take the cash, buy a house in University Hills, or use the cash to buy elsewhere in the area—or, for that matter, use the cash to buy a boat or car. If they bought houses in the surrounding communities, they could also gain from the ongoing housing appreciation in the area.

As it happened, the housing subsidy was and remains an inducement for faculty members to buy bigger houses and lease bigger lots than they would have bought had they been required to pay market prices for their square footage. Of course, this means that the available land has not likely accommodated as many faculty members and their families over the years as it could have accommodated were market pricing used.

The embedded housing subsidy has also likely caused faculty members who bought the larger houses to hang on to them longer than they otherwise would. Outside of the subsidized University Hills development, many parents whose young adult children move to places of their own do what comes naturally: they downsize their housing. The downsizing process not only reduces the housing costs of the homeowners with contracting family sizes, it also frees up the stock of larger houses to be bought by younger parents with growing families.

In University Hills, however, that process has been abated for two reasons:

- First, the large houses owned by downsizing families are cheaper than they would otherwise be. So, the downsizing families can be expected to continue to retain their "excessive" square footage, as has been the case. (There was one notorious case of a wife of a deceased prominent faculty member who held onto her five-bedroom/three garage house for years until she died in her eighties, in spite of the fact she lived only in the downstairs part of the house.)

- Second, since appreciation of the faculty housing has been capped by the rise in the consumer price index, faculty members with contracting families often have limited equity in their houses and, hence, have less to gain (than they would if their houses had been market priced) by moving to smaller and

> cheaper houses and diverting their equity to other asset forms, for example, stocks and bonds.

One unfortunate, and unanticipated and unintended, result of the rules of ownership and resale is that the university has begun to lose younger faculty members to other universities because they can't move to larger houses in University Hills and can't afford to buy larger houses in the surrounding Orange County communities, where housing price increases have hardly been restricted to the rise in the consumer price index. The annual rise in the price of housing in Orange County since 1990 has been one of the highest in the country.

Indeed, between 1990 and 2007, the median housing price in surrounding Orange County communities appreciated by more than four times the rise in the consumer price index. This means that the professor who bought the $200,000 house in University Hills in 1990 could only sell the house for $318,000 in 2007, but if the professor did sell out, he or she would have to shell out in 2007 perhaps $1.2 million to $1.5 million to buy a comparable house in the surrounding Irvine community. The implied housing subsidy has, accordingly, jumped dramatically. Assuming a comparable house in the surrounding community is only $1.2 million and an interest rate of 6%, the price differential between inside and outside University Hills, in round numbers, is $900,000, or $54,000 a year in 2007—a subsidy, I might stress, that is collected year after year *only if the faculty member stays put.*

The growing disparity between the prices of houses in University Hills and the surrounding communities has resulted in many faculty members holding onto their houses after they retire. With the shortage in housing in University Hills, the university has used the available housing stock strategically, often offering the available houses to much sought-after distinguished professors on the so-called "priority list" who tend to be in their late forties and fifties, if not sixties. Many such faculty members can expect to spend more years in their houses retired than they spent in their houses during their active teaching and research year at UC-Irvine.

Because of the growing spread between the prices of houses in University Hills and in surrounding communities, the housing deals offered years ago have been described as "golden handcuffs." Many faculty members have no choice other than to stay put. Other faculty members who relocate after retirement to other parts of the country have an added incentive to use their University Hills homes as second homes (although they have to make sure that they follow the letter of the definition for "primary residence"). After all, their capped resell prices make their houses cheap places to own and to use on trips back to Southern California to enjoy the close-to-perfect weather no more than five miles from the Pacific Ocean, as well as the virtually bug-free environment (factors that help explain why housing prices are so much higher in Southern California than in most other parts of the country).

When I retire, you can bet my wife and I will hold onto our University Hills house for as long as either of us are alive. Why? First, my wife and I have a college-age daughter who thinks of our house as her homestead, a place to which she wants to return as she goes through her adult life. Second, when we decided to buy the house in University Hills, we freed up funds that were invested in securities. These financial assets have appreciated so that we could cash them in and buy another place in the community, but why should we? We would then be narrowing our investment portfolio with a larger portion being invested in housing, which implies added risk. More importantly, the shift of our assets from financial securities to housing would mean a shift of "income" from cash that can be used to buy many things to a single-purpose in-kind benefit, housing. We deem the cash from our investments more valuable.

The university now realizes it is in a housing bind, one that could have been anticipated with a little hard-nosed economic thinking, but, of course, wasn't. University Hills is "graying" as more and more faculty members retire and do what I plan to do—retire in place. Indeed, some faculty members jokingly call University Hills a retirement community—an academic "Leisure World" of sorts—because of the growing number of aging faculty in the neighborhood with canes and walkers. For the time being, the university has been able to bring younger faculty into the neighborhood, but only by building more houses. However, the available land for additional University Hills homes will soon run out—perhaps in as little as 5 years, long before the university expects to stop the growth of students and faculty—after which the graying of University Hills can be expected to accelerate, especially since the housing program will by then have been in place for 30 years, a tenure of service often sufficient to achieve maximum benefits from the university's defined-payment retirement plan.

What can be done to relieve the growing housing shortage (there are over 600 people on the waiting list at this writing)? Unfortunately, not much—short of allowing current homeowners to sell their houses at prices above the current pricing caps. If faculty members can only sell their houses well below market, where will they go? How will they pay for houses in the community?

If the university allows faculty members to sell at market (so that they can move out), then it might have a public relations problem of some magnitude, given that current homeowners would be allowed to pocket the capital gains associated with living on state property. But I don't see why such would be considered any more unfair or inappropriate than the current system that allows identified faculty to garner the value of state property by continuing to live where they are.

Then, what other options does it have, once it uses the last acre of its "free" land—if it truly wants to continue to build the quality of its active faculty, not its retirees? One course the university has taken has been to elevate reminders of the "primary residence" requirement by investigating several supposed violations. Faculty members have also become neighborhood police squads, reporting on retired neighbors who do not appear to be meeting the residency requirement on the grounds that they don't seem to live in their houses very much, arguing that that it is "unfair" that unused houses are denying young faculty cheap housing. In other words, the price controls will make more and more faculty members neighborhood snoops and nannies, hardly an anticipated and intended consequence. But there is a question the nannies will have to ask themselves: Are the faculty members who use their houses only a few months of the year depriving young faculty members housing any more than the aging retirees (and their spouses) who continue to squat in their houses for decades after they retire? Did they not pay for the right to use their houses on a limited basis through their active years by suffering salaries below what they would have demanded, absent the housing benefit?

The solutions may now be limited. One possible solution might be to allow faculty members to rent their houses to other faculty members for long stretches of time. At least such rentals would make more houses available to more young faculty members for longer periods of time. That is, such greater leniency of the rental rules can result in greater use of the available housing stock.

In the end, the university might simply have to use donated or state funds to buy out professors from their University Hills houses at something above capped rates just to free up houses for the (supposedly) higher goal of continuing to expand and upgrade its faculty through the coming years. And why shouldn't it? The university has demonstrated that it will use an extraordinarily valuable university resource—land—to build its faculty. Why not use its donated real dollar resources to continue to do the same? Certainly there will be a cost. But the land used for housing was hardly ever "free," because the university could have leased the property (and any commercial units built on the land) and used the rents collected to pad faculty members' salaries (or do any number of other great things).

Now, if the university wants to free up houses, it will have to incur a cost of some magnitude. No escaping that fact of economic life. However, the cost of faculty buyouts will not likely have to be as great as the differential between housing prices in University Hills and surrounding communities. This is because some unknown number of retiring faculty members will want to retire elsewhere in the country, perhaps in places like Utah and North Carolina where housing prices can be higher than in University Hills but lower than in Orange County, or the rest of California. The university simply can offer a buyout price equal to a comparable house in the faculty members' retirement destinations. Granted, some retiring faculty members can be expected to game the buyouts system by proposing to retire in places with high housing prices, but such problems can be overcome with contractual provisions, at least to a degree, that payment will only be made if the faculty member relocates to where he or she indicates (and remains there for some specified period of time).

Alternately, the university can use a buyout auction system similar to the one airlines regularly use when they are overbooked. When the airlines need passengers to release their seats to people on the wait list, flight attendants will usually announce a "low" buyout price (say, a seat on the next available flight to the person's destination plus another roundtrip ticket to any of the destinations served by the airline within the continental United States). If an insufficient number of passengers accept the flight attendants' offer, then the deal can be sweetened (say, to two tickets to any destination in the world flown by any airline). The university can simply gradually up its buyout premium until the desired number of houses is freed up. Faculty members thinking about moving will be put into something of a competitive quandary that can cause them to reveal something close to their true minimum sellout price. When faced with the initial offer, you can imagine a faculty member thinking, "Should I take the offer on the table now or wait for a better one? If I wait for a better one, I could be left out in the cold, not able to get a premium price at all, because others have taken all available buyouts."

Okay, you don't like to apply market solutions to universities. Can you give me a better one? Renege on past-signed contracts and force aging faculty members to downsize their houses? That's a surefire recipe for lawsuits that can cost the university dearly. Suppose we limit by contract the years that newly arriving faculty members can stay in their houses. The university could also force new hires to accept a contractual provision that requires them to sell out when they retire. All you have done through such provisions is lower the value of the housing fringe benefit, which smart prospective faculty members should surely be able to figure out—*if university administrators making the rule change can figure it out.* Contractual limitations on the use of houses will have a way of feeding into new faculty members' starting salaries (or other fringe benefits) that will be higher than they would be without the housing forced-resale restrictions.

If only the university had thought through these pricing issues 30 years ago—if it could have.

CONCLUDING COMMENTS

There is a theme running through the discussion of various pricing puzzles in this chapter: "You can't fool Mother Nature, and you can't fool market forces" (at least not for long). Market prices for tradable goods, especially those with some durability like cars and houses, have a way of capturing the goods' disadvantages and advantages—and *changes* in those advantages and disadvantages. So it is that new car prices drop substantially when the cars leave the dealer's lot for the first time, partially because of the inability of the buyers (relative to dealers) to make a resale market for the cars they just bought. And new-car buyers need to understand that used-car buyers won't be fooled systematically into believing that used cars available for sale, as a group, are likely to have the same risks of repairs as new cars sold by dealers. If they are fooled, the pain of their purchases will no doubt lead them "as by an invisible hand" (Adam Smith's pat phrase) to correct the error of their buying ways. That is to say, the price differential between new and used cars can be expected, at least eventually, to reflect not only the wear and tear that goes with the normal use of cars, but also risk cost that goes with the prospect of used cars being lemons (or more defective than cars that people keep).

Similarly, if house buyers see value in views, that value will be reflected in the prices of houses with views. Prices, in other words, will absorb some (not necessarily all) of the value of the views, which is a solid explanation for why many people who value views don't seek properties with views (and often seek properties with big negatives, for example, an occasional natural disaster).

Also, this chapter has sought to drive home an easily overlooked lesson: when we try to help victims of natural (or even unnatural, for example, workplace) disasters through public aid, some, if not all, of the value of the help will be captured by hikes in the prices of assets owned by the victims. The aid that policymakers provide can also constrict future changes in public aid policies. Once the aid for natural or manmade disasters is captured (or capitalized) into the prices of property, then any withdrawal of the aid can give rise to a "disaster" of its own, given that the aid withdrawal can undermine the value of property as surely and as completely as the natural and manmade disasters that gave rise to the aid in the first place.

NOTES

[1] Akerlof (1970).

[2] Hecker (2003, p. 2).

[3] For more discussion on how policymakers can become constrained in their policy options by the so-called "Transitionary-gains trap," see an insightful article by Gordon Tullock (1975).

[4] Aguiar and Hurst (2004).

[5] Friedman (1957).

[6] Becker (1965).

[7] Aguiar and Hurst (2004).

[8] Aguiar and Hurst (2004).

[9] Dale and Krueger (2002).

[10] This data comes from reports made to the Graduate Council when I was a member.

Chapter 3

WHY SALES

Why do retail stores use seasonal (after-Christmas) and intermittent ("manager's blowout") sales over the course of the year? Answers to such questions are no doubt many, given the diversity of researchers and practitioners in economics and marketing who have worked on them. However, almost everyone is agreed that many sales (and other forms of promotions covered in following chapters) are founded on two economic lines of argument, "price discrimination" (charging different consumers different prices for different units) and "peak-load pricing" (a special category of price discrimination that involves charging higher prices during hours and days of heavy demand and lower prices at other times).

As will be argued, the economic theory of price discrimination presumes that retailers who use sales (and all other firms that price discriminate) must have some degree of monopoly power that they are exploiting via differential prices (a claim I accept in this chapter to explore key points but dispute for some market conditions discussed later in the book). More importantly, as we will see, an investigation of the economics of price discrimination can provide an explanation for a host of other differential pricing strategies, including (but hardly limited to) scholarships provided by universities and colleges, airline fares, soft drink prices at fast-food restaurants, adult/children prices at the movies (a topic to be considered in Chap. 4 on popcorn prices from which this book draws its title), and coupons (considered in Chap. 5), as well as annual and seasonal sales at department stores. First, we need to lay out the economic foundations of price discrimination methods in general. (Readers who are steeped in price-discrimination theory can skip the following section.)

CHAPTER 3

PRICE DISCRIMINATION THEORY

Economists and marketers have long argued that firms can be more profitable by charging different consumers different prices rather than charging one uniform, market-clearing price.[1] That is, firms should price discriminate wherever possible. Here, we will seek to understand the underlying economic logic of that position.

NECESSARY CONDITIONS FOR PRICE DISCRIMINATION

Price discrimination has a commonly understood definition: It involves setting multiple prices for the same good across consumer groups and across time periods. Economists and marketers have understood that price discrimination requires that two conditions be met. First, a firm interested in price discrimination must have some degree of market (or monopoly) power, or the ability to choose among various price-quantity combinations (a claim, as noted, I will consider critically for some forms of price discrimination, most notable in the printer/ink cartridge market). As noted in our earlier discussions, there is a much-heralded "law" in economics, the *law of demand*, or the assumed inverse relationship between the price of a good and the quantity buyers are willing and able to buy of the good (assuming all other market forces remain unchanged). If the price of the good rises, then less of it will be bought, and vice versa. Buyers might be willing to buy 1 unit at a price of $9, 2 units at a price of $7, and 3 units at a price of $5, and so forth. The seller must be able to search through the available price/quantity combinations with the goal of choosing that combination that maximizes profit. If the seller cannot do that—or must take the price dictated by competition, say, $5, as is the case in the wheat market—price discrimination is obviously not possible (except in some identified cases).*

Second, the product sold must not be easily resold (or resold at low costs). If a product can be resold with relative ease (or at low or zero cost), then buyers who are offered the product at a low price can turn around and resell the product to buyers charged a higher price, with the reseller pocketing a profit in the (arbi-

* For a more detailed discussion of the law of demand, and its graphical representation, see my textbook written for MBA students with Dwight Lee (2006) or consult video module 2.1 I have done on the law of demand at http://media.merage.uci.edu/McKenzie/Modules.html.

trage) process. If a publisher were to try to sell its economics textbooks to students at one university for significantly less than to students at another close-by university, students at the first university would soon learn that they could buy more books than they need and resell them at the second university. If students don't discover the profit opportunity from arbitrage, then surely one of the many used textbook buyers who prowl the hallways of faculty office buildings buying up "comp copies" of textbooks would not likely hesitate becoming textbook arbitragers among school bookstores, provided, of course, the price differential were sufficient to cover the resell costs. Alternatively, the students at the second university could walk or drive to the first university to buy the textbook. (We will return to the issue of price discrimination in the textbook market after we have developed a way of thinking about how firms can charge different consumer groups different prices.)

PRICE DISCRIMINATION AMONG BUYERS

If different buyers are willing to pay different prices, then the seller can make more profit by charging different buyers different prices. For example, following through with our earlier example of buyers willing to pay prices of $9, $7, and $5 for the first, second, and third units of the good, then the seller can take in only $15 in revenues if a price of $5 is charged for all 3 units. However, if each buyer of the first through third units is willing to pay the prices indicated—$9, $7, and $5—each for a unit, then the seller can obviously make more money by charging the individuals those prices. Total revenue will then be $21 ($9 + $7 + $5), and profits will rise by the same amount as revenues, $6. This is the case because under both pricing strategies, the production run is the same, 3 units, which means production costs do not change with a switch from a pricing strategy of a constant price, $5, for all 3 units to a strategy of price discrimination among buyers, different prices for the different buyers.

Again, if the firm has no choice over price to be charged and/or if the good can be readily resold, then price discrimination will not work. If the firm offers the good to one buyer for $5 and another for $9, then the $5-buyer will buy two at $5 and resell one unit to the first buyer for something less than $9. Ditto for resells to the buyer charged $7. Of course, I must add the caveat that the cost of resell in each case has to be less that the difference between the buyer's purchase price, $5, and the resell price. This type of price discrimination—different prices charged different buyers—abounds in the world we all encounter on a daily basis.

College and University Scholarships. Colleges and universities are renowned for providing students with "scholarships," supposedly all distributed for "merit" and "need." Without questions, some undetermined amount of scholarship money is allocated for those intended purposes. However, private colleges and universities often charge extraordinarily high prices (now, often more than $40,000 for tuition and fees, and room, and board), and many of them often grant more than half of their students some form of "scholarship."

For example, in 2007, Amherst charged $41,600 in tuition, fees, and room and board and provided need-based grants to 78% of incoming students that averaged $29,400. Duke charged $41,200 and provided need-based grants to 86% of incoming freshmen that averaged $24,000. In a survey of 107 private colleges and universities, the *New York Times* found that 95% gave more than half their students need-based grants.[2] We have to wonder why so many students are meritorious and needy with their scholarship awards being handled on a case-by-case basis.[3]

If merit and need explained their prices and scholarships, why don't the universities just lower their prices and save the administrative costs? The fact of the matter is that colleges and universities, especially private ones, use scholarships as a method of price discriminating and "maximizing revenues," a buzzword often used by admissions officers (not just economists). They post a high price for all, and then grant scholarships based on the universities' estimate of the difference between their posted prices and the amount students are willing and able to pay. Indeed, the spreading acceptance of price discrimination among colleges and universities helps explain, as we shall see, the dramatic increase in the average tuition at four-year private colleges and universities during the last half century. During any 17-year period between 1958 and 2001, tuition at the nation's colleges and universities rose 1.2 to 2.1 times the rate of inflation.[4]

Indeed, colleges and universities often determine the allocated scholarship by asking prospective students (or, perhaps, more accurately, their parents) on the financial aid applications exactly what price would cause them to matriculate. The universities then simply send out a congratulatory letter, announcing the "scholarship," which happens to be close to the difference between their posted prices and the prices the prospective students indicated would cause them to matriculate. And admissions officers are willing to negotiate on price, as indicated by the sentiment in a letter that Carnegie-Mellon University's admissions office sent applicants, "If you received a financial-aid package from us that was not competitive with other offers, let us know."[5]

Admissions officers, of course, love to have students apply for "early admission," which means that students can be accepted as early as, say, the November before their following fall enrollment. To validate their early admission request, early-admit students must agree to turn down all future acceptances. Such an argument means that early-admit students are less likely than other students to be offered scholarships—*because* they have declared themselves to be willing to pay the posted price if admitted and have cut off later lower-price options. Effectively, early admit students give up any bargaining power they might have, and consequently likely pay a higher price, than students who do not ask for early admission.

Admissions officers have also found that prospective students, especially in-state ones, who visit their campuses before applying or who affirm their desire and intentions to attend once they have been put on "wait lists," are less price sensitive than others. As a consequence, they are less likely to be offered scholarships.[6]

Price Reductions over Time. The pricing strategy used by universities in their scholarship allocations can also be found in the sale of electronic gadgets, for example, the Apple iPhone and USB thumb drives. We noted at the start of the book that Apple introduced two models of the iPhone in late June 2007, a model with 4 gigabytes of storage for $499 and one with 8 gigabytes at $599. By the first week of the following September, Apple terminated sales of the 4-gig model and lowered the price of the 8-gig model by $200, or to $399.

Thumb drives have regularly been introduced over the last few years with each successive generation having a greater storage capacity (128 megabytes, 256 megabytes…1 gigabyte and so on). Generally speaking, a given size drive has been introduced at a relatively high price, for example, several hundred dollars for the first 1-gig drive, only for the price to decline precipitously over following months. In April 2006, the online price of a 1-gig Imation Clip Flash Drive had fallen to just under $60. By August, the price had fallen to about $35. At the time of this writing (August 2007), the same drive could be bought for $19.86, less than a third its price 16 months earlier.[7]

Without question, some of the price decline in USB flash drives over time can be attributed to cost savings from technology improvements, growing production runs, and growing competition in the industry. Still, it makes sense for producers in markets with any degree of competitive imperfections, to introduce their products at a high price, and sell to buyers who are willing to pay high prices, and then lower their prices over time to appeal to people who won't pay the initial high prices (and are further down the demand curve for the product). Seen from this perspective, the price reductions are not the major source of increased expected profits from charging declining prices with time. Rather, the major source of expected added profits is the initial high prices that could not be extracted if sales started at lower price levels, because those willing to pay the higher prices would buy at the lower introductory prices. The price reductions with time can be planned and scheduled when the firm's production plans and release date are set.

Admittedly, this form of price discrimination is necessarily imperfect. This is because buyers can begin to *expect* price reductions with time. Some buyers *willing* to pay the initial high prices will learn to hold on their purchases, but their shifts in purchases will leave the impatient buyers willing to pay the initial high prices all the more exposed to high initial prices, because of their impatience and price insensitivity. The delays in purchases of some buyers with moderate price insensitivity can cause the posted price to rise, with the seller taking advantage of buyers who, by their failing to delay their purchases (or to fail to pay careful attention to firms' pricing strategies over time), reveal their high price insensitivity.

PRICE DISCRIMINATION WITH INDIVIDUAL UNITS BOUGHT BY BUYERS

Our earlier discussion of the market demand—with prices of $9, $7, and $5 for the first, second, and third units, respectively—can be the demand covering three different buyers, as discussed—or it can be the demand for a single buyer who sees the value of successive units of the good falling as more of the good is bought and consumed. That is, the value of the first unit is $9 to the buyer; the second, $7; and the third, $5. If the firm can structure sales so that the buyer pays those prices for the different units, then the firm can, of course, earn more profits than would be earned if the three units were sold for one price, $5. The decline in prices of additional units sold is often seen as giving the buyer a "break" on the price for additional sales. Alternatively, this pricing strategy can be viewed as a hike in the initial price, from which the price "breaks" can be given.

Drink Prices at Restaurants. Stores charging different prices for different units sold can be seen everywhere. Jack in the Box, a fast-food restaurant, offers customers three sizes of soft drinks: small (12 ounces), medium (20 ounces), and large (32 ounces)—with prices of $1.39, $1.85, and $1.95. This means that the cost per ounce for the small drink is 11.6 cents. The cost per *additional* ounce (over the ounces received in the small drink) on the medium drink is 5.8 cents, and the cost per *additional* ounce on the large drink is .6 cents.

Note that the restaurant is not giving high school football players who buy the large drinks a price break, which the drop in the marginal price per ounce might suggest. Even when they buy the large drinks, the players are still paying $1.39 for each of their first 12 ounces. They get the price break on the additional ounces, but only then because the restaurant believes that it has to lower the marginal price to entice the players to guzzle more ounces. The restaurant makes more profit off the large drinks than the small ones. This is because restaurant makes the rather large profit off the first 12 ounces, plus some smaller profit off the additional 20 ounces in the large drink.[8]

Other Products. The kind of pricing strategy for drinks (hamburgers, fries, and virtually all other menu items) employed by Jack in the Box is found throughout the fast-food and sit-down restaurant industry, and the same strategy can be found in pricing of popcorn and candy bars whether sold at movies or in airports. The pricing strategy is no less common in grocery stores whether the items are cans of beans, rolls of paper towels, packages of candy, or cartons of milk. Some of the price differences for various sizes can be chalked up to differences in cost of packaging (a widely recognized explanation), but some of it can also be chalked up to the fact that stores are "walking their customers down their demand curves" (a not-so-appreciated explanation). Just how rapidly the price declines is, of course, dependent upon how responsive—or *elastic*—customers' demands are. The lower the responsiveness of buyers to a price cut—the lower the *elasticity* of demand (or the higher the *inelasticity*) of demand—the more rapid the expected decline in the price for the marginal units of the good.

MARKET SEGMENTATION

Our discussion to this point has a theme: sellers would love to be able to figure out the demands of individual customers, carefully crafting their prices so that each and every customer pays the maximum price he or she is willing to pay for each and every unit. The last thing a seller wants to do is charge everyone the same price. Often, however, sellers must do that, but only after no other pricing or promotion strategy can be devised. The finer the price discrimination among buyers and the units sold, the better for the seller.

It might be easy to view price discrimination as a strategy *option* that can be taken or set aside—and it is that to a degree, but only a degree. This is because price discrimination can add to firm profits, as explained, and therein lies a compelling reason firms can be *pressed* (if not *forced*) to price discriminate. If firms that can price discriminate don't do so, their stock prices will be suppressed below what they could be because profits will be lower than they could be with price discrimination. In the absence of price discrimination, savvy investors can buy the stocks at low stock prices by way of a friendly or hostile takeover, change the firms' pricing structures to include price discrimination, and sell their stocks at a capital gain as the firms' stock prices rise to reflect the greater profit with the installed price discrimination strategies.

Many firms will not be able to adopt the kind of finely tuned pricing structures implied in our foregoing examples. The problem is that figuring out the demands of individual customers and charging each customer a different price for different units can be costly, or the costs can exceed the greater revenue potential from price-discrimination strategies. It is very tough for many restaurants, for example, to identify the price sensitivity of individual customers (say, by their looks or dress) as they walk through the doors. This doesn't mean that they can't price discriminate; they only have to develop a less ambitious strategy than charging every customer a different price for each unit bought.

One such strategy can be to recognize that different *groups* of buyers have different demands, with different price sensitivities, which means that firms can charge the different groups different prices according to the groups' different price sensitivities. Consider the problem of price discrimination through so-called *market segmentation* in the simplest of possible cases, with buyers being divided into only two groups. One group of buyers—Group A—is highly *insensitive* to price changes. They will buy more when the price falls and less when the price rises, but the changes in both directions will not be all that great. The other group—Group B—is highly *sensitive* to price changes, meaning that a price change, up or down, will lead to a relatively large change in the quantity purchased.

To start, suppose that the producer of widgets is selling a total of 130 units at the same price, $1.50, to both Groups A and B. Members of Group A buy 60 widgets, and members of Group B buy 70 widgets. If the price of widgets is raised by a third to $2 to Group A, sales will go down by only a sixth, or by 10 widgets, from 60 to only 50 units. The seller gains revenue in raising the price when the price increase is relatively greater (in percentage terms) than the quantity reduction. The producer had initial revenues from Group A of $90 when the price was $1.50 and the quantity sold was 60 ($1.50 × 60 = $90). When the price to Group A is raised to $2, revenues from that group rise to $100 ($2 × 50 = $100).

Suppose the producer takes the 10 widgets not sold to Group A and sells them to Group B by lowering the price for Group B from $1.50 to $1.45, a drop of slightly more than 3%. Sales, however, rise by 14%, from 70 to 80. Sales revenue drawn from Group B also rises from $105 ($1.50 × 70 = $105) to $116 ($1.45 × 80 = $116).

The producer has increased profits by shifting the 10 widgets from Group A to Group B since production costs must be the same, given that output remains steady at 130 widgets. However, total revenues from both Groups A and B have risen—with total revenues going from $195 ($90 + $105 = $195) to $216 ($100 + $116 = $216). Since production costs are unchanged, the shift in sales from A to B increases firm profits by the increase in revenues, or by $21 ($216 – $195 = $21).[9]

The producer should obviously raise the price charged Group A and shift sales to Group B so long as revenues from both groups rise. A little less obviously, the producer should go further and continue to shift sales of widgets from Group A to Group B so long as the combined revenues from both group rise—which is to say, so long as the revenue from both group rises *or so long* as the rise in revenues from either group is greater than the fall in revenues from the other group.

The working rule for the price discrimination is probably now transparent: The greater the price sensitivity of the group—or the higher the elasticity of demand—the lower the price. Conversely, the greater the price insensitivity of the group—or the lower the elasticity of demand—the higher the price. If customers can only be put into two groups—or *market segments*—then the more price-sensitive customer group (Group B) should be charged a lower price than the less price-sensitive group (Group A). The difference in prices charged the two groups will reflect the difference in the groups' price sensitivity. The greater the difference in price sensitivity of the two groups, the greater the price differential.*

Tailoring prices to the price sensitivity of buyer groups is commonplace. Consider these examples:

- The prices of McDonalds' drinks and hamburgers are higher in airports (as illustrated for members of Group A) than in places around town (for members of Group B).[10]

- Continental (and every other airline) charges passengers who book their flights early (three or more weeks in advance) and who have Saturday-night stayovers (Group B) less than it charges passengers who book their flights just before they leave (Group A).

* If you need a more intense discussion of price discrimination, see Chap. 11 in McKenzie and Lee (2006) and the video module 11.5 I have done on price discrimination by market segments at http://media.merage.uci.edu/McKenzie/Modules.html.

- In 2006, Apple charged buyers of its all-white MacBook (Group B) $1,399. The company charged buyers of its laptop configured the same way but in solid black $1,499 (Group A), a difference of $100. In 2007, we can only surmise that the relative price-sensitivity of white laptop buyers increased as white laptops became commonplace, given that Apple lowered the price another $100, to $1,299 with no price change in the black model and only modest improvements in the model's specs.

- Ralph's (and virtually all other) grocery stores provide customers with "frequent-buyer" or "Club member" cards that entitle holders (Group B) discounts not provided customers without cards (Group A), on the argument that people who tend to buy frequently (and/or buy in large quantities) have good reason to comparison shop and to obtain the frequent-buyer cards: they can prorate their search costs over a large number of purchases.[11]

- Tully's Coffee Cafes sell a tall drip coffee for $1.75. They sell a café latte for $3.20. Granted, café latte may take more expensive ingredients and more labor than a tall drip, but hardly close to $1.45 in added price of the café latte. The customers who buy the tall drip tend to be price sensitive (Group B), whereas buyers of café lattes tend to be price insensitive (Group A), with the two groups selecting into the two groups when they order their drinks at the counter.

- In 2007, Whole Foods Markets sold organic bananas to shoppers (Group B) for $.99 a pound. They sold nonorganic bananas to shoppers (Group A) for $.59 a pound. The price sensitivity of the two groups of banana buyers goes a long way toward explaining the price differential. If the price differential only covered the cost differential, Whole Foods would not be as eager to carry, and provide shelf space, for both types of bananas.

A TEXTBOOK CASE OF TEXTBOOK PRICE DISCRIMINATION

A classic example of price discrimination is the international differential in the price of college and university textbooks. The ability of the publisher to charge different prices at different universities is understandably a function of the cost of moving books between two university markets: The higher the cost of moving books between university markets, the higher the price differential between the markets can be. This means that a lower cost of moving textbooks should lead to greater arbitrage opportunities, given price differentials, and to a narrowing of the price differential over time as the arbitrage opportunities are discovered and exploited by students (and used book buyers).

In the past, textbook publishers have sold many of their textbooks in the USA for much higher prices than they have charged in the UK. One study found that after adjusting for the length of textbooks and their formats (hardback versus paperback), the prices of 268 textbooks (outside of economics textbooks) at the Amazon site in the USA averaged 31% higher than at the Amazon site in the UK. The prices for 204 economics textbooks on Amazon-US averaged 49% higher than on Amazon-UK.[12]

Consider one vivid example of the price differential in textbook pricing at Amazon-US and Amazon-UK. On the day these words were typed, Robert Pindyck and Daniel Rubinfeld's textbook on *Microeconomics* (6th edition)—which, by the way, carries a lengthy discussion of price discrimination—was "on sale" on Amazon-US for $159.33. The same book was listed "on sale" on Amazon-UK for the pound equivalent of $77.95—half the US price![13] The price differential cannot be chalked up to cost differentials, given that almost all textbooks sold on Amazon-UK and covered by the study mentioned above were printed in the USA.[14] Indeed, if there is a cost differential, the UK textbooks costs were higher because of the cost of shipping the books from the USA to the UK.

For years Amazon has been preventing US students from buying books on Amazon-UK (or other European Amazon sites). Now that restriction has been abandoned, students have gradually been discovering that they can buy their textbooks from Amazon-UK, and have been doing so in growing numbers, mainly because the price differential is often substantially greater than the trans-Atlantic shipping costs.[15] The expected growing shift in textbook purchases between the USA and UK sites can be expected to increase the demand for textbooks on Amazon-UK and decrease the demand for textbooks on Amazon-US, causing a narrowing of the price differential toward a differential that reflects the shipping (and any other reselling) costs. You can imagine that textbook publishers will see a need for raising their UK prices because, otherwise, they will end up forgoing higher-priced US sales for lower-priced UK sales (and perhaps incurring the added costs of dealing with international transactions).

Obviously, the difference in textbook prices on the two Amazon sites is a textbook example of price discrimination, attributable, as suggested by our foregoing analytics, to differences in the price sensitivity of students in the UK and the USA. But that observation raises the question, why might UK students be more price-sensitive than US students? It's hard to give a complete answer, because of the multitude of differences between the British and American students and their markets. But I can offer tentative observations that might provide a partial explanation for the difference in price sensitivity. The differential might be explained in part by income differences. Incomes in the USA are generally higher than in the UK, which could result in US students not caring as much as UK students about the prices they pay, especially since textbook expenditures come on top of the relatively higher costs of public and private higher education in the USA than in the UK (which, for public universities, comes free of tuition for students who are admitted).[16] Textbook prices are, therefore, more salient, and constitute a higher percentage of students' out-of-pocket educational costs in the UK. It could also be that the used book market is more developed (because the textbook market is far larger) in the USA than in the UK that could lead to lower resale costs and higher used textbook prices in the USA than in the UK. Indeed, the lowest "used price" on Amazon-US for the Pindyck/Rubinfeld textbook mentioned above was $98 (at the time of this writing). The lowest "used price" on Amazon-UK was $35.16 (at the going exchange rate). Hence, US students could recoup about 62% of their new book purchases, whereas UK students could recoup only 45%, which helps explain why US students might be less sensitive to new book prices than UK students. Still, there is money to be made (or, perhaps more accurately, saved) by US students buying their books on Amazon-UK for $77.98 and selling as used in the US used market.

There are obvious potential (and real) interplays between new and used book markets. If textbook publishers hike their new book prices, then student demand for used books can be expected to rise, driving their prices up. A good working rule is, the higher the new book price, the higher the used book price (all other considerations equal[17]). Of course, the development of the used book market means that the elasticity of demand for a given textbook should be expected to rise *after the first year of adoption*. Without the used book market, publishers might have good reason to hold their prices down on the first year of sales of a new edition, because they can imagine that the lower initial price can stimulate future adoptions (to the extent that some, perhaps only a few, professors consider the prices of the books they adopt) by creating "market buzz" about their text. With the emergence and development of the used book market, publishers have less reason to hold their prices down for future sales. Hence, publishers can be expected to exploit whatever inelasticity of demand they have in the first year of a new edition, meaning that the used book market can drive up the prices of new textbooks. In addition, publishers can be expected to try to kill off the market for used books by bringing out revised versions of their textbooks with shorter sales cycles. The added cost of more frequent editions can feed into higher prices for the first year of new editions. (I admit that I remain puzzled why publishers don't systematically drop their prices after the first year of an edition to better compete with used books.)

Of course, another explanation for the differential in US/UK textbook prices can be that US students care relatively less about the prices of their books, because a higher percent of their book expenditures are covered by their parents than is the case in the UK. To the extent that parents pay for books, students have less incentive to find out what texts are required for their courses early enough to order their books online, much less from a foreign website from which the texts may have to be shipped with a delay. The longer students wait until they learn of their assigned textbook, the less price-sensitive they will be, and the higher the prices publishers can charge.[18]

One study suggests that one of the more prominent reasons for the US/UK price differential is that textbooks in the USA are the focus of courses of study and are generally required. In the UK, textbooks are far less frequently required. UK students are more frequently assigned a variety of readings than is the case for US students. To the extent that textbooks constitute a less important component of course assessments, UK students can more easily forgo textbook purchases with less damage to their grades and standing in their classes. Hence, they can be more responsive to textbook prices than US students.

Given ongoing changes in educational technology, including the delivery of printed material, and the lowering of international transaction costs (via Amazon and other online booksellers), we have to expect a significant narrowing of the US/UK price difference over time. Digital versions of many textbooks can now be downloaded by both US and UK students to laptop and desktop computers chapter by chapter at modest prices per chapter ($1.99 each). That's only $59.70 for a thirty-chapter textbook. At iChapters.com there are at least a couple of dozen microeconomics textbooks for download at such prices, drawing into question the long-term viability of the $159 price for a new, printed version of the Pindyck/Rubinfeld microeconomics textbook.[19]

Textbooks can be easily pirated already, especially when they can be converted (through downloading and scanning) to digital formats. Pirated textbooks, whether in paper or digital form, represent a serious competitive threat to textbook publishers, perhaps a more serious threat than used book buyers (who must also be threatened by the emergence of the pirated copies). Pirated copies of books should be expected to impair publishers' ability to charge high prices and to price discriminate.

Perhaps the biggest competitive threat to textbook prices, especially in the US market, is the option of nonpurchase that students have. The National Association of College Stores found that nearly 60% of college students surveyed nationwide do not buy all of their course materials. The State Council on Higher Education in Virginia found that 40% of the students surveyed in the state did not buy the textbooks for their classes, attributable in part to a tripling of textbook prices between 1986 and 2004 (which now come with many more auxiliary materials).[20]

Even with technological advancements that depress textbook sales and prices, a price differential could still persist. While the elasticity of demand might be expected to increase in all markets, differences in market elasticities can remain, which can leave a price gap between markets. However, again, we should expect the price gaps across markets to narrow. That is, the price gap can be expected to move toward (but not necessarily to) the added cost of transacting across markets.

Before leaving the book example, it needs to be noted that publishers can also segment their market and price discriminate by producing different formats—hardback and paperback versions—of books with identical content. They might reason that people who buy paperback editions might not have good reason to hold their books for as long as do buyers of hardbacks. They might also reason that people who buy hardback books when the books are first released are more eager, and less-price sensitive, than buyers who can wait months, or more than a year, to buy their paperback versions. Understandably, paperback editions of books that are released after hardback editions almost always carry a lower price than the hardback editions, and the price differential between hardback and paperback editions has been found, as expected, to be greater than the difference in production costs.[21]

THE LOGIC OF AFTER-CHRISTMAS SALES

After going through models of price discrimination with my MBA students, I apply the analysis to several of the topics covered in the foregoing pages. I also invariably ask (maybe after a mid-class break), "Why do so many online and offline retailers have after-Christmas sales?" The students generally are quick to respond something to this effect, "To get rid of all the unwanted winter and Christmas merchandise" or "To reduce inventories for tax purposes." The students might explain that after-Christmas sales are a consequence of store buyers' misjudgments on the market demands for various products and mistakes in ordering.

I then ask if there are other reasons, a question that is typically met with palpable silence and a look of puzzlement on students' faces, as if there could be no other explanation. I grant them that misjudgments and mistakes can explain many things that happen in business, but I then ask if store-wide sales—year after year after year—could possibly be chalked up solely to misjudgments and mistakes. I point out that they could have provided such an explanation for sales before ever setting foot in my class—or listening to the earlier lecture. If after-Christmas sales can be chalked up to misjudgments and mistakes, I have to ask why are the store buyers at Nordstrom's (or any other prominent department store chain) retained—year after year after year? Should they not be fired and replaced with people whose misjudgments and mistakes are not as pervasive and persistent? After all, we are talking about stocking "errors" at Christmas that are systematic, that is, extend across the stores and result in table after table and rack after rack of "excess inventories" that are discounted by 50% or more. Indeed, many stores announce "storewide" after-Christmas sales with price cuts of "25% or more."

By my raising the puzzle of after-Christmas sales in this chapter, you might rightfully conclude (as my MBA students do—eventually!) that the logic of price discrimination, which we have developed for trans-Atlantic textbook sales, is also intimately linked to after-Christmas sales and perhaps all other seasonal and intermittent retail sales, to lesser or greater extent. That is to say, retail stores have after-Christmas sales (often deep ones) because the price-insensitivity of their customers takes a plunge between the day before Christmas and the day after.

Before Christmas, many customers *need* the goods they buy to be able to stand witness to the considerable (often only imagined) joy of their love ones and friends on Christmas morning receiving their gifts. Before Christmas, many customers are working and have high opportunity costs of their time; they also might have low storage costs. They have not yet filled their cabinets and closets with countless gifts, most wanted but some kept only out of respect for the givers. After Christmas, many buyers are often fully stocked with more goods than they need, or want. Many are often on holiday breaks at Christmas time, with low opportunity time costs.

More to the point, before Christmas, buyers' demands are highly inelastic. After Christmas, they are highly elastic because they have time to consider more carefully the prices charged by any number of sellers, and they have to see significant price reductions to stuff their cabinets and closets with more products. As pointed out earlier (in our discussion of the price responsiveness of Groups A and B), firms can maximize profits only by playing to the different elasticities of demand, which means that they should charge relatively higher prices before Christmas *in anticipation* of charging relatively lower prices afterwards.

Stores should be expected to order earlier in the year with both market—pre- and post-Christmas—demands in mind. After all, buyers often can't wait until the week of Christmas to place their orders for after-Christmas sales, especially when the goods have to be produced in remote corners of the globe. Seen from this perspective, after-Christmas sales on most items are *planned*. That is, many storewide sales are not matters of misjudgments and mistakes. The so-called "price cuts" after Christmas are not that at all, at least not in the sense that they are unanticipated and unplanned. The higher before-Christmas prices fit the higher demand and lower price elasticities of demand that stores then face. The after-Christmas prices fit the then lower demand and higher price elasticities of demand. Christmas allows stores to segment their markets with the prices charged before Christmas being higher than it would be if a constant price for both market segments had to be charged.

Of course, the elevated before-Christmas prices, followed by expected after-Christmas sales, can cause many price-sensitive shoppers to postpone as many purchases as they can until after Christmas. But such postponements are not necessarily all bad for stores, since the postponements further segment their markets into price-insensitive and price-sensitive shoppers. Purchase postponements can leave the before-Christmas market dominated by highly price-insensitive customers, giving rise to some additional price increase tailored to the demands of the before-Christmas shoppers. Shoppers who delay their purchases can increase the after-Christmas demands for goods, thus tempering the extent of the after-Christmas price cuts.

One reason for the growing popularity of gift cards at Christmas is that gift givers understand that the gift card recipients can get greater value from a given dollar amount on the cards because the cards can be used in after-Christmas sales.[22] Are gift cards advantageous to the givers and recipients? It's not easy to say, given the crosscurrent of market forces cards can put in motion. Still, it might be helpful to highlight a few of the forces.

Gift-card givers can avoid the difficulty (cost) of honing individual gifts to recipients when the recipients' preferences are not known very well. Many gift givers might give more in terms of dollars on gift cards than they would in "real gifts," since givers don't have to incur the search costs of finding real gifts and might want to assuage recipients hurt feelings from not having real gifts to open on Christmas morning.

On the other hand, if givers are themselves price sensitive, they might give fewer dollars on the card than they would spend on real gifts, given that the givers can anticipate that the recipients will be able to buy merchandise at lower prices after Christmas. The recipients might miss the joy of having real gifts under the tree on Christmas morning, but they can be more than compensated by the knowledge that the gift cards allow them to buy what they know they want and by the knowledge that the gift cards hold more real (price-adjusted) dollars than would have been spent on more expensive real gifts before Christmas.

SALES AND THE ECONOMICS OF INFORMATION

We can now address a more general question: why do stores have intermittent sales, some of which are as predictable as after-Christmas sales (for example end-of-summer sales)? Why do stores have other sales that are less predictable than after-Christmas sales (for example, "managers' blow-out mid-season sales")? Again, without question, some sales of some items can be the consequence of sellers misjudging market demands for goods. Consumers are often fickle in what they will buy.

However, following the late George Stigler's "economics of information",[23] Hal Varian has argued that many sales across the year are devices by which store managers can (again!) separate price-sensitive from price-insensitive customers.[24] Many buyers are price sensitive because they have low opportunity costs, both in time and storage. These buyers have ample time to monitor newspapers and television programs (and other media) for the sales announcements.

In short, Varian takes note of an unheralded fact of market life: there always exists in markets a dispersion of informed and uninformed buyers, with the degree of information shoppers have on prices related, as we have explained, to their search and storage costs. Informed buyers can be expected at stores when sales are announced and to load up on the goods that are on sale. Once informed price-sensitive shoppers have loaded up on goods, there will be time between the sales for stores to hike their prices for the buyers who are (rationally) uninformed about sales and who are willing and able to pay higher, nonsale prices.

This line of argument also helps explain why in any geographically spread market—say, a city—the same good can be sold at different prices with nontrivial differences in the prices. Some sellers face shoppers who are informed about prices across the market, because they have low search costs, and are, hence, price sensitive. Other sellers face shoppers with high opportunity costs of becoming informed shoppers and, hence, face inelastic demands, which means their relatively higher prices can persist (a line of argument developed by Steven Salop and Joseph Stiglitz[25]). One might conclude that the uninformed price-insensitive shoppers are being "ripped off." They are, but only in the sense that they are charged more than their informed counterparts. The price-insensitive shoppers can still *be better off than they would have been had they incurred the search costs.* They can rightfully believe that their *effective* prices paid—lower sticker prices paid plus the search costs—are greater than what they actually pay for higher sticker prices because of their lack of attention to comparative shopping.[26]

At the start of Chapter 1, I noted that as I was finalizing this book, Audible.com had announced a "Summer Clearance Sale" of an extra 25% off its low prices for its audio book titles. I also suggested that Audible's "clearance sale" was something of a puzzle because Audible doesn't have an inventory, aside for the digital master copies of its more than 30,000 audio books. It would hardly want to get rid of its masters because that would greatly limit its sales to one copy per book. It could obviously do far better by keeping its masters (which cost precious little to inventory) and selling digital copies (which cost little to nothing to reproduce). Then, why did Audible announce its clearance sale? Maybe there is some marketing gimmick to the use of the word "clearance," but my guess is that other Audible subscribers are no less savvy than I am. They realize that "clearance" is irrelevant to the announcement (and to similar word usage by other brick-and-mortar stores); what is important and eye-catching is the "25% off." Our discussion of sales in this chapter reveals the most likely explanation for the announced "summer sale:" Audible has either detected a difference in the elasticity of its demand in the summer vis-à-vis other seasons of the year or it has detected that some Audible site visitors pay attention to its sales announcements and respond to them. Others not seeing them leave themselves open to higher prices when they return later to the Audible site to download audio books.

CONCLUDING COMMENTS

Economics can be a fascinating subject in one unheralded regard: a single simple model of market behavior can explain much of what we observe in the world about us. You really don't need to know a lot (in the ways of principles) to do some hard-nosed economic analysis. I hope readers will agree that our model of market segmentation and price discrimination is very elementary, but explains many observed price differences.

Firms can obviously make a lot of money by creatively designing "better mousetraps." From the perspective of this chapter, I hasten to add, they can also make a lot of money from creatively designing ways of segmenting their markets.

Readers should not deduce that the resulting price discrimination is simply a means by which sellers can take advantage of buyers. Sometimes consumers can be worse off, but sometimes such is not the case, especially in the long run. In creative methods of market segmentation and price discrimination are economic (above-competitive) profits that can stimulate the development of more products than would otherwise be available for consumers.

In chapters to come, we will often repeat with variation and amplifications the lessons learned in this chapter. Having stressed the gains to be had from matching prices with buyer price sensitivity, we need to end with a caveat. Just as competition can cause producers to improve their products' quality and features, competition can also undercut, with time, the profit potential from creative methods of price discrimination. As noted in our discussion of textbooks, with time students can be expected to learn that they can lower their out-of-pocket textbook costs by buying online and then by going to sites for textbook sellers in far-removed markets, thus undercutting the ability of publishers to price discriminate. As more and more informed shoppers begin to delay purchases until they can take advantage of after-Christmas sales and then give store gift cards on Christmas morning (so that the recipients can take advantage of after-Christmas sales), sellers will have growing reasons to extend their sales backward to before Christmas. Sales (at times other than after Christmas) by a few sellers can lead to a proliferation of sales by a growing range of sellers, and eventually to the emergence of some sellers who dispense with sales altogether. Such sellers can be expected to promote "everyday low prices." Welcome to Wal-Mart!

NOTES

1. For early discussions of price discrimination, see Pigou (1962) and Robinson (1965). For modern textbook discussions of various forms of price discrimination taught university and college students, see Becker (1971), Pindyck and Rubinfeld (2004), and McKenzie and Lee (2006)
2. *Education Life* (2007).
3. Consider Richard Vedder's discussion of the use of scholarships for purposes of price discrimination (2006).
4. Lifson (2004). See also Vedder (2004).
5. As reported by then-*New York Times* business columnist Peter Passell (1997).
6. Passell (1997).
7. As reported on the web site NexTag Comparison Shopping on August 14, 2007, http://www.nextag.com/imation-1gb-usb-flash-drive-clip/search-html.

[8] The marginal cost of an ounce of drink must be lower than .6 cents. Otherwise, there would be no reason for a profit-maximizing firm to systematically charge so little for the marginal ounces in the large drink. Restaurants are not in the business of doing their customers a favor any more than the customers go to restaurants to do their owners a favor.

[9] Granted, price discrimination by market segments might give rise to higher sales costs. However, the central point of the discussion is that price discrimination can raise profits. The increase in profits is only reduced by the added sales costs. It goes without saying that if the increase in sales costs from price discrimination exceeds the increase in revenues from price discrimination, then price discrimination is not a viable firm strategy.

[10] It might be thought that McDonalds' prices have been pushed up by airport rental rates, and such may be the case, but only to a degree. It might be more appropriate to say that airport rental rates can be relatively high because price insensitivity of travelers, given their time constraints and inability to look for eating options outside of the airport.

[11] Indeed, "quantity discounts" given by a great variety of stores, middlemen, and manufacturers are often viewed as representing economies stores achieve from selling in bulk and reducing the transaction costs of associated with multiple purchases, which can explain the price differences—in part. Another explanation can be that the retailers, middlemen, and manufacturers are simply tailoring their prices to the price sensitivity of different groups of buyers, with the discounts going up with the quantity bought.

[12] One study involving a comparison of 268 textbook prices on Amazon.com (USA) and Amazon.co.uk in May 2002 revealed that after adjusting for the length of the textbook and format of the textbooks (hardcover versus paperback) in a regression analysis, textbook prices on Amazon.com were 31% higher than on Amazon.co.uk (Cabolis et al. 2005).

[13] The sale price of *Microeconomics* in British pounds was £39.91, with the price of a pound equal to $1.9532.

[14] Cabolis et al. (2005).

[15] Lewin (2003a and 2003b).

[16] Higher incomes in the USA can push students' demand curve for textbook further out to the right, meaning that any given textbook price can be lower down their demand curve for textbook, which can (but not necessarily will) put US students in a range of their demands with a lower elasticity. To see this point more clearly, see the discussion of the elasticity of demand along a given demand curve and at different demand levels in McKenzie and Lee (2006, pp. 270–276). The presumption of a generally higher elasticity of demand in the UK could be expected to show up in

prices for trade books higher in the USA than in the UK, and the study relied on for differences in textbook prices has also shown that trade books carry a 13% premium in the UK vis-a-vis the USA (Cabolis et al. 2005).

[17] We should, of course, realize that the used book price will be determined, in part, by how widely the text is adopted: The greater the adoptions, the more fluid the used book market can be, the greater the demand of used-book buyers, leading to a relatively higher used-book price.

[18] I present this argument with some hesitation on accepting it without reservation. This is because many (if not most) students can figure that money spent on textbooks will, to a nontrivial extent, be money that parents can't spend in other ways (for instance, as an increase in student allowances).

[19] You can sample the availability of downloadable chapters at iChapters, http://www.ichapters.com/comsite5/bin/comsite5.pl.

[20] As reported by Kinzie (2006).

[21] Clerides (2002).

[22] According to one report, Christmas gift card sales were expected to jump nationwide 32% in 2006 over what they were in 2005. In Indiana, 69% of Christmas shoppers were expected to buy gift cards with an average value of $117 (up from $88 in 2005) (Knight 2006).

[23] Stigler (1961).

[24] Varian (1980).

[25] Salop and Stiglitz (1977).

[26] Varian (1980).

Chapter 4

WHY POPCORN COSTS SO MUCH AT THE MOVIES

Going to the movies and downing a tub of popcorn and an oversized soda is as American as… well, going to a baseball game and getting several hotdogs and beers. Both outings can now put a nontrivial dent in any family's entertainment budget.

There are two notable features of family trips to movie theaters:

- First, theaters charge nonelderly adults and children (generally, under the age of thirteen) and seniors (generally, no younger than fifty-five) different prices for admission tickets but not for popcorn (and other concession items). Why?

- Second, theaters' charge for large tubs (containing only 7 ounces) of popcorn $7 (in my area of the country), or close to a dollar an ounce (with an ounce of popcorn equaling about 3 cups in volume).[1] In addition, the price of a tub is nearly three-quarters their charge for a (non-senior) adult ticket and over 90% the price of a child or senior ticket. Again, why so much when a bag of popcorn kernels is so cheap? At this writing, popcorn costs $.85 a pound in two-pound bags at local (Southern California) grocery stores, with a pound of popcorn kernels making (according to my rough estimates) slightly more than three theater-size tubs of popped popcorn. Add in the cost of vegetable oil, and the cost of materials for a theater-size tub of popcorn made at home is only $.55. [2] This means that a theater-size tub of popped popcorn bought in theater lobbies is nearly thirteen times the materials cost of home-popped popcorn and that the profit margin for theaters on materials alone must be well over 90% (especially since theaters can buy their popcorn and oils with substantial quantity discounts).

The easy and most frequently cited explanation for these pricing strategies is that theaters are taking moviegoers to their monopoly cleaners by their discriminatory pricing on tickets for adults and children (along with senior citizens). Indeed, discriminatory pricing is prima facie evidence of monopoly market power, or so economists have conventionally argued.[3] Moreover, theaters effectively trap consumers once they go through their ticket turnstiles, thus permitting extortionist pricing on popcorn (and other concessions).

Without question, movie theaters often have a measure of monopoly pricing power. After all, some theaters are the only theaters in a town or an area of a city, at times because of zoning restrictions and at other times because of shopping malls' interest in reducing competition in order to increase their rental payments. Also, distributors license movies to theaters within identified "clearance zones," with one theater in each zone getting a particular film, for example, any one of the movies in the *Harry Potter* series.[4]

While there is a measure of truth in claims that theater prices reflect an equal measure of monopoly power, we will see in this chapter why that easy answer is hardly the whole truth of theaters' pricing strategies. What we will find, among other things, is that popcorn prices are high in part because of the reduced prices for children. In addition, because theaters cannot be owned by movie producers and distributors (because of a series of court orders that date to the late 1940s), theaters have an incentive to hold down (relatively speaking) all ticket prices in order to increase the demand for popcorn (and other concessions), thus allowing theaters to hike their prices on popcorn and other concessions and their profits. Along the way, we will find that theater popcorn is actually pretty cheap—on the margin!

DIFFERENTIAL THEATER TICKET PRICES

At the time of this writing, the Regal Theater chain in Southern California charged non-senior adults $10.50 for tickets and charged children, 12 and under, and seniors $7.50 for tickets. The differential in ticket prices for adults and children has been easier for economists to explain than the high price of popcorn (and other concession items), mainly because of the several lines of available standard monopoly arguments economists can and have tapped, no one of which is likely to provide a full understanding of theater pricing.

One line of argument is well worn among economists: The differential pricing for adults and children can simply be chalked up to price discrimination by market segments introduced in the last chapter. To review that earlier discussion, adults are (supposedly) less price sensitive—or have more inelastic demands—for going to the movies. That is, adults don't change (in percentage terms) the number of movies they see in theaters as readily as do children when their ticket prices are hiked. Why?

REASONS FOR ADULT–CHILDREN PRICE DIFFERENTIALS

One plausible (albeit partial) explanation may be that adults' time is more valuable (given their paying work opportunities), which has a threefold consequence.

- First, adults' higher incomes can hike their demand for going to the movies, which *can mean* (but doesn't necessarily mean) that they are not as pressed to respond to a ticket price increase. This can mean that any given increase in the ticket price can have a lower percentage reduction in sales to adults and that theaters experience an increase in box-office revenues and profits, given that the costs of providing the theater seats will not be materially affected by attendance.[5]

- Second, (non-senior) adults incur greater (opportunity or time) costs than children to search out alternative prices for different movies at different theaters, which implies that adults may be less aware of lower prices of movies elsewhere and alternative forms of entertainment and, therefore, are less able to respond to a price hike out of simple ignorance (albeit a level of ignorance rationally sought).

- Third, because of adults' much greater time cost, any given hike in the movie ticket can represent a lower (percentage) increase in the *total cost* of going to the movies for adults than for children. And marketing research does show that a given dollar change in the price of a good can affect the willingness of buyers to respond to a low-price product relatively more than a higher-priced product.[6] To see this point, consider an adult who earns $40 an hour (or the equivalent of $83,200 a year) and is typically asked to pay $10.50 a ticket for a two-hour movie—$3 more than the child's ticket. If (for purposes of explanation) the adult's wage is a rough approximation of his or her opportunity for going to a two-hour movie, the adult experiences less than a 4% increase in the total effective cost of seeing the movie when the adult ticket price is raised by $3, from $7.50 to $10.50. How is that? The total cost of going to a two-hour movie for an adult earning $40 an hour is $87.50 when the ticket price is $7.50 [($40/hour × 2 hours) + $7.50 ticket price = $87.50]. A hike in the ticket price by $3 to $10.50, or 36%, raises the total cost of the movie experience to $90.50, or by a mere 3.4%.

The same $3 increase in the admission price for children, whose opportunity cost of time is far lower—say, $2 an hour (a generous pay rate for young children, which I use only for purposes of illustration)—than the opportunity cost of the adult, would represent, in our illustration, more than a one-fourth increase in the total cost of seeing the movie. The total cost of a child going to a two-hour movie is $11.50 ([$2/hour × 2 hours] + $7.50 = $11.50). A $3-increase in the child's ticket price represents a 26% increase in the total cost of the child going to the movie. This means that, everything else being equal, we should not be surprised if young children are more sensitive to any price increase than adults, given that the price increase for children is larger in percentage terms and more salient in terms of their reference cost.

I grant you that people's wage rate is not always a good measure of opportunity cost. People tend to go to movies in their off-work hours, *because* their opportunity costs can then be (but not necessarily will be) lower. We shouldn't allow the particulars of our example to deny the larger points at issue: The cost of going to the movie can be some multiple of the ticket price, because of the value of time involved. The opportunity costs of people's time can rise with age, because of their growing skills and experience and job opportunities. As the opportunity cost of moviegoers' time rises, their sensitivity to a ticket price increase can fall (everything else being equal). One explanation for theaters setting an upper age limit for children's tickets at 12 is that by such an age, children's opportunity time costs have risen to the point that they, too, have become significantly more price-*insensitive*, which is reason enough for the theaters hiking the ticket prices of children above 12.

The differential pricing for adults and children can also be explained by the fact that, like it or not, many parents value seeing movies themselves more than they value their children seeing them (especially when movies contain rough language and violence). In such cases, the theaters have to lower children's ticket prices in order to encourage parents to take their children to the movies or to send them off to the movies by themselves. In this latter regard, movies have to compete with babysitters who often charge less per hour than the federal minimum wage, and who sometimes charge less for two hours of babysitting than the adult ticket price for two parents.

Of course, I recognize that studios produce movies solely for children, with

Finding Nemo, an animated film from Pixar being a grand example (especially since it won the Oscar for being the best animated film released in 2003). In such cases, parents care more about their children seeing the films than they, the parents, care about seeing them. This might suggest that parents' tickets should be lower than their children's tickets. Perhaps so, but only in some cases. As all parents know, children have ways of pressuring their parents to take them to the movies, and to feed them at the concession counters. That pressure can translate into a reduced price sensitivity on the part of the parents. Besides, the crucial issue to parents in such cases is not so much the *relative* prices of adult's and children's tickets, but with the overall cost (including all ticket and concession expenditures and their time costs) they incur from going to the movies. It's no accident that family/children's films are released during times (for example, Thanksgiving or Christmas) when many parents are off work and their children are out of school. The total cost of family trips to the movies is then lower than at other times of the year, because of the lower opportunity costs of all family members, which means that studios and theaters can charge more for tickets and concessions than at other times of the year, or fill more seats at constant prices.

PEAK-LOAD PRICING

Another alternative explanation for the difference in the price of adult and children tickets starts with the proposition that the main goal of movie theaters is to fill as many seats as possible at all times of the day. Seats that go empty at various times of the day represent revenues that can never be recaptured and theaters' costs vary little with how many seats are filled when the projector is turned on. As economist Steven Landsburg has pointed out, the lower price for children may have nothing to do with the form of pure price discrimination (and the implied monopoly power) just discussed.[7] Rather, the price differential may have everything to do with the fact that children (and senior citizens) tend to go to the movies during periods of slack demand, in the afternoons and early evenings, when the cost of providing the added seats for children (and seniors) is relatively low, if not zero. The gain in revenues from the added seats sold to children (or seniors) from their lower ticket prices more than offsets the reduction in revenues from the lower ticket prices for children who would have gone to the movies at the adult prices. The net increase in revenues goes largely to theater profits, again, given that virtually all costs (other than clean up) are not materially affected by the extra seats that are filled.

From this perspective, the adult and children's ticket prices are a rough form of "peak-load pricing." The prospect of this explanation having validity can be seen in the fact that at Regal Theaters children's ticket prices ($7.50) are often close to, if not identical with, adult ticket prices ($7.50) for matinee showings (before 5:30 P.M.).

CONCESSIONS SALES

Perhaps an even more incisive explanation for the difference in adult and children's ticket prices is that children buy more concessions—popcorn, sodas, and candy—or that they cause their parents to buy more concessions than they, the parents, would otherwise be inclined to buy. Given that the profitability of the concessions can be crucial to the overall profitability of the theaters, theaters have an added incentive to lower the price of children's tickets. The lower children's price can be seen as a way the theaters can increase the demand for and price of concessions. What the theaters lose on children's ticket prices (much of which would have gone to the movie studios, as we will see) they can recoup on concession revenues. The lower the cost of the concessions and the more theaters can charge for them, the more the theaters should be willing to cut the price of admission. From this perspective, we have a partial explanation for why theater popcorn costs so much: The exorbitant cost of popcorn can be chalked up in part to the cut in the price of admission for children.

UNIFORM POPCORN PRICES

Why don't theaters also charge children less for popcorn and other concessions than they charge adults?

One explanation could be that, in contrast to their demands for tickets, adults and children are, more or less, equally responsive to changes in the price of popcorn (they have the same *elasticities* of demand). This means that theaters have nothing to gain from using a lower price of popcorn to lure children to buy more popcorn.

Perhaps an even better explanation is that a lower children's price on popcorn would only cause parents to send their children to buy the popcorn instead of going to the concession counter themselves. We can imagine that if there were a significant price discount for popcorn for children, enterprising children would buy up extra tubs at the concession counter and then hawk them in the aisles to adults, splitting the price differential with their older customers. For example, if a large tub of popcorn costs adults $7 and children $4.50, children could more than cover their cost of admission by buying several tubs of popcorn and reselling them to adults for $5.50.

Price discrimination works for ticket prices because theaters can post ticket takers at their turnstiles. The ticket takers can ensure not only that everyone who enters has a ticket, but that only children (or those who *look* to be age twelve and under) are admitted with children's tickets.

THE HIGH PRICE OF THEATER POPCORN

A large tub of theater-popped popcorn (which, by the way, has close to 1,700 calories and up to 130 grams of fat when buttered!) sold for $7 in Southern California at the time these words were typed (and probably more by the time these words are read). As noted, if a comparable size bowl of popcorn were popped at home, the popcorn would cost, in terms of out-of-the-pocket expenditures, a little more than half a dollar, with the raw materials for commercially popped popcorn costing substantially less than home-popped popcorn because of the price breaks commercial vendors can get from their quantity purchases.

One commercial popcorn machine vendor estimates the cost of a popper full of popcorn made on its poppers to be no more than a nickel. Add in 8 cents for the paper tub at the theater, and the theater's profit margin from material costs alone is obviously very high. [8] Assuming that a theater-quality commercial popper can make about a tub of popcorn, the material cost represents less than 2% of the cost of a $7 tub of theater popcorn, leaving a profit margin from material costs alone of over 98%.

However, like so many other goods, the material cost is hardly the most consequential cost consideration for theaters offering popcorn. The labor required to make popcorn is far more consequential. If it takes fifteen minutes (on average) for a worker to make, fill, and sell a tub of popcorn and the typical worker behind the concession counter makes $9 an hour (which is close to the entry-level retail pay rate in Southern California), the popcorn costs the theater upwards of $2.40 in material and direct labor costs (not accounting the cost of plant and equipment and indirect labor costs for doing all the other things that need to be done in theaters in order for moviegoers to want to buy the popcorn: taking tickets, running the projector, and cleaning the theaters, just to name a few labor costs). Still, the profit margin on popcorn appears high no matter how the costs are calculated, and the *marginal* cost of making and selling additional tubs of popcorn is much lower than the *average* cost, which makes popcorn sales very profitable *on the margin*, which is the key reason theaters push popcorn sales.

Still, why so much in terms of price and profit margin for in-theater popcorn? One transparent response is that home-popped and theater-popped popcorn are not the same products. They taste and smell differently. Indeed, theaters have an incentive to offer a different product from what moviegoers can make at home. They also have an incentive to accentuate the smell and sound of the popping popcorn in their lobbies, thus increasing moviegoers' assessments of the value of the theater-popped popcorn (and their market demand) and, at the same time, reducing theater goers' price sensitivity.

When popcorn was first sold in movie theaters in the 1930s, it was trucked in, after having been popped in remote locations, primarily because of the fire hazards the available popcorn poppers then presented. Sales of popcorn in theaters didn't take off until the late 1940s when the popper technology improved, reducing the fire hazard, and moviegoers' senses were teased by the sound and smell of popcorn in the lobbies. Theaters deliberately sought to enhance the smell and sound of the popping popcorn, making them "audible and smellable edibles." They further sought to increase the demand for popcorn (and other concession items) by using yellow popcorn that pops to a greater volume than white popcorn, but also gives the appearance of having been buttered (which means theaters butter costs could be curbed).[9] If theaters couldn't offer a different product worth more than home-popped popcorn or could not manipulate the demand in their lobbies by the smell, theaters would be unable to charge so much because far more moviegoers would sneak in their own popcorn with them to the movies.

Another transparent answer to the popcorn pricing riddle is that the profit on theater popcorn is not nearly as high as the above-cited figures suggest—when all costs are considered. After all, unlike other retail establishments, the floor space and equipment dedicated to popcorn popping are expensive and used only a few hours of most days (largely in twenty-minute segments between film showings). Very likely more important costs incurred by the theater are the costs of labor (wages, fringes, and taxes) involved in both making the popcorn, standing around doing nothing when no one is at the concession counters, and cleaning up after movie patrons who take tubs of popcorn to their seats, only to occasionally spill them (and almost always leaving crumbs behind).

If labor were not an important cost factor, surely more moviegoers would make their own popcorn at home, bag it, and sneak bags in with them to the movies. If our moviegoer who earns $40 an hour (and values his time at home and in theater by that amount) were to make a large bag of home-made popcorn to take to the movie and if the time involved were as little as twenty minutes (it took me twenty-three minutes to fill a theater-size tub, taking two rounds of popping in my sizable home popcorn popper), the labor (opportunity) cost alone for the home-popped popcorn would be $13.33, which means labor costs alone would be more than the cost of a tub at the theater's concession counter!

Granted, theaters have prohibitions against bringing outside food into their theaters (for the obvious purpose of increasing the demand for and prices of their concessions). However, if the true full cost differential between theater-popped popcorn and home-popped popcorn were as stark as appears to be the case in a comparison between the purchase price of theater popcorn and the *materials* cost of home-made popcorn, then surely many moviegoers, especially children, would take full advantage of opportunities to hide their bags of home-popped popcorn on the way into the theaters. And hiding is hardly difficult, as some movie patrons can attest! All one needs is a jacket in cold weather and a large purse or just a shopping bag that gives the appearance of being filled with purchases in warmer weather. Even when stopped occasionally by the ticket taker, smuggling home-produced popcorn could still be a highly paying proposition over a sequence of trips to the movies—*if popcorn were as excessively priced for the value provided as popularly lamented by moviegoers.*

As it is, one researcher found that a major reason theater popcorn may appear more expensive than popcorn sold elsewhere is that the theater portions are larger.[10] The average price of buttered popcorn *per quart* sold in 21 suburban and metropolitan theaters in the Mid-Atlantic states was actually close to 10% *below* the average price of popcorn per quart sold in 18 large shopping malls.[11] However, the researcher also found that the average price per ounce of medium-sized soft drinks sold in theaters was 37% higher than the average price per ounce sold in 24 convenience stores. The average price per ounce of four different candy bars sold in theaters was nearly double the price per ounce in convenience stores.[12] Given the relatively greater price and smaller size of candy, no one should be surprised if more candy is smuggled into theaters than popcorn.

THE MISGUIDED ENTRAPMENT THEORY OF OVERPRICED POPCORN

When asked why popcorn costs so much in movie theaters, many people who believe they understand the problem with full clarity have a pat answer: "The movie theaters lure moviegoers into their lobbies with hit movies. They are then trapped and effectively forced to buy what the theaters offer at their counters, since there are no competing sellers allowed in the lobbies. Hence, the theaters are, for all practical purposes, monopolists, which necessarily means the theaters can charge anything they want for popcorn."

Surely there is at least a grain of truth to such a line of argument. Like almost all businesses (other than grain farmers in the Midwest), theaters have some control over their prices. Markets are hardly perfectly fluid (as economists' market model of perfect competition, idealized in all microeconomic textbooks, suggests[13]). If there were as much truth to this argument as its proponents think, we have to wonder why the theaters stop at charging $7 (or whatever) for a tub of popcorn? Why not $10 a tub? For that matter, why not $20 or $50 a tub?

PRICING LIMITS FOR MONOPOLISTS

One of the more fundamental errors in the argument's logic is that moviegoers do have a choice over whether to buy or not buy popcorn. This fact alone suggests another fundamental error suggested by the claim that monopolists "can charge anything they want for popcorn." That simply is not true, and never has been. Monopolists are, like all firms, constrained in the prices they charge by their products' market demands. As stressed throughout this book, the demand for any good is an inverse relationship between the price of the product and the quantity sold: the higher the price a monopolist charges, the lower the quantity that the monopolist will sell—a rule of market behavior we have deferred to in every chapter to this point in the book.

Granted, at a very low price, a monopolist can raise its price and sell less, but with revenues (achieved from the price times quantity sold) and profits rising. For example, suppose that a monopolist charges $1 for its "widgets" and sells a hundred of them. If the monopolist raises its price to $1.25 (or by 25%) and the quantity falls to 90 units (or by 10%), revenue will rise from $100 ($1 × 100 units) to $112.50 ($1.25 × 90 units). Profits will rise by *more* than the $12.50 increase in revenues. This is because (in most production processes) there is bound to be some reduction in production (materials) costs with the drop in sales from 100 to 90 units.

However, as the monopolist raises its price, there is bound to be some higher price beyond which any further increase in the price will lead to a drop in revenues. We know this will be the case simply because we know there is some extraordinarily high price (for an extreme example, $500 for a tub of popcorn) at which point even the most powerful of monopolists will sell absolutely nothing because even a monopolist can't force consumers to buy its good at such a ridiculous price—especially not popcorn. At that very high price, the monopolist will then have absolutely zero revenue ($500/popcorn tub × 0 sales = $0)—and, necessarily, zero profits. If an increase in price can initially lead to greater revenues and eventually some very high price will yield zero revenues, then as the price is gradually raised, there has to be a price beyond which an increase in price will lead to a decrease in revenues that exceeds the reduction in costs from the curb in sales. Profits must then decline.[14]

The entrapment theory of high popcorn prices is flawed in another important way: People don't have to go to any particular movie theater. They also don't have to go to the movies, or they can eat a bowl of popcorn before going in the theater. They can do any number of other things with their time. People have choices, plenty of them. To this extent, movie theaters are hardly the monopolists they have been made out to be. Theaters must face the fact that their prices both on admission and on popcorn can affect how many people go to the movies and then buy theater-popped popcorn. As with so many other businesses, theaters clearly must be mindful of their costs and what they charge on all fronts, as evidenced by the fact that in recent years, several major movie theater chains that haven't been so mindful of their business basics have filed for bankruptcy.

MOVIES AS BUNDLED EXPERIENCES

Most theaters understand that they aren't simply in the business of selling seats to watch particular movies. Movie theaters are selling "experiences" or "entertainment bundles" in one-and-a-half to three-hour segments. These bundles include several components, with, perhaps, the movie and popcorn being two of the more important components.[15] For these *bundles*, theaters can charge some overall price. Our law of demand applies again: The higher the price of the bundles, the lower the quantity bought. Assuming the theater is pricing its bundles in accord with what the market will bear, this means that if a theater decides to raise the price of popcorn, it must lower the price of admission to hold attendance constant. It also means that theaters can manage their demand for and price of popcorn through the price they charge for admission: The lower the price of admissions, the greater the ticket sales, and the greater the demand for and price of popcorn. Needless to say, theaters can be expected to seek to optimize on the overall price of their entertainment bundles, and the prices of the bundles' separate components—all with the goal of maximizing their profits.

As an aside, I am certain some readers will object to my assumption that theaters will charge for their bundles all that the market can bear. I make that assumption for the patently obvious reasons that theaters' high prices on tickets and popcorn suggest that's what they are trying to charge: what the market will bear, within limits of what they can know about their market demand. Moreover, there are market pressures that encourage theaters to charge as much as they can and make as much profits as they can. If they systematically charge significantly less than what the market can bear, the theaters' profit streams into the future will be undercut. Their stock prices will also suffer on financial markets. As noted before, savvy investors who believe they know better what the market can bear can be expected to buy controlling interests in the companies, charge what the market will bear, and raise the companies' profit streams. The investors can then sell out with a handsome capital gain as the stock prices rise to reflect the greater future profit stream.

Why is the price of theater popcorn so high? If theaters could easily tell who among the people who reach the ticket windows loved popcorn, they should be willing to let those people in with a price break on tickets (which they do for children, as we have noted). However, among adults, it is not always easy for people at the ticket windows to spot the heavy popcorn eaters (although overweight people might prove to be good candidates for popcorn sales). By charging a high price for admission, theaters could be excluding from their lobbies many potential popcorn lovers, and denying themselves profit on their popcorn sales. In addition, theaters would have to lower their price of popcorn to compensate for the higher ticket prices in the overall price of their entertainment bundles they have for sale. Economists David Friedman and Steven Landsburg have argued, (apparently) independently of one another, that theaters' best pricing strategy is to try to hold the price of tickets down and raise the price of popcorn on the grounds that the popcorn lovers get more benefits from their movie experience than the non-popcorn lovers.[16] They should be willing to pay more, and do pay more through such a pricing arrangement. By holding down the price of tickets and elevating the price of popcorn (and other concessions), theaters are able to increase the number of potential popcorn buyers, with popcorn (as we will see) having a higher profit margin than tickets. If theaters were to do the reverse, raise ticket prices to lower popcorn prices, then they could not only curb ticket sales, but also popcorn sales.

Economists Luis Locay and Alvaro Rodriguez have an alternative way of explaining the high price of popcorn.[17] They reason that a film is a fixed good in the sense that moviegoers can't buy more or less of it. They buy their tickets and watch what comes on the screen. The ticket price is an admission price to do two things: 1) see the film and 2) buy popcorn (and other concessions). Moviegoers' demand for popcorn varies greatly. Theaters could sell popcorn to all who pay the admission fee. Why not simply focus on those moviegoers who have high demands, charging a very high price for the "small" bag of popcorn and then "walk those buyers down their demand curves" by lowering the additional price for additional increments of popcorn (a pricing strategy we discussed in Chap. 2 when talking about the pricing of drink at fast food restaurants). Seen from this perspective, the high price of popcorn can be attributed, not so much to the market power of theaters, but to the intense demand for popcorn of a segment of all moviegoers. Moviegoers with a less intense demand for popcorn might as well blame their fellow movie patrons, not the theaters, for pricing them out of the popcorn market in the theater lobbies.

MOVIE SCREENING CONTRACT

A point that Friedman and Landsburg and other economists have missed is that theaters have an added incentive to lower the price of admission and hike the price of popcorn built into their contracts for the movies they show. Theaters often bid for movies in terms of the percentage of their box-office receipts. Theaters regularly bid 70% of their box-office receipts—with their bids sometimes reaching 95% of their box-office receipts—for the rights to show a movie.[18] Theaters could, and have, bid a fixed amount—say, $100,000—for the rights to show a movie for a multiple-week engagement. However, because, as entertainment economist Arthur De Vany has argued, the success of a movie is very unpredictable (even when a movie has star power and is a sequel to a successful movie), a fixed amount bid means that the theaters would assume a great deal of risk, which explains why fixed bids alone are rarely used in contracts negotiated between theaters and studios.[19] Making the bid in terms of a percentage of box-office receipts increases the incentive studios have to make popular movies, which can give rise to greater ticket *and* popcorn sales and which, in turn, can give theaters a reason to hike their bids for movies. The percentage take way of buying movies thus shifts the theaters' risk costs, thus allowing for gains to both the theaters and the movie producers and distributors.[20]

Obviously, because of contractual provisions that cause the theaters' to fork over a major share of their gate receipts, theaters have a built-in incentive to keep their ticket prices low in order to raise their popcorn and other concession prices. If the theaters cut their ticket price by $1, they reduce the box-office receipts they get to keep by as little as 5 cents per seat sold, and usually by 30 cents.

But they can then raise the price of popcorn by $1 (to keep the overall price of their entertainment bundles constant). By cutting the price of admission, theaters will not only sell more seats, they will gain the high marginal profit on the greater popcorn sales due to greater ticket sales. The profit margin on additional popcorn sales is substantially greater than the relatively few cents they would have gotten to keep on a lost box-office ticket sale.

You can bet that there is a constant struggle between movie producers (and distributors) and theaters over admission pricing, with the producers understandably wanting higher admission prices than are optimal from the standpoint of the theaters. When movies are released for showings, movie producers' costs are pretty much fixed. This means that the movie producers whose receipts are a percentage of the theaters' box-office receipts want the movie theaters to charge that price that maximizes theater revenues (not theater profits). Given that the producers are paid a percentage of box-office receipts, that one price that maximizes box-office revenues would therefore maximize the producers' revenues and profits (again, given that their costs are more or less sunk costs, which means they have been incurred and can't be changed). On the other hand, theaters have an incentive to charge less than the revenue-maximizing ticket price because more seats sold means more popcorn sold.

Of course, the conflict between theaters and producers can be ameliorated in two basic ways. First, producers can be given a share of the theaters' revenues on concessions.

Second, movie producers and movie theaters can form what are called vertically integrated firms (meaning the production, distribution, and theater components of the movie industry would all be controlled by a single firm organization). Such firms could then juggle their ticket and popcorn (and other concession) prices to maximize their organizations' collective profits. Such vertically integrated firms would not have to deal with the so-called transaction costs involved in producers and distributors negotiating rental prices for their movies with theaters. All parties would not then have the hassle—meaning incur transaction costs—of dealing with the pricing conflicts, given their different objectives as separate firms. The integrated firms would not then have to incur the monitoring costs that studios have to incur to make sure theaters accurately report their box-office receipts (and with theaters having to fork over 70 or more percent of every dollar reported, the temptation to falsify reports is obviously substantial). Not only would integrated firms have lower costs and greater profits, but because of the cost savings, ticket and popcorn prices could also be lower.

THE SUPREME COURT AND THE HIGH PRICE OF THEATER POPCORN

Indeed, before the late 1940s all major movie studios—for example, Paramount, Fox, and Warner Brothers—owned chains of movie theaters, very likely in part to minimize the hassle factors or transaction costs we have noted. However, because in the 1940s the studios required their theaters to charge customary admission prices and restricted showings in non-owned competing theaters, the U.S. Department of Justice took the studios to court for monopoly price fixing, arguing that the studios were clearly violating the nation's antitrust laws (specifically, the Sherman Act).

After a series of lower court decisions, the studios were required by the U.S. Supreme Court in 1948 in *the United States v. Paramount* to divest themselves of their theater chains. The presumption underlying the ruling was that divestiture would lead to greater competition in the theater market and lower ticket prices. However, the exact opposite occurred: In the two decades following the divestiture decision, movie ticket prices rose substantially relative to the general price level.[21] To be exact, between 1948 and 1958, movie ticket prices rose by more than 36% (despite the incentive theaters had to try to substitute concession revenues for ticket revenues), while the consumer price index (CPI) rose by only 20%. Between 1958 and 1968, movie ticket prices rose by almost 69%, while the CPI rose by between 15 and 16%.[22] In short, the *Paramount* decision probably increased industry costs that showed up in ticket prices that spiraled upward.

The decision also gave rise to tension between the producers and theaters on ticket and concession pricing discussed above: freed of direct studio control, the theaters sought to curb the rise in ticket prices in order to elevate their popcorn prices. Put another way, the price of popcorn is probably today higher than it needs to be, or should be. However, a measure of the inflated popcorn prices can be chalked up to an ill-conceived antitrust ruling back in the late 1940s and to continuing legal restrictions on the ability of studios to hold ticket prices up.[23] This means that were studios allowed to freely reinvest in theater chains (and organize their contracts with theaters as they did before the *Paramount* decisions), the price of popcorn would likely fall relative to the price of tickets (with the overall real price of the movie bundle going down). The popcorn lovers would no longer be subsidizing (albeit indirectly) as much as they now do the ticket prices of the non-popcorn lovers.

CHAPTER 4

THE COST OF THEATER POPCORN—ON THE MARGIN!

In the foregoing analysis, I have calculated the cost of popcorn the way many moviegoers are inclined to do so, in terms of *average* price, say, per ounce. Such a take on the price of theater popcorn is instructive, but it still misses a key insight about the price of theater popcorn, that the price of theater popcorn is not all that expensive *on the margin*. Consideration of the marginal price of additional ounces of popcorn can tell us much about moviegoers' responsiveness to the price of popcorn (or their elasticity of demand) and also something about theaters' marginal cost of popcorn production.

As it happens, Regal theaters sell three sizes of popcorn, "small," "medium," and "large" (with the large being the "tub" I have used in calculations to this point). The prices of the three sizes are structured the way we might expect, given the analysis of price discrimination covered in Chap. 3. The prices of the three sizes are $5.50, $6, and $7, respectively.

According to my own rough estimates (developed from my actually buying several containers of each of the three sizes from local Regal theaters), the small bag of popcorn contains (on average) close to 4 ounces of popcorn (not counting the weight of the bag), which means that the average price per ounce is $1.375.[24] If you buy the medium size, you will spend 50 cents more, but you will get about twice the ounces of popcorn (about 8 ounces). The price of the *marginal* ounces is therefore about 12.5 cents—which makes those ounces pretty darn reasonable, at least on the margin (don't you think?). This way of looking at the popcorn pricing structure also suggests that Regal must be figuring that its actual *marginal cost* of producing additional ounces of popcorn in the medium bag is something less than 12.5 cents. It is unlikely that Regal would sell additional ounces of popcorn if its production cost were not less than 12.5 cents. If the additional costs of the additional ounces were, say, 22.5 cents, then Regal would be losing a dime on every additional ounce sold. No profit-maximizing theater would want to sell more popcorn to lower its profits.

What makes Regal popcorn pricing strategy really interesting is that while the tub filled with popcorn is actually heavier than the filled medium bag (because the tub itself is more than twice the weight of the bag), the tub of popcorn contains 12% fewer ounces of popcorn (7 ounces for the "large" tub versus 8 ounces for the "medium" bag, again, according to my rough estimates). However, the fewer ounces do not mean that the tub is a worse deal for *all* moviegoers—*because the tubs are refillable while the medium bags are not.*

I can attest that the tub looks as though it holds more popcorn even when the tub is positioned side by side with the medium bag. However, from my samples of containers, the tub is clearly a worse deal for those moviegoers who buy a tub and believe (wrongly) that they are spending an additional dollar to get more popcorn. The tub is also a bad deal for those moviegoers who do not know that the tub is refillable.

The tub can be a great deal for groups of hungry teenagers and large families who have learned to share, and don't mind trotting off, in the middle of the movie, for refills. For the groups that refill the tub twice, the marginal cost of the additional ounces of popcorn is really quite low, perhaps as low as 5 cents an additional ounce (which, again, suggests that the marginal cost of popcorn popping is very likely lower than 5 cents).

Even moviegoers who buy tubs of popcorn intending to go after one or more refill but who never avail themselves of the opportunity can still look on the large tub of popcorn as a better deal than the medium bag of popcorn because they view the value of having the *option* of refills is worth more than the additional dollar cost.

You shouldn't infer that the groups that refill the tubs are avoiding paying the high price for popcorn embedded in the small bag. Everyone who buys popcorn by the medium bag or the large tub pays that price for those first four ounces in their containers, and everyone who buys the tub pays the 12.5-cent marginal price for each extra ounce embedded in the medium-size bag of popcorn.

Again, what the theater is doing is walking its patrons down their proverbial demand curves. They aren't so much lowering the marginal price of the additional ounces as they are hiking the price on those first few ounces. And this kind of pricing structure allows theaters to effectively charge all popcorn buyers some "admission price" for concessions, which can be used to cover their many overhead costs in providing concessions and cleanup. The pricing structure, which has a rapidly declining price for the marginal ounces of popcorn indicates that theaters are convinced that moviegoers are relatively insensitive to marginal price charges (or they have fairly inelastic demands for popcorn), or else the drop off in the price would not have to be so great to induce moviegoers to move to the next larger size. Alternately, the pricing structure for the small and medium sizes suggests that Regal can hike its price per ounce eleven times—from 12.5 cents per ounce for the added ounces in the medium bag to $1.375 per ounce for the small bag—before moviegoers will cut their consumption of popcorn in half. This observation, in turn, suggests that moviegoers' major problem with the high price of movie popcorn is not that they are dealing with a seller that is trying to earn as much profit as they can from buyers; sellers do that all the time. Rather, moviegoers get hit with a high price on the first few ounces because they, as a group, are relatively price insensitive. Whose fault is it for the high price of popcorn at theaters? I lay the blame more on fellow moviegoers than the theaters, if "blame" is appropriate in such matters.

CONCLUDING COMMENTS

Popcorn is, supposedly, a cheap product to make at home, but only because most people think only of the few cents the kernels of corn cost. They overlook the opportunity cost of their making a bowl of popcorn, and that is not a bad oversight for home-produced popcorn when popcorn is typically made in the evening when all family members are settled in for an evening of, say, watching a movie from a DVD—that is, when people who make the popcorn have few other opportunities, and their time typically has little monetary value. When people are preparing to go to a movie, they may have an array of alternative activities, including continuing to work at the office. Then, the time cost of popcorn can escalate such that, as explained in this chapter, home-produced popcorn can be quite costly, which leads to a lesson from this chapter: One reason theaters can charge a lot for popcorn (at least on the first few ounces) is that home-produced popcorn is expensive to make. And we can extrapolate: to the extent that people's time costs of making popcorn at home increases, theaters can hike their popcorn prices. Moviegoers might feel a sense of entrapment at the movies when they notice the price of popcorn, but any sense of entrapment can probably be chalked up more to the constraining force of people's time cost than their being physically inside the theater with no popcorn sellers other than the theater.

One portion of the ounce of truth, not well recognized, is that consumers in many (if not most) product markets rarely ever consider (or even think to consider) their own costs of producing the goods they buy, because such consideration would be a waste of time. Their personal cost of producing a good they seek to buy (for example, a laptop computer) is usually far removed from the price that they are charged by someone else for the good. Consumers have grown accustomed to comparing prices of producers (other than themselves) and picking the best price.

In theaters, when moviegoers go through the turnstiles they don't usually have a choice of alternative suppliers. That means, at the point of purchase, moviegoers are left without a choice. Their beef with the price of theater popcorn is probably that they see themselves as more than competitive on cost (without considering their opportunity-of-time cost), something that is not usually the case. They may think they should be able to get the same kind of deal on theater popcorn as they get on so many other goods and services they buy. In this regard moviegoers may see theater popcorn as a "bad deal" only because it is not a far better deal than it is. But then I suspect moviegoers don't think the matter through, to see where the "truth" of the matter ends, abruptly. Suppose on going through the turnstile, the walls of the lobby were lined with popcorn vendors, all after your business. They would clearly compete on price, and the price would likely fall to competitive levels, as it is supposed to do, which would be somewhere close to the marginal cost of popcorn production. That price would mean that vendors would not be able to recover some nontrivial costs of popping popcorn, not the least of which would be the lobby space, much less marketing and administrative overhead, a consequence that could lead to no one selling popcorn. In short, the restriction on alternative sources of supply inside the lobby is probably a policy that enables theaters to cover overhead costs, and then some, all of which can be welfare enhancing for moviegoers in the long run.

That digression aside, the main point of this chapter remains that the entrapment theory of movie popcorn pricing leaves much to be desired, mainly because almost all (other than brain-dead) moviegoers are aware that popcorn (and other concessions) are higher (on the first few ounces) at movies than elsewhere. If popcorn prices were truly higher than the cost a lot of moviegoers would incur to make popcorn at home, we would observe them finding creative ways of sneaking home-produced popcorn into the movies. The fact that such is very infrequently observed (even among moviegoers who complain about the high price of popcorn at the movies), I've got to believe that the price of theater popcorn is not all that far out of line, and is cheaper to buy than home-produced popcorn is to make.

Having said that, there has been a legal force in the theater/movie industry that has probably inflated the price of theater popcorn somewhat, the Supreme Court's forcing theaters to divest themselves of their theaters, which has given theaters a profit incentive to suppress their price of movie ticket prices (as much as they can) in order to inflate popcorn prices. As studios are gradually given greater freedom to reacquire theaters, or vice versa, we might expect the price of popcorn to fall, but the fall in popcorn prices will likely be at least *partially* offset by higher ticket prices. I've italicized "partially" because if movie studios do acquire rights to buy and sell theaters freely (which they are gradually acquiring), the industry will likely operate more efficiently. The greater efficiency in the industry can translate into, on balance, lower prices for the bundled experience of having a night out at the movies (which includes the cost of both the tickets and the popcorn).

NOTES

[1] I actually bought four tubs of popcorn from a Regal Theater in Irvine, California. The average weight of the popcorn in the tubs was 6.75 ounces, not counting the weight of the paper tub (1.75 ounces).

[2] These are my cost calculations from having made at home enough popcorn to fill a tub from a Regal Theater three times over. However, my tub of popcorn was significantly (47%) heavier than the popcorn made at the theater (for reasons I don't understand). My reported cost of a tub of home-produced popcorn, $.55, is the cost, assuming equal weight. Assuming equal volume, the cost of the home-produced tub of popcorn was $.81. To make the calculations, I weighed the bag of popcorn and bottle of oil before and after the popping was complete to make the necessary cost calculations. The reported cost of a home-produced tub of popcorn is the average of the three tubs made at home.

[3] As economists have conventionally argued, if movie theaters weren't protected monopolies, to one degree or another, new entrants to the theater markets would provide adult seats at the price of children's seats. Since the adult/children's pricing structure has persisted for a long period of time, existing theaters must be protected from new competition by market entry barriers, or so the conventional argument is developed.

[4] The clearance zones are determined by the potential box-office receipts of movies, as well as the population of the area. The clearance zones can have radii of from 3 miles in major cities to 15 miles in small towns (Orbach and Einav 2001, pp. 10–11).

[5] For example, if one low-income group of moviegoers is willing to buy 100 tickets at a price of $6.50 a ticket and 60 tickets at $9.50, the theaters revenues decline from $650 ($6.50 × 100) to $570 ($9.50 × 60). If the group's income goes up and the count of tickets demanded goes to 150 at $6.50 and then drops to 110 at $9.50, then the theaters' revenues increase from $975 ($6.50 × 150) to $1,045 ($9.50 × 110). Notice that both examples have the exact same price increase of $3 and the exact same decline in the number of tickets sold of 40.

[6] Kahneman and Tversky (2000a).

[7] Landsburg (1993 Chap. 16).

[8] FunFoodZ of Evansville, Indiana, maker of commercial popcorn poppers and carts, estimates that the raw materials for popcorn popped on its machines to be a nickel (2 cents for the raw popcorn, 2 cents for oil, and 1 cent each for the oil and salt), as found at http://www.hi-profit.com/funfoodz/nfppcproft.asp, accessed March 2, 2004.

[9] See Smith's discussion of the "popcorn boom" (2001, Chap. 6).

[10] Harris (1996).

[11] However, I hasten to add a word of caution in accepting the conclusion that popcorn sold in theaters is lower than popcorn sold in malls: The *average* price of theater popcorn could misrepresent the relative price of theater popcorn and popcorn sold elsewhere because the theater popcorn is sold in larger portions (Harris 1996). The added popcorn could be sold at low marginal prices, thus pulling down the average price of theater popcorn.

[12] Harris (1996, p. 44).

[13] Perfect competition is a market in which there are numerous producers of an identical product with completely free (costless) entry by producers. Consumers are also fully aware of all prices charged by all of the numerous producers. In such a market setting, producers have no control over price, and the prices charged will only enable producers to recover their production costs (including risk and opportunity cost). However, clearly real-world markets do not match well with economists' perfectively competitive model. Moreover, as the Dwight Lee and I have argued elsewhere (2007), perfect competition is not a market setting that is likely to maximize growth and economic wellbeing over time, simply because producers have little to no incentives to develop new and better goods and services.

[14] More formally, economists reason that the profit-maximizing monopolist will certainly increase its price so long as its revenues go up. However, the monopolist will not stop raising its price when revenues begin to fall. This is the case because the decrease in costs from lower sales can be greater than the decrease in revenues. Hence, profits can still rise under such conditions. For example, if the monopolist raises it price from $6 to $7 and sales decline from 100 to 80 tubs of popcorn, revenues fall from $600 ($6 × 100) to $560 ($7 × 80), or by $40. The monopolist would still raise the price so long as the drop in costs were greater, say, $50. The monopolist's profits would rise by $10. The rule economists proffer for monopolist's (or any other firm's) pricing strategy is that price should be raised so long as the reduction in revenues is less than the reduction in costs. It should stop raising it price when the reduction in revenues equals the reduction in costs.

[15] See Friedman (1990, pp. 28–29, 90–93, 249–250.) for a more formal discussion of how theaters price movies and popcorn taken as bundles.

[16] Landsburg (1993, Chap. 16).

[17] Locay and Rodriguez (1992).

[18] De Vany (1991 and 2004).

[19] De Vany (2004).

[20] For more details on how movies are priced and distributed, see Tyson (2000).

[21] See Conant (1960) and Crandall (1975).

[22] Crandall and Winston (2003).

[23] For an extended discussion of the perverse economic consequences *Paramount* decision, see De Vany (2004, Chaps. 7–9).

[24] My weight estimates are necessarily "rough" because there was a nontrivial amount of variability in weights of the popcorn in the several containers I bought. You can also imagine that the variability in the weights is dependent on the clerk filling the containers, whether he or she fills them to overflowing and packs the popcorn in the containers, as one clerk did to the point of crumbling the popcorn.

Chapter 5

~

WHY SO MANY COUPONS

Coupons—those slivers of papers that offer price breaks on so many of the products we buy—seem ubiquitous. They fall out of Sunday newspapers like confetti. They stare at us on almost every page of magazines we peruse. They cover the wrappings and boxes, inside and outside, of foods and other products we buy. Postal workers stuff our mailboxes with them. And they line the shelves of grocery store aisles. Many families have organized banks of them.

Indeed, coupons are a major worldwide business. During the first half of 2006, although down from the year before by 6%, there were 153 billion coupons distributed to Americans, or close to 42 million a day, with one-fourth of all coupons requiring the purchase of two or more products. The worth of the all *distributed* coupons during the first half of 2006 was an average $1.27 (with, generally speaking, the value of the coupon rising with the price of the product).[1] In the second quarter of 2006, half of surveyed households reported redeeming at least one coupon. However, those households that used coupons redeemed close to a dozen during the 3 months prior to the survey. The mean value of the *redeemed* coupons was $1.01.[2] The total count of coupons redeemed had been by 2006, it should be noted, in a steady decade-long decline, with the fall over 40% between 1997 and 2006.[3]

Coupons come in a variety of forms, several major categories of which include the following:

- **Free-standing inserts**, which are coupons that are unattached to publications;

- **Package coupons**, including
  - *Peel-off coupons*, which must be used at the time of purchase;

- *On-package coupons*, which can be seen on a purchased product, but must be used with a future purchase (of the same or different product),
- *In-pack*, which are similar to on-package coupons, aside from the fact that buyers may not be aware of them until they use the products purchased;

- **Online**, which are coupons that can be printed from web sites set up to distribute coupons for various sellers;

- **In-ad**, which are printed in advertisements in newspapers and magazines; and

- **On-shelf** or **shelf-pad**, which are coupons that can be found along store aisles, often just below the couponed product; and

- **Electronic checkout and discount** and **instant redeemable**, which are coupons that are automatically redeemed at the time of checkout.

However, the overwhelming majority of distributed coupons (88%) are the free-standing inserts. The second most widely used form is the in-ad coupon, constituting a distant 3% of all coupons distributed.[4]

As is perhaps transparent from household trash bins, the redemption rate on coupons is meager (and falling), a scant .8% for all distributed coupons during the first half of 2006.[5] As might be expected, the redemption rate generally rises with the dollar value of the coupons.[6] And, as might also be expected, the redemption rate for peel-off and on-shelf coupons is, on average, several times the redemption rate for all coupons, and sometimes above 50%.[7] Frequent-shopper discounts, given to shoppers who have their store cards scanned, have begun to supplant coupons at many stores, especially grocery stores.[8]

Why so many coupons, if so few are actually used? One nonconsequential answer is that coupons are a relatively cheap form of product promotion, costing firms less than a penny ($.007, according to one report[9]) per distributed coupon, but such a small price per coupon results in a total cost of more than $1 billion for manufacturers. Obviously, the relatively few coupons redeemed must generate a lot of value for manufacturers. But how can firms generate value—profits—from cutting their prices in a consequential way to shoppers? Why don't they just cut their prices and avoid all the printing and redemption costs associated with coupons?

If coupon distributors make a lot of money from coupons, why would they ever collude (as they have) to suppress couponing? Why would consumer groups and the antitrust enforcers oppose (as they have) collusive arrangements among coupon distributors? If you have no idea why those questions involve serious pricing puzzles, then you need to read on.

Without doubt, coupons serve many business purposes. They can, and have been used, for market research, to assess the price sensitivity of buyers in different parts of the country (by sending out coupons with different dollar values to different groups of buyers), to determine how "deal prone" different consumer groups are, to determine the appropriate prices firms should charge in the future and to induce trials and repeat customer business.[10]

Coupons that are received on one purchase and must be used to buy the same product on the next purchase can increase buyer "switching costs" and can foster brand loyalty, which is another way of saying they can increase the inelasticity of consumer demand, permitting a rise in the before-coupon, posted price. To the extent that competitors follow with similar coupons that increase the switching costs of their buyer base, the market becomes more segmented and the demands facing all manufacturers can become more inelastic, making price reductions by all less profitable.[11] However, the most common lines of argument developed by economists to explain the pervasive use of coupons are twofold: First, coupons allow for price discrimination. Second, they allow for peak-load pricing.

COUPONS AND PRICE DISCRIMINATION

Our discussion of price discrimination in Chap. 3 permits us to quickly lay out a prime economic reason for coupons: Coupons are an ingenious marketing invention that enable sellers to segment their markets into different buying groups with divergent price sensitivities and then to price discriminate, charging the price-insensitive group more than the price-sensitive group.

Coupons may be rightfully viewed as *ingenious* because they enable sellers to hide their role in hiking the price to the price-insensitive buyers. Sellers simply set a posted price that is higher than it would otherwise be absent the distribution of coupons. The higher posted price is the price that will be paid by people not redeeming coupons who, presumably, are relatively price-insensitive buyers (and must be if coupons are to work their magic on profits and to continue in use). The seller might not be able to tell price-sensitive from price-insensitive customers as they walk in the door, but the seller need not do that. The seller can simply count on the price-sensitive buyers to self-identify themselves by presenting the coupons. We can expect buyers presenting coupons to be relatively price sensitive because of the time and effort they devote to finding the coupons, clipping, storing, and retrieving them, and then presenting them for redemption. By virtue of their going to such trouble, "couponers" declare their interest in getting price breaks, and the lowest prices possible. They also demonstrate, by presentation of coupons, which might be worth only a few cents, that the opportunity cost of their time is minimal, which means that they likely have time on their hands to engage in extensive comparison shopping on price from online and brick-and-mortar sources. Just being aware of alternative products means that they can be relatively responsive to price breaks.

By identifying themselves as price-sensitive customers, couponers reveal, inadvertently, the rest of buyers as being less concerned with price (at least not sufficiently concerned to develop a bank of coupons). In a sense, buyers with coupons effectively enable sellers to stick price increases to buyers without coupons. Many buyers without coupons must, indeed, be really unconcerned about finding price breaks, given that they often go through checkout counters without coupons, even when the coupons are on multicolor peel-off pads just below where couponed products are shelved. According to one study, something over half of the units of products with coupons on pads just below the products are bought by people who do not go to the trouble of peeling off a coupon and taking it to the counter.

There are several good reasons for expecting buyers without coupons to face a posted price when a coupon promotion is going on that is higher than the posted price prior to the coupon promotion.

- First, the coupon can increase the demand for the couponed product, even among buyers who do not use coupons, because the coupons can
  - Draw attention to advertisements,
  - Create "market buzz," especially for new products that, by their newness, have not been experienced by buyers who can, enticed by the "trial price" after coupon, can use the couponed product and pass along their assessment to friends, colleagues, and family members, and
  - Give rise to future purchases, especially when in-packaged and on-packaged coupons are tied to future purchases or to the purchases of other products.[12]

- Second, once buyers with coupons have been identified, then the old prior price, founded on some average of the elasticities of the price-sensitive and price-insensitive buyers, will be lower than the profit-maximizing price that is appropriate from the isolated price-insensitive buyers.

No one should be concerned about buyers without coupons, I hasten to add. Those buyers *can* have—and *do* have, according to research—relatively high opportunity costs.[13] The monetary value of their time that they would have to devote to couponing over a range of products could be greater than the monetary value of the coupons clipped, stored, retrieved, and redeemed. They are simply better off taking the higher prices, just as the buyers with coupons are better off by redeeming coupons the monetary value of which is greater than the monetary value of their time devoted to couponing.

COUPONS AND PEAK-LOAD PRICING

Coupons enable stores to engage in another form of price discrimination, "peak-load pricing" (a generally unrecognized argument among economists, but briefly explained by economist Steven Landsburg[14]). Grocery stores are usually very busy during the week in late afternoon and early evening. In those peak shopping hours, time-constrained, price-insensitive shoppers on their way home from work dominate store customers. Customers who shop in mid-morning often do not have jobs and are more likely than late-afternoon shoppers to be price-sensitive because they have time on their hands to search for the lowest prices on the products they buy. Coupons are a device for cutting prices for mid-morning shoppers, which means coupons are also a device for hiking prices (before coupon redemptions) for relatively price-insensitive shoppers during peak hours, reducing prices during off-peak hours, thus allowing a given number of customers to be served with a smaller number of checkout counters than would otherwise be required.

EVIDENCE ON COUPONING

Marketing and economics researchers have spilled a great deal of ink on the market and bottom-line effects of coupons. One of the strongest empirical findings is, as noted, that buyers who use coupons tend to have lower opportunity costs of time and, as a consequence, have higher elasticities of demand.[15] More concretely, working parents, who are often seriously time constrained from the demands of work and family, tend to be less frequent users of coupons than nonworking women. Senior citizens tend to use coupons more frequently than younger adults.[16] Buyers with cars can take advantage of coupons—and, in general, can be more "deal prone"—than those buyers without cars because buyers with cars often have lower costs of getting to stores with "deals" (with or without coupons).[17]

It also follows that buyers with low opportunity storage costs (in areas of the country with low housing costs) can be expected to be more responsive to coupons. They can stock up on products when coupons are offered.[18] Hence, we might expect coupon distributions and redemptions will be lower in places like Manhattan, New York City that has high land and building space costs than in places like rural Grundy, Texas that has low land and building space costs. Indeed, because of difference in land and space costs, we might expect coupon distributions to be greater in the USA than in Japan.

If coupon redemption is negatively related to opportunity costs, it follows that, from both economic theory and evidence, redemption rates should be positively affected by the dollar value of the coupons and the shelf prices of products, and this is precisely the general conclusion from empirical research on coupon redemptions.[19] Not surprisingly, buyers who were most likely to buy products, before coupons, were most likely to redeem their coupons for those products.[20]

Any number of researchers have indeed found that coupons have been used to segment markets with the end result being what theory predicts. The price-insensitive buyers are charged a higher price than was charged absent the coupon promotion. The price-sensitive coupon redeemers are charged an after-coupon-redemption price that is lower than the price before the coupon promotion, just as the theory of market segmentation predicts.[21] The coupons do what they are supposed to do, not so much as raise total firm sales as to bolster profits, although coupons could do both, which researchers have found to be the case.[22]

If a chief aim of coupon promotions is to get price-sensitive buyers to self-identify themselves, then it is understandable why some retailers will happily take coupons issued by competitors. Indeed, they might prefer to accept the coupons of competitors than distribute their own, because doing so allows the retailers to free ride on the promotional costs suffered by their competitors (an economic force that can be expected to lead to "too few" coupons issued, just as "too little" will be reduced in the presence of "external benefits"). Also, once the price-discrimination logic of couponing is understood, there is no reason retailers (independent of what dollar value manufacturers place on their coupons) should not start offering "double (or even triple) coupon" deals (meaning the cents off the price is multiplied by two or three).[23] Whether retailers double coupon depends, as might be supposed, upon exactly how price-sensitive coupon redeemers are. By offering double coupons, retailers can further segment their markets by first redeeming the coupons of buyers who accept the enticement of the original value of the coupons. Once those customers have been served, double coupon offers can then be used to appeal to buyers who need a greater price enticement to incur their higher opportunity and storage costs. Double and treble couponing, in other words, enable sellers to charge different buyers at different points on the sellers' demand curves different prices (a form of price discrimination first discussed in Chap. 2).

COUPON COLLUSION

The foregoing analysis of coupons is built around one theme: Coupons are a creative way for firms to exploit their market power to generate extra profits. Coupons may for some firms in some markets be promotional devices for extracting extra profits, and only that. However, we should not slide down the slope of assuming that the distribution of *all* coupons in *all* markets is a promotional device that serves no competitive purpose. I stress that caveat because real-world coupon strategies seem to suggest coupons can be founded on good old market competitiveness.

In the mid-1990s major coupon distributors began trying to curb their coupon distributions, with any such curb being inconsistent with the theoretical presumption that coupons allow everyone to increase profits. General Mills announced in 1995 that it intended to do away with coupons in favor of "everyday low prices." However, when other cereal manufacturers did not follow General Mills' lead, General Mills abandoned its termination of coupons.[24] The following year, Procter & Gamble and nine other major coupon distributors agreed to terminate the distribution of all coupons in three cities in upstate New York. Consumer groups protested, going so far as to organize boycotts of P&G products. P&G terminated its coupon cartel when antitrust prosecution was threatened, paying out $4.2 million in penalties to close the antitrust investigation.[25]

This case is interesting because the particulars do not square with the type of monopoly, price-discrimination theory of coupons developed to this point. If all coupon promotions do nothing more than enable manufacturers to generate monopoly profits, why would manufacturers want to suppress their distribution? The transparent answer is that suppression makes no economic sense—*if coupons are not used in competitive market environments*. The antitrust enforcers should not want to break up a cartel, because such a break-up would send the conspirators back to extracting monopoly profits through price discrimination embedded in their coupon distributions.

Again, the P&G coupon cartel case makes no sense from conventional monopoly, price-discrimination theory but does make sense from a different theoretical perspective, the economics of information. The late George Stigler argued in his 1961 seminal paper on "The Economics of Information" (briefly mentioned in Chap. 3) that one of the most unrecognized but widespread features of markets is "price dispersion," which means that product prices (and qualities) can differ across markets.[26] The extent of price dispersion can be influenced by, among other economic considerations that affect buyers' search costs, the information consumers collect on prices, a point that led Stigler to several important deductions:

- The greater the size of the market (in terms of geographical breadth and count of products), the greater the search cost and the greater the degree of price dispersion across the market;
- The more buyers spend on a good, the more incentive they have to incur search costs and the lower the price dispersion; and
- The greater the number of repeated purchases, the more extensive buyers' searches and the lower the price dispersion.[27]

According to Stigler, consumers will necessarily be driven to acquire some optimal amount of information on prices (and other product features), given search costs, which means they will remain uninformed about some prices in their markets. Buyers can also be expected to acquire more information on the prices of more products when search costs fall.

Stigler's argument suggests that search costs can fall for any number of reasons, not the least of which might be the advent of new and more effective means of advertising product prices (and features). Why? Because advertisements can contain easily accessible information on prices, which can ease the search costs of buyers, causing them to know more than they would know without advertisements about where to find the best buys. Advertisements might be costly, but they can still lead to lower (average) prices because they induce price competition as buyers move to the lower price sellers in their markets.

Coupons can be seen not only as means of competitively lowering prices, but also as one more effective form of advertising product prices, which means that more consumers are alerted to more prices across their markets that, in turn, can intensify price competition among all firms (both those that distribute coupons and those that don't). Coupons might still result in a gap in the prices paid by relatively price-sensitive and relatively price-insensitive consumer groups, but the gap can emerge at a price base that is lower than would have been realized if coupons were never created and widely used. In short, the profits of firms in given markets can, because of coupons, be lower than they would have been without coupons, or if coupons, as a form of spreading price information, were suppressed.

From the perspective of Stigler's information economics, coupons can increase market efficiency in two ways: First, they make advertisements more cost effective. Second, by increasing consumer information of the existing price dispersion, coupons can foster greater price competition among manufacturers. From this perspective, P&G's coupon cartel makes economic sense, for P&G and its conspirators, but not consumers. P&G, no doubt, would like nothing better than to suppress any force that encourages price competition.

This perspective makes understandable an array of research findings on an important effect of coupons on many shelf prices, namely that shelf prices have often gone down—not up—with coupon promotions.[28] That relationship becomes more understandable when it is realized that coupon promotions can be used when manufacturers are faced with a lower market demand for their products, necessitating a price cut of some sort.[29] Coupons can be used to make sure that the full price cut is not received by all consumers. Coupons also tend to be used most heavily at the end of manufacturers' fiscal years, when they may be trying to lower their inventories and to boost revenues to improve their profit picture.[30]

Coupons can also add to store efficiency by allowing stores to expand their customer bases and engage in peak-load pricing. The customers induced by coupons to shop at stores in off-peak hours enable stores to spread the cost of their plant and equipment over more sales. As noted earlier, stores might even be able to reduce their employment of plant, equipment, and checkout counters. Such efficiency improvements can show up in increased market supply of available outlets and greater downward competitive pressures on prices, which can lower posted shelf prices, as well as lower prices after coupon redemptions at times other than off-peak hours.

Sellers (both manufacturers and retailers) may, in short, be using a two-prong approach to cutting prices and improving sales. Instead of cutting prices across the board, they cut prices to price-insensitive buyers somewhat, but then offer price-sensitive buyers an even greater price reduction through coupons. In effect, the price charged price-insensitive buyers is still higher, *relatively speaking*, than the price charged price-sensitive buyers.

Shelf prices could also fall with the issuance of coupons because of market reactions of competitors. As marketers Aviv Nevo and Catherine Wolfram have presented the argument, let's suppose that Kellogg wants to increase its sales of Raisin Bran with a narrowly targeted coupon promotion among relatively price-sensitive college students. General Mills might try to protect its market share in breakfast cereals with a lower shelf price for Cheerios, or with a coupon of its own for college students just to match Kellogg's coupon promotion effort.[31] But then Kellogg might respond by broadening its coupon distribution to professors, and then to students' parents, and so on. The end result can be that coupons are spread so widely that Kellogg gives up on coupons and decides to lower its shelf price.[32] This is to say, efforts to price discriminate to bolster profits can, under some market conditions, lead to across-the-board shelf-price reductions and to lower, not higher, firm profits.

CONCLUDING COMMENTS

For a long time economists have told their students that coupons are a creative mechanism by which price-sensitive consumers notify sellers of their price sensitivity, enabling sellers to segment their markets and to charge consumers without coupons more than they charge consumers with them. The presumption has always been that coupons elevate sellers' profits. While I have developed the standard argument in this chapter (and still believe that it has a place, albeit limited, in monopoly/price-discrimination theory), I caution that competition has a way of nullifying the profitability of the most creative pricing and promotion schemes, including coupons.

NOTES

1 The value of coupons distributed during the first half of 2005 was $1.23, an increase of 3.3%, below the inflation rate of 4.3% for the 12-month period ending June 2006 (an unusually high inflation experience for the USA in recent years). This means that the average real value of coupons fell slightly more than 1%.

2 CMS (2006). The real, inflation-adjusted average value of redeemed coupons fell by 2% from June 2005 to June 2006.

3 Daniel (2007).

4 CMS (2006).

[5] The 2006 redemption rate for all coupons was 14% lower than the redemption rate for 2004, which was 1.2%. Two possible explanations for the decline in the redemption rate are that both the real value of coupons and the average length of time consumers had to redeem their coupons fell.

[6] Coupons worth $.01 to $.24 constituted less than .1% of all coupons distributed and only .2% redeemed. Coupons worth $.50 to $.75 constituted 19% of coupons distributed and 19% of coupons redeemed. Coupons worth $.75 to $1.00 accounted for 52% of all distributed coupons and 41% of coupons redeemed. Surprisingly, however, coupons worth more than $1.00 constituted 17% of all distributed coupons, but only 8% of all redeemed coupons (CMS 2006).

[7] CMS (2006). *Advantage Update*, no. 3, accessed on February 14, 2007 at http: //www.retailwire.com/Downloads/AU_3-06.pdf

[8] The count of frequent shopper discounts was not counted before 2006, perhaps because of their minimal importance. During the 12 months ending in mid-2006, frequent-shopper discounts constituted .1% of all coupons but 1.23% of all redeemed "coupons" (CMS 2006)

[9] As estimated by Santella & Associates for 2001, accessed on February 16, 2007 at http://205.212.176.204/coupon.htm.

[10] Nielson (1965).

[11] Banerjee and Summers (1987).

[12] See Levedahl (1984), White (1983), Narasimhan (1984), and Sweeney (1984).

[13] Narasimhan (1984).

[14] Landsburg (1993, p. 164).

[15] Narasimhan (1984).

[16] Bawa and Shoemaker (1987a).

[17] Blattberg et al. (1978).

[18] Blattberg et al. (1978) found that buyers who owned their own houses were more deal prone, and more price sensitive, than buyers who lived in apartments, since house owners have more storage space and lower inventory costs than apartment owners.

[19] See Ward and Davis (1978); Reistein and Traver (1982); Shoemaker and Tibrewala (1985); Bawa and Shoemaker (1987b); Bawa, Srinvasan, Srivastava (1997); and Vilcassim and Wittink (1987).

[20] See Shoemaker and Tibrewala (1985), Bawa and Shoemaker (1987b), Neslin and Clarke (1987), and Krishna and Shoemaker (1992).

[21] Gerstner and Hess (1991).

[22] Vilcassim and Wittink (1987) and Dhar and Hoch (1996).

[23] See Krishnan and Rao (1995).

[24] As reported by Nevo and Wolfram (2002, p. 337).

[25] As reported in Nevo and Wolfram (2002, p. 337).

[26] Stigler (1961).

[27] Stigler (1961).

[28] See Corts (1998) and Nevo and Wolfram (2002)

[29] Sobel (1984) and Aguirregabiria (1999). Alternatively, coupons may cause retailers to use couponed products as "loss leaders," enhancing the amount of foot traffic they experience in their stores by giving buyers a break on the shelf prices in addition to the break they get from the coupons (Lal and Matutes 1994).

[30] Oyer (1998).

[31] Nevo and Wolfram 2002.

[32] See also Corts (1998).

Chapter 6

WHY SOME GOODS ARE FREE

Economists spend a great deal of time explaining how market prices are determined, and almost all of that time is spent explaining why prices are positive (above zero). Their price analyses almost always reinforce an often-repeated quip: "There is no such thing as a free lunch."

Economists' emphasis on positive prices is understandable because most goods cost something to produce, and most production processes are constrained at some point by the old and familiar *law of diminishing returns*, which simply means that when more and more of a variable resource like labor is added to a fixed resource, like an individual plant or parcel of land, beyond some point the additional output from the additional labor must diminish. If the additional output didn't begin to diminish beyond some point, then the world's production of a good such as tomatoes could be grown on a single acre of land (or really in a flower pot). All that would be needed is for the number of workers to be continuously expanded. Since we know that growing the world's tomato supply on an acre of land is not possible, it follows that for most production processes additional output from each additional unit of labor added will begin to diminish beyond some point. It follows that beyond some point, the additional or marginal cost of production will begin to rise, at least for most goods and services. The positive and increasing marginal production costs will place a lower bound on the price that can be charged.

Granted, the plant and land size do not have to remain fixed for all time. All resources can be expanded with resulting *economies of scale*, or falling production costs, at least over some initial range. However, firms can become so large that they run up against organizational and communication constraints. Workers' and managers' incentives to contribute as productively as possible to firm efficiency and profits can be undercut by the fact that their individual contributions can, beyond some size firm, become hard to measure. Their lack of contribution can become obscured by the number of employees and size of the firm's output. Hence, *diseconomies of scale* can be expected beyond some point in firm growth, which raises again the prospect of positive rising marginal production costs and a lower (positive) bound to the price that can be charged.[1]

We would not normally expect a business to sell a good for a price below its marginal cost of production. If the price for the good were $5 and the marginal production costs were $6, the firm would be losing $1 on the last unit produced. Hence, the positive and rising cost of production will usually ensure that the price charged will be positive.

Having learned these lessons with care, many students might understandably be puzzled on leaving their introductory courses by the prevalence of so many goods that have *zero* prices, or are "free." All they have to do is look around for free goods, from parking to internet access at their universities and in coffee shops, to any number of sources of information on the Web. Microsoft has for more than a decade given away its browser, Internet Explorer. For several years, Dwight Lee and I gave away one of our textbooks over the internet, until a publisher asked to publish it, at which time we removed it from the Web site. Even now that the textbook has been published, we allow anyone to download, free of charge, the more than sixty video modules, in which we briefly review key components of the textbook.[2]

Free goods: what a good puzzle to face! I won't be able to explain zero (or even below-zero) prices of all goods that carry them, but I can present arguments other economists have developed to explain some of them, and add some new arguments, as well. The discussion is necessarily wide ranging, starting with an explanation for free wireless access in coffee shops and ending with a discussion of why some "piracy" can be good for producers. In between, I explain why the pricing strategies of Microsoft and street-drug dealers have much in common.

PROFITS FROM ZERO PRICES

Puzzlement over zero prices can be relieved often by a simple fact of business life: many firms can increase profits by providing customers a valuable service and not charging them for it, at least not directly. A good example of zero pricing is the wireless internet access provided in many coffee houses. This access makes it possible to enjoy a cup of coffee while catching up on e-mail or the news from a favorite web site. The wireless access is obviously costly to provide, but it is also a valuable service to coffee house customers, one for which many would no doubt be willing to pay more than enough for the coffee house to recover its cost. But because of the nature of wireless internet access, not charging for this service can benefit the customers and, at the same time, increase the profits of the coffee house.

First, consider the customers' benefit from cups of coffee. Its consumption is said to be "rivalrous." That is, when one person benefits from a particular cup of coffee, someone else is denied those benefits. Charging for the coffee by the cup makes sense because the charge ensures that the person buying the coffee by the cup places a monetary value on the cup of coffee that is at least as much as the value that someone else who could also drink that cup would place on it.

In the case of internet access, on the other hand, once wireless internet access is provided, it is simultaneously available to everyone in the coffee house. There is no rivalry in consumption. When one person is "consuming" internet access, her consumption doesn't reduce the internet access available to others. In this case, charging a customer for internet access would reduce her use, and benefit, without benefiting another customer and without reducing the cost to the coffee house. So, once the access is provided, charging for it directly will reduce the total value it provides consumers. Customers are better off without a charge.

It's tempting to think that this is such an obvious point that it is silly to make it. Aren't customers always better off getting things they value without paying? Actually, no, they aren't. Clearly, people are better off not paying for goods if those goods are still made available. But how many hamburgers, shirts, and cars would you get without paying enough to cover their costs and provide suppliers with a reasonable profit? We are better off paying for the goods we want than not paying and going without.

But, in the case of internet access, the local coffee house can profit by giving away the access. Indeed, it can actually profit more by giving away access than by charging for it (with a slight exception considered in a moment). Since the more consumers who use the access, the greater value they realize from patronizing the coffee house, the more the coffee house can charge for coffee (and whatever else it sells). Also, once internet access is provided, there is no additional cost to the house when another customer logs on. So, if the coffee house charges for internet access, it reduces the use of the access, the value to consumers, and therefore the total amount it can charge them without reducing its cost. The best strategy is then to make the access free of charge and let the customers pay for the value received in the price of the coffee.

There is a parallel here between a service like internet access in a coffee house and the decorations in, and general ambiance of, restaurants and stores. It is costly (easily running into the hundreds of thousands of dollars) to provide an attractive look and feel to a restaurant, but when done well, customers value it by more than the costs. It would not pay for the restaurant to charge for ambiance directly, however, since once it is available there is no extra cost to the enjoyment another customer receives from it. The better approach is to charge for the ambiance in the price of the meals, which people are willing to pay because of the pleasant surroundings.

There is a qualification to the above pricing strategy that leads to considering situations that make it profitable to completely reverse the strategy by charging for admission into an establishment and then giving away what is served inside. To see this, let's go back to the coffee house and internet access. When stating that once the access is provided, it costs nothing when another person uses it, we ignored an important limitation—space. Coffee houses have only so much space and if they have an attractive feature like internet access, then some will come primarily for the access, linger excessively, and crowd out others who are also interested in the coffee. Obviously, the more popular a coffee house, the more of a problem space is likely to be. This may explain why Starbucks charges for using its wireless "hotspot" service, which is as much a charge for the use of a table as it is for the use of the internet access.[3] Whether Starbucks will continue this charge is debatable, however, since some coffee houses are providing wireless access at no charge, at least in the college towns of Athens, Georgia and Irvine, California. Clearly, from my personal travels, more and more hotels are providing web access at no charge—more accurately, no direct charge.

Space limitations are important in the pricing of many goods and services. For example, the fee universities charge for taking courses commonly depends on the number of units taken.[4] Students are obviously paying for the right to sit in class and benefit from the knowledge and lectures of their professors with their tuition payments. Use of the university facilities such as the library is made available at no additional charge. But, there are other facilities on campus that are likely to be more popular than the library, and more subject to space limitations, such as the recreation center, the parking decks, and the campus movies. Not surprisingly, students are typically charged extra for the use of these facilities.

Space limitations also provide part of the explanation for why the prices on dinner menus are higher than those for the same, or almost the same, meal at lunch. People typically don't linger over lunch as long as they do over dinner, so at least some of the higher dinner prices are charges for the extended use of the limited restaurant space.

When the facility itself is the main attraction because, for some reason, it is special, if not completely unique, then it may be appropriate to charge for admission to the facility and give away many of the things consumed in it. Few people go to a restaurant or coffee house just to enjoy the décor, even when very nicely done. The food and coffee are the dominant attractions, and so it makes sense to let people enter the restaurant for nothing and charge only for the food. This is not true of cruise ships, however, even though they are occasionally thought of as floating restaurants. The main attraction of cruise ships are the cruises, not the food. If people were allowed to board cruise ships free of charge, they would quickly be full of passengers, with, no doubt, a long line left at the gangplank as the ships pull away from the dock. It would require outrageous prices for the food served on board to cover the cost of building, operating, and maintaining the ship. Under such a pricing arrangement, cruise ships would be overcrowded with dieters and provide less value to the typical passenger than they now do, and therefore generate less profit. Far more value and profit are created by charging people on the basis of the quantity and quality of the space they want (more for a large cabin with a view than for a small cabin without one), and including the food in the price of the cruise.[5] Of course, there are services on board, like massages, haircuts, and skeet shooting that are paid for separately.

Disney World and Disneyland are also good examples of facilities that are major attractions in their own right, quite apart from what patrons do once they get inside. Many people would enjoy walking through Tomorrow Land and along Main Street in Disneyland without going on any of the rides, so it makes sense to charge admission to the park but not for the individual rides. This is in contrast with a run-of-the-mill amusement park where there is often no admission fee, but charges for the individual rides, simply because there is no value to being in the park apart from going on the rides. There are, of course, plenty of things for sale in Disneyland and Disney World, including food. As opposed to a cruise ship, it makes no sense to include the price of food in the admission fee to an amusement park because people remain in the park various lengths of time. Some people stay long enough to get breakfast, lunch, and dinner for their admission fee, while others get only lunch. If food were covered in the admission price, those who got only lunch would be effectively subsidizing those who downed three meals.

The point of this section is that firms charging for everything they sell, at least directly, is not always a good idea. It is often more profitable to provide services "free" since that increases the value of complementary services on which firms can set prices at levels that generate more revenue than would be achieved by charging directly for everything. Charging one price for a group of related services also eliminates the expense of collecting fees, while increasing customers' convenience and reducing their transactions costs.

CHAPTER 6

THE NATURE OF PRODUCTS AND PRICING STRATEGIES

When economists talk about positive prices for goods, they typically mean what might be called "regular goods." In order for a good to be considered "regular" consumers must know its value; its value must be unaffected by how many other consumers are buying and using the good, and current consumption of the good will not affect future consumption.

There are three classes of goods that don't fit the usual theoretical mold economists use:

- Experience goods,
- Network goods, and
- Addictive goods.

The inherent characteristics of these goods can provide producers with an incentive to lower their prices, if not give them away or even *pay* prospective consumers to buy them, at least for an introductory period of time. Of course, producers can be expected to make price concessions in the short run, but *only* if they can reasonably count on future payoffs that more than cover the initial below-cost pricing, which they can rightfully view as a part of the required investment in developing the market for a new good.

THE PRICING OF EXPERIENCE GOODS

Experience goods are goods whose value cannot be fully known before using it. When we contemplate buying something new—say, a new laundry detergent or the first published work of a budding novelist—by definition we have precious little information on the quality and usefulness of the good, and may have even less of a basis on which to judge the good's subjective value to us, a fairly obvious point that economist Phillip Nelson brought to economists' attention nearly four decades ago.[6] True, we may have used products from the manufacturer of the laundry detergent or the publisher of the novel, but the value of such information can be limited since a substantial majority of new detergents and books introduced in any year are disappointing in one or more regard, and many fail miserably before the end of their first year on the market.

Hence, trying new products of any kind can be a gamble for consumers just as for producers. Producers might introduce ten products in the hopes that one or two of them are sufficiently successful to cover their own costs, plus the development and production costs of the eight or nine products that are poor financial performers, or that fail altogether. Similarly, consumers might have to try ten products in order to find one or two products they like sufficiently to make all ten purchases worth their prices plus the search costs incurred. Because of the gambles involved, many producers put only the most promising products into production—those that seem to be a quantum leap ahead of any available products or those that pass the assessment tests of focus-groups or reviewers. Consumers often do the same thing by staying with products they know or similar products. This means that producers often have to go with an even more restricted set of new products: those that can cause consumers to change their entrenched buying habits (patterns or rules) because they are perceived to be only marginally different from established products. Given that consumers often confront an array of "new" products touted as "improved," which all too frequently are not, consumers will understandably be guarded in the products they test, given the low probability of finding a "new and improved" product that lives up to its billing. For this reason, "For new technology to replace old," the late management guru Peter Drucker is widely reported to have once quipped, "it has to have at least ten times the benefit."

Consumers' ignorance of new products is, however, not an insurmountable barrier to consumption, however. Rather, it is just another economic (cost) barrier for both producers and consumers to overcome. And, there are gains to be had by both consumers and producers from overcoming consumers' ignorance barrier.

Consumers can, of course, diminish their own ignorance of the intrinsic value of new products by experimenting with an array of new products and by searching out media reports of the experiments and objective and subjective evaluations of experts, such as product reviews from the laboratory technicians at publications like *Consumer Reports*. Consumers can also seek the advice of friends, family members, and colleagues about their new product experiences. Indeed, many family, collegial, and friendship groups, who have grown to know each other's preferences and who share their acquired information from experiencing new goods, can serve two economic functions:

- First, information-sharing groups can reduce the number of new products each group member needs to experience, thus reducing each member's search costs.

- Second, an information-sharing group can increase the diversity of the group's "portfolio" of new goods, because of the enhanced information provided through members' objective and subjective evaluations. Just as selecting a diverse portfolio of financial securities can reduce the risk investors face and can increase the rate of return for the risk takers, so groups who share information about the value of new goods can reduce the risk members face in buying them.

With people accustomed to obtaining product information from others, it should be no surprise that television ads, especially "infomercials," rely heavily on "testimonials" about products. Consumers might rightfully fear that the testimonials have been corrupted by payments made to the people who testify to products' worth, and therefore can be expected to discount the testimonials' value, but that hardly means that they can or will *totally* dismiss them. After all, testimonials can be remembered in compressed form without specifically remembering the people giving them.

Consumers might also give a testimonial some credibility since the spokesperson did think enough of the product to endorse it for a fee. Nevertheless, consumers' reluctance to give credence to aired evaluations for unknown products from unknown people helps explain the value of brands and "star power" on consumer purchases. Well-known brands can corrupt their own values by asking people/actors to provide fraudulent evaluations. Established "stars" can likewise undermine their own credibility (and future income stream) if they endorse seriously flawed products.

That is to say, consumers are likely to give more credence to product endorsements for established brands from celebrities (or anyone else whose reputations are a significant source of their future incomes) than to unknown people because the stars have more to lose from misrepresenting their true assessments of the products they are endorsing. Consumers can also reason that producers are paying "big bucks" for celebrities' endorsements, which can suggest that the producers have confidence that the products being endorsed are superior to others and will measure up to the stars' claims.

Naturally, the value of information about new products depends on the prices and variety of similar products, as well as the frequency of purchases and durability of goods. Consumers may seek little information about the experiences of others with, say, penny gumballs, since it is cheaper for most consumers to try several different gumballs than to take the time to ask others for evaluations. On the other hand, if the options are expensive—say, plasma or LCD screen televisions—then experience sharing is likely to be common. If there are few options in a product category—say, cans of mixed nuts—consumers might find trying all options to be less costly than seeking others' evaluations. However, when there are many options, as is the case for new fiction and nonfiction books (a market in which tens of thousands of new books are released each year), then product reviews and word-of-mouth information sharing can be expected to be important to the success (and failure) of books.

If consumers expect to purchase a product frequently or if the product is highly durable, then consumers have a strong economic motivation to engage in information searches, including obtaining the personal assessments of others in their relevant groups. For a good that is bought frequently, such as bread, there are greater gains to be had from finding the right product, and substantial reason for incurring search costs. Or, if a good is durable (and especially if the good is expensive, as is the case with the purchase of a car), then an extended search can alleviate substantial costs of making mistakes.[7]

Regardless of how they seek to overcome their ignorance, we can expect consumers to extend their search for information about prices and the objective and subjective values of new products so long as the additional gains from searches exceed the additional search costs. And, we should expect the additional gains from an extended search for information on prices and product quality to fall, at least beyond some point, since consumers will initially focus their attention on the most productive avenues of search. Additional costs of searching for information will probably escalate when a search is extended, since consumers will usually start their searches by giving up their least valuable activity. Since, by definition, the cost of searching is the value of what could have otherwise been done in that same amount of time, consumers make the cost of the initial search as low as it can be. To extend their searches, consumers have to give up more and more time to do other things, which means that the additional cost of extending their searches (and the value of what is given up) will rise.

The point is that as a search for information is extended (and the additional gains fall while the additional costs rise), there is some rational limit to how much people will do to allay their consumer ignorance, which means that there are economic limits to how many goods people will experience. This also means that consumers will remain, to some degree, rationally ignorant of the prices and the objective and subjective worth of many products, which implies that buying mistakes will abound, but the costs involved in these mistakes are expected to be less than the costs of avoiding them.

Since the cost of searching out pricing information is typically far lower than the cost of searching out information about product qualities (given that experiencing a good can be far more time consuming than reading and comparing prices), we would expect consumers to be far more knowledgeable about the prices of an array of new (and old) goods than their objectively and subjectively assessed qualities. To the extent that consumers restrict both their searches and their experiences with new goods, we should expect consumers' search costs and the costs of experiencing new goods to somewhat limit the entry of new competitors. Limits on entry should give producers of established, long-experienced goods a market advantage, meaning a monopoly edge, or the ability to charge more for these products than if search costs were lower.

Put another way, the higher the search costs for information on the worth of new goods (with bad experiences with new goods being part of the search costs), the more producers of established products can charge. This is because consumers can reason that it is less costly for them to continue to consume a known good and pay a higher price for it than to incur the search costs necessary to find alternative products that are better deals.

However, consumer search costs can also create an upper limit on the prices that producers of established products can charge. Producers of established products must understand that price hikes can lead to extended consumer searches and more consumers defecting to new products that prove to be "improvements" over the established (overpriced) products.

The internet (along with other forms of media) has been a boon for consumers seeking pricing and product quality information, mainly because the internet has lowered search costs for comparative pricing and for objective and subjective assessments of product quality. Many web sites now provide comparative prices for just about any good or service. Product reviews by experts and users are also easily accessible on the internet. As a consequence, the internet has undercut the strength and duration of any monopoly pricing position that established products might have had.

How can producers of new products get around consumer inertia grounded in risk aversion and search costs? For a growing array of producers, the solution has been to allow consumers to "experience" the good by giving away the product initially or by passing out samples of new products bundled with Sunday newspapers or offered at "taste booths" in stores. Costco now has so many sample booths in its warehouses on Saturdays and Sundays that shoppers can practically eat lunch by sampling foods as they move about the aisles. Car dealerships offer extended test drives that may last for days. Newspapers often offer free trial subscriptions to new residents. Music buyers can go online and stream samples of performers' music. Many studios offer movie trailers both in theaters and online. Publishers now allow prospective buyers to free downloads of the first few chapters of new books.[8] I noted earlier how Dwight Lee and I have taken sampling of textbook materials to a new level by allowing anyone to stream or download the accompanying video modules (with the running time of all 65 video modules exceeding nine hours).

Why the "freebies"? For producers, freebies can have both competitive and monopoly intentions. The competitive intention is perhaps obvious. Producers of new products use freebies to lower consumers' search and experience costs, thus encouraging consumers to move away from established brands, and sampling does increase sales.

According to one study, 92% of in-store shoppers prefer free product samples to cents-off coupons.[9] Another study found that 70% of shoppers will try a sample when asked and 37% of those who try the sample will buy the product. In-store samples can boost sales on the day the samples are given out by as much as 500%.[10] This explains why stores and manufacturers spent in 2002 $1.2 billion providing free product samples.[11] While another study did not find a difference between the increase in product sales to samplers and nonsamplers, it did find that sales of sampled products to the samplers goes down as the number of samplers increases at any one time.[12] Costco store managers I've casually and briefly interviewed have an easy explanation for the growing number of sample booths they have scattered throughout their stores. Daily sales for almost any product that is sampled can rise 30% above the sales bases for days when the samples are not provided. For meat samples, daily sales can "easily" more than double.[13]

However, sampling can also have monopoly intentions and effects. Any increase in demand can translate into the producer being able to charge more for purchased products than they would otherwise. Indeed, some producers may hike their prices during the time they are handing out freebies. Producers can also reason that by giving consumers free samples, they will cause consumers to truncate their searches for objective and subjective evaluations of other products, which suggests that producers can hike their prices somewhat because their sampling encourages consumers to remain ignorant of other products. The sampled product can then become the established product, which means that producers can hope that sampling lowers consumers' sensitivity to a price increase. Of course, the sampling advantage might not last for long, given that other producers will have reason to follow suit and provide samples or otherwise offer free trials.

Producers' use of freebies is necessarily limited by the ability of other producers to benefit from the experience consumers get from the freebies. If consumers can sample a new product—say, a new set of earphones—and then buy the exact same product from someone else, then producers are going to be very reluctant to provide the samples, for the simple reason that producers providing the samples incur costs that producers not providing samples don't have to incur. Producers not providing the samples still reap gains from greater sales, which can be further expanded because these producers can charge a lower price than those providing samples.

Hence, one market condition that helps to explain the prevalence of "free goods" is, ironically, restrictions on competition. Because branding is one market entry restriction, branding (or at least the potential for branding) can encourage the distribution of freebies (or just samples). Brand loyalty can restrict consumers from switching to other producers and can restrict the entry of potential competitors (or duplicators).

Such entry restrictions should not be viewed as all bad, if they encourage freebies and sampling—and, for that matter, encourage the development of new products and their markets. We have copyright laws that restrict market entry precisely to provide requisite economic inducements for the development of products and their markets. Publishers would not be likely to release nearly as many new books each year and provide for sampling on the web if, once the books and their markets are developed, anyone could pirate the books and sell copies more cheaply than the prices that must be charged by the originating publishers who incur the book and market development costs.

Indeed, pirating of digital (or electronic) goods—digital books, digital music, digital movies—is a major threat to the development of such goods precisely because the reproduction (marginal) cost of digital goods is either zero or close to zero. That means that pirates can make money at prices slightly above zero. The problem of piracy is compounded by the fact that all buyers of digital goods can potentially become distributors by giving away, via the internet, numerous free copies to friends, family members, and colleagues who themselves can become relay pirates. This piracy kills off original producers' incentives to develop digital goods in the first place. This means that "free (digital) goods" could come with a huge societal cost, the nondevelopment of goods that, if they were developed, could improve human welfare far more than free goods.

THE PRICING OF NETWORK GOODS

Consumers obviously receive value from the candy bars they eat. The value of their candy bars is not consequentially affected by the fact that other consumers may (or may not) be buying candy bars. A *network good* is categorically different. It is a good the value of which is affected by how many other people are buying and using the good. The greater the number of users of a network good, the greater is its value to all users.[14] The classic example of a network good is the telephone. Telephones require a real-world, physical network through which calls can travel. A telephone is of no value to the owner if the owner is the only person with a phone. If someone else owns a phone, then the value of the phone goes up for both phone owners because they can call each other. As the sales of phones increase, the value of the individual phones can increase because the growing number of phone owners have an expanding array of calls they can place.

The operating system for desktop and laptop computers is also a network good with "network effects" (or benefits to users from the prevalence of other users) that could rise even more rapidly with the number of users than is the case for the telephone. Unlike the telephone, anyone with a personal computer can get some benefits from owning an operating system even if no one else owns one, because the computer owner doesn't need to involve anyone else to use his or her computer. However, if other people use the same operating system, then all users can share their work and perhaps more effectively develop projects together. Thus, the value to all users can rise with the number of users.

In addition, with a rising number of users of a given operating system, software developers have a growing incentive to write applications for the operating system, which increases the value of the operating system to all users, thereby increasing the demand for the given operating system. The increasing demand for the operating system can stimulate the development of even more applications for the operating system, which can further increase its sales.

If a given operating system shows signs of becoming dominant, then the market can "tip" toward the system as everyone starts buying it in *anticipation* that application developers will write more applications for it. Application developers can write more applications for the operating system in *anticipation* that users will all want to use the operating system, all because people begin to believe that the system will be dominant and will have a greater array of more valuable applications than alternative operating systems. The developer of the operating system to which the market tips will see its demand escalate, with its market share expanding *because* its market share is expanding, while other operating system developers will see their demand and market shares contract as their operating systems' value for their users drops with the contraction in the number of their users and available applications.

For a network good such as an operating system, there are economies of scale on both the supply and demand sides of the market. There are scale economies on the supply side since an operating system is a "digital good," which means its reproduction costs are close to zero, if not zero, because software is (largely) nonmaterial, made of nothing more substantive than 1s and 0s, or electrons. A digital good can be duplicated by pressing a few buttons on a computer.

There are scale economies on the demand side because, as noted, the value realized by users escalates with the growing number of other users and applications. Under such market conditions, we should initially expect the competition among existing operating system developers to be aggressive, if not fierce, because the payoff can be so big: dominance of the entire market. The loss is also potentially large—elimination from the market—as consumers and application developers move to the dominant operating system.

Producers of regular goods, such as the candy bar mentioned earlier, have the usual reasons to lower their prices. They face the ever-present law of demand, or the inverse relationship between price and quantity, considered in all previous chapters. If a regular-good producer lowers its price, it can sell more units to more consumers. The consumers who buy the regular good individually gain because of the lower price, but not because of the greater sales to more people. The demand, along with buyers' value of the good, stays put with a reduction in price. Hence, when considering regular goods, economists stress, in effect, a rule to their students: "Price doesn't affect *demand* (or the functional relationship between price and quantity). Price affects the *quantity demanded*. Other market considerations—for example, income and weather—affect demand (that is, the position of the demand curve when graphically illustrated)."*

Producers of network goods face the usual incentives to lower prices in the near term since lower prices can lead to greater near-term sales. However, they have an additional incentive to lower prices: the greater current sales can increase the value of the network good to all consumers. Therefore, greater sales can hike future demand. Moreover, an even lower current price for the network good can lead to even greater current sales, which can lead to an even greater hike in the future demand.

Following the inherent market logic of network goods, there is no reason why producers of a network good such as the operating system should stop lowering its price to something that is "low." Why not "charge" a zero price? For that matter, why not "charge" a below-zero or negative price (which means the developer pays the users to buy the operating system)? Such lower prices can also stimulate initial, short-run "sales," raise current use, increase the array of applications, and

* For a more detailed discussion of network good, graphically illustrated, see McKenzie and Lee (2006, chapter 7) and video module 7.3 at http://media.merage.uci.edu/McKenzie/Modules.html.

hike future demand even further than future demand would be with only a "low" price.

Of course, zero and below-zero prices can't be expected to last forever because the operating system developer must ultimately be able to cover its development costs. Indeed, the developer can be expected to charge zero and below-zero prices only because such prices enable the developer to eventually raise its prices with the expansion in the future demand. With the higher future price, the developer can more than cover its current and future production costs and any initial outlays made in the form of below-zero prices.

That is to say, to justify initial zero and below-zero prices, the developer must anticipate some monopoly or market power that will enable it to charge above-competitive prices going forward. In the case of an operating system, the developer might acquire an ability to charge above-competitive future prices because users can become "locked in" to the operating system, but only to the extent that users confront the prospect of incurring "switching costs" to move to another operating system. To switch operating systems, users may have to buy and learn another operating system and maybe even a new computer. Users will also have to forego the benefits of belonging to the established operating system network with all other users and with the array of available applications. Because of such switching costs, alternative operating systems may have a tough time entering the market and attracting users.

While restricting entry, switching costs can have benefits not only for the established operating system developer, but also for consumers. First, the switching costs can hold the network together, with the network benefits continuing to flow to all users. Second, the prospects of the operating system developer being able to charge above-competitive prices and to reap monopoly profits in the future can heighten the operating system developer's incentive to lower its price initially for the purpose of developing the network and to aid application developers in writing programs for the operating system. A reasonable working pricing rule could be: the greater the expected future profits, the lower the initial price—and the more likely the current price will be zero or below zero.

Third, switching costs can be expected to impose an upper bound on the price the established operating system developer can charge in the future—and, consequentially, a lower bound to the initial price. If users perceive that the price that will be charged going forward in the future is greater than the perceived switching costs, then users can be expected to make the switch to another operating system.

The established operating system developer's ability to charge a high price can also be checked by new entrants proposing to cover some of the users' switching costs. Why would any new entrant do such a thing? The answer has already been laid out in the above discussion of the interaction between the current price charge and future demand: A new operating system entrant might cover some switching costs with the intent of building its own network, thus enabling the entrant to charge above-competitive prices in the future. This line of argument means that any established operating system developer might dominate its market—indeed, it might be the only operating system developer—but can still face strong competitive pressures to contain its future price that, again, can restrict the incentive the operating system developer has to lower its initial price.

And, the operating system developer has to recognize user fears that he will in fact charge an exorbitant future monopoly price, making users reluctant to take the initial bait in the form of low, zero, or below-zero initial prices, and won't join the network. That is, user fears of exorbitant future price hikes can make it difficult for the operating system developer to build the network and become the dominant, or only, operating system developer, all of which can, again, check the ability of the developer to charge a monopoly price in the future, which can also check how far it can lower its initial price.

NETWORK EFFECTS AND THE MICROSOFT ANTITRUST CASE

The foregoing line of argument obviously helps to explain why Microsoft's Windows now dominates the operating system market, with more than 80% of the world's personal computers running Windows. In fact, Microsoft's chairman Bill Gates laid out the forgoing pricing logic for a network good like an operating system in 1981 in a conference talk. Gates asked his audience of Microsoft executives at a retreat,

> Why do we need standards? It's only through volume that you can offer reasonable software at a low price. Standards increase the basic machine that you can sell ... I really shouldn't say this, but in some ways it leads, in an individual product category, to a natural monopoly: where somebody properly documents, properly trains, properly promotes a particular package and through momentum, user loyalty, reputation, sales force and prices, builds a very strong position with that product.[15]

In mid-1985, Gates wrote John Scully, then CEO at Apple, asking if Apple would consider licensing the Mac operating system to Microsoft. Gates explained to

Scully that the Mac system needed to be disconnected from a particular computer and then sold to *all* computer manufacturers at a low price in order to build a network of users and application developers. Scully turned down Gates, and Gates followed the strategy of offering Windows at a low price and, to overcome switching costs, easing the problems new and established application developers faced in writing for Windows by freely giving away application development kits.[16] The rest is history. The Mac operating system lost its market dominance and IBM was never able to get market traction with its OS2 operating system, while Windows took over the operating system market.[17]

Consequentially, Microsoft became what the Justice Department viewed as a monopoly. In its antitrust complaint filed in 1998, the Justice Department charged Microsoft with being the "sole entry point" to the operating system market and having "no viable competitor,"[18] all founded, the Justice Department attested, on the special economics—network effects and switching costs—of the operating system market.[19] The federal district judge presiding over the antitrust case concurred totally.[20] Moreover, Microsoft was protected from competition by the "application barrier to entry," the tens of thousands of applications that had been written for Windows, which "would make it prohibitively expensive for a new Intel-compatible operating system to attract enough developers and consumers to become a viable alternative to a dominant incumbent in less than a few years."[21]

Economic legal consultant Franklin Fisher gives more details of the Justice Department's network theory of market dominance in his testimony for the government:

> Where network effects are present, a firm that gains a large share of the market, whether through innovation, marketing skill, historical accident, or any other means, *may* thereby gain monopoly power. This is because it will prove increasingly difficult for other firms to persuade customers to buy their products in the presence of a product that is widely used. The firm with a large market share *may* then be able to charge high prices or slow down innovation without having its business bid away (emphasis added).[22]

Fisher adds later, "As a result of scale and network effects, Microsoft's high market share leads to more applications being written for its operating system, which reinforces and increases Microsoft's market share, which in turn leads to still more applications being written for Windows than for other operating systems, and so on."[23]

When Microsoft began giving away its browser Internet Explorer and "paying some customers for taking IE [Internet Explorer],"[24] the Justice Department charged Microsoft with "predatory" pricing, a strategy that could only be designed to destroy Netscape, the then-dominant browser on the market, and to protect Microsoft's monopoly in the operating system market.[25] The Justice Department argued without qualifications in its filing of facts with Judge Jackson that Microsoft's business practice, including its pricing strategy, "makes sense *only if* there is a monopoly to protect."[26]

What is really baffling about the Microsoft antitrust case is that both the Justice Department and district court judge failed to understand that Microsoft's zero price for Internet Explorer could be justified by the network-effects arguments on which the lawyers and economists at the Justice Department had founded their original antitrust complaint. First, unlike in the traditional definition of monopoly, consumers could benefit not only from getting Internet Explorer free of charge, but also from having Internet Explorer integrated into their Windows operating system and not having to install a separate program. Microsoft's aggressive marketing strategy could also yield benefits to Windows users by holding the network together and, consequently, by having application developers continuing to write for Windows.

By integrating Internet Explorer into Windows free of charge, Microsoft was trying to maintain and expand its market for Windows, and it was trying to take over another adjacent market: browsing. A *monopoly* is expected to do much the opposite by restricting market supply in order to raise its price. There was no evidence introduced at trial to indicate that Microsoft had acted like the monopoly it was charged with being, but these points are only the tip of a host of arguments I've considered in my earlier book on the Microsoft antitrust case (and, hence, need not dwell on any further here).[27]

OPTIMUM PIRACY

With all of the hullabaloo surrounding the free downloading of all digital goods—music, books, movies, software—via Napster in the 1990s and the internet ever since, you might think that piracy is an unmitigated scourge in the digital era. But, might piracy be a mixed blessing for firms, especially those that produce digital goods with potential network effects? Should such firms not seek some *optimal* level of piracy?

Without question, piracy is much more problematic for modern digital goods than for old, material-based industrial goods. Additional units of industrial goods like cars are very costly for buyers to reproduce, mainly because their production requires a mammoth investment in plant and equipment. On the other hand, as already noted, digital goods can be reduced to 1s and 0s (or electrons). Once the first unit is produced it is very cheap for buyers to reproduce their own units for personal use and resale. Indeed, every copy of a digital good sold has the potential for being a master that, with no more equipment than a personal computer and an internet connection, can be used to produce and distribute an endless number of exact replicas at little or no marginal production cost. Every user of a digital good, in short, is a potential pirate—and a potential competitor.

Hence, not surprisingly, the Business Software Alliance found that "thirty-five percent of the packaged software installed on personal computers (PCs) worldwide in 2005 was illegal, amounting to $34 billion in global revenue losses due to software piracy," with the median piracy rate among the 97 countries studied estimated at 64%.[28] Central and Eastern Europe had a piracy rate of 69%. The Asian/Pacific region had a piracy rate of 54%, while North America had a piracy rate of a "mere" 22%.[29] Vietnam had the highest piracy rate of 90%, while China ranked fourth among countries in its piracy rate, which was 86%.[30]

Before Napster was declared illegal in 2001, the file-swapping, internet-based company had 50 million users freely swapping songs. The count of CD albums sold rose by a scant .4% between 1999 and 2000, after rising at a compounded rate of 14% a year from 1991 until 1999. Between 1999 and 2000, sales of CD singles fell by 39%, after rising at a compounded rate of 33% a year between 1991 and 1999, according to the Recording Industry Association of America.[31]

Does it follow that piracy should be altogether stamped out? Of course not, and no business would ever try to do so—because of the enormous cost that would be incurred in even trying for a zero piracy rate. At some point, as piracy is reduced, the cost of reducing piracy even further would be higher than the added revenues from greater legitimate sales. Perfection on any economic front is simply not, and cannot be, optimal, much less a viable option.

In addition, for many goods, some piracy can actually add to legitimate sales. This is the case partially because piracy can create network effects. For example, people might start buying a particular computer program because they want to be compatible with others who are using pirated, as well as purchased, copies of the program.

Piracy can also generate its own form of "marketing buzz." The "buzz" can convince some consumers that the pirated software will be widely used, raising the demand for legitimate copies. Indeed, some consumers might reason that if the good is not subject to at least some piracy, then it is not likely to be sufficiently popular to become the industry standard. For example, one explanation given for WordPerfect's rise to the top of the word processing market in the

1980s was that the program could be more easily copied, illegally as well as legally, than other word processing programs. Back then, when most word processing programs could not read the files of other word processing programs, having a lot of pirated copies around very likely stimulated sales and increased WordPerfect's market share.

Another example comes from researchers who reported in the *Journal of Marketing* that with the elimination of copy protection for spreadsheet and word processing software programs in England, sales went up between 1987 and 1992 by one copy for every six copies that were pirated. These researchers also concluded that when the software was introduced, pirating was very limited (as expected, given that there were few copies to pirate). Eighty percent of the copies actually bought were very likely attributable to the growing network effects of the pirated copies. Over time, the count of pirated copies decreased to 15% of all available copies.[32] Other researchers have argued that "counterfeiters" help producers identify useful technologies.[33]

Similarly, other researchers have found that illegal copying of printed publications has actually increased publishers' profits, mainly because the publications (journals, for example) that can be copied are more valuable and because the publishers can price discriminate between individual users (who might have limited needs for copying and therefore are charged a low price), and libraries (which have a demand for allowing their patrons to copy their holdings and which are therefore charged a much higher price to offset that copying).[34]

Piracy can, no doubt, present real problems for producers of digital goods. Beyond some point, piracy can eat into sales. Moreover, the potential for piracy can constrain legitimate firms' price increases, given that their price hikes can increase the demand for and price of pirated copies—and the incentive the pirates have to generate more copies. In addition, a firm that lets it be known that it looks upon piracy as an acceptable business expense, not a moral wrong, can expect to have more piracy problems. At the same time, firms must understand that their objective should be to regain control of pirated copies, not to stamp out piracy altogether. Some piracy can be good for any number of businesses.

While it is widely recognized that too much piracy can be bad for business, it can also be bad for the pirates themselves as well as the users of pirated digital goods (not the distributors of pirated goods, although too much piracy can be bad for their business, also). The problem with unchecked piracy is that the developers will have a tough time competing with pirated copies, since their effective price is close to zero. To compete for users, developers might have to lower their prices so much that, while they can cover their reproduction costs, they won't be able to cover their product and market development costs. This means that developers, seeing the prospects of rampant piracy and close-to-zero prices or close-to-zero sales, will curb the array of digital goods they produce (and will curb the continued development of new editions). Digital-goods users will then suffer the value they could have had from products that go undeveloped and underdeveloped.

Granted, with protections against piracy (in the form of copy-protection technology-based "locks" on digital goods or in the form of patents and copyrights, which means that pirates can suffer legal penalties), digital goods developers can make a lot of money, as have Microsoft, Oracle, and (lately) Apple. Pirates and users may rightfully reason that their copies cost the developers nothing. Besides, the prices developers charge are far too high, a point pirates might surmise (mistakenly) from their own low copying costs. It may in fact be true that *successful* developers make far more than a competitive rate of return. Hence, their "exorbitant" profits are in some sense "unjustified."

At the same time, I have to stress that "exorbitant" profits for *successful* developers may be a necessary precondition for a continuing flow of innovative digital (or nondigital) products. The problem developers face is that coming up with *successful* products is a major crap shoot, meaning that developers need the prospect of exorbitant profits on the few successful products in order for the *expected* profits (with potential profits discounted for risk of failures) to be large enough to spur development.

To clarify the point, suppose that the total product and market development costs for a digital product total is $9.9 million. Suppose also that only one in a hundred digital products can be expected to yield an "exorbitant" profit of $1 billion. The problem developers face is that they do not know which one of their products will be successful. Hence, the *expected* profits on the development venture will be $10 million (.01 × $1 billion), or slightly more than is required for the developer to incur the upfront $9.9 million in development costs for a given digital product. Put another way, if the profit *potential* for a successful product were a "mere" $900 million, the product would not be developed because the *expected* (discounted-for-risk) profit of $9 million would be less than the development cost of $9.9 million. Therefore, piracy can undercut product innovation because it can wipe out the "exorbitant" profits that are essential for product development in risky market environments—and, I stress, all markets have elements of risk and uncertainties precisely because they are evolutionary processes with most everything important that emerges being the product of a gazillion interactions, only a small portion of which can be under the control of any firm, even a dominant firm in its industry.

CHAPTER 6

THE PRICING OF ADDICTIVE GOODS

Because they give rise to a chemical, bodily dependency, *addictive goods* inspire, to varying degrees, their own continued and expanded consumption. That is, the consumption of the good today creates a need for the consumption of the good into the future, perhaps, for some people, at an increasing rate.[35] Classic examples of highly addictive goods (for many people) include heroin, cigarettes, and alcohol. The list of mildly addictive goods (for many people) might include chocolate and television shows.

The analysis for addictive goods can follow our analysis of experience and network goods in several important respects, because current consumption of all three types of goods can lead to increases in future demand. A reduction in the initial price of the addictive good can increase sales initially, but because the good is addictive, future demand for the good can be expected to rise. The lower the initial price, the greater the future demand. The more addictive the good, the greater the future demand for any given reduction in the price. This implies that the more addictive the good, the more responsive (or elastic) the long-term demand.

With the tie-in between current and future consumption, zero and below-zero initial prices should be open pricing options for producers of addictive goods (as is the case for producers of experience and network goods). Producers might want to initially give away their products, or pay consumers to use them, in order that they can become chemically hooked. Once they are chemically hooked, users' (subjective) switching costs can rise. Indeed, users can become *locked in*, unable to switch out of the good's consumption. Once users are hooked, producers can jack up the price, which, of course, is a strong motivation for giving the product away in the first place. Accordingly, the stronger the addiction, the lower the initial price that can be justified.

Cigarette companies in the 1960s and before followed the give-away strategy. They hired college students to walk their campuses passing out packs of cigarettes. Heroin dealers are renowned for giving users their first hit (or hits). Stores that sell boxes of chocolates often give away samples, partly to allow prospective buyers to *experience* the good, but also to create a *need* among buyers to eat more of their chocolate.

Giveaways can't be expected to be prevalent in highly competitive markets, ones with a large number of sellers and with virtually open entry into the market. In such markets, sellers who provide giveaways can cause consumers to become hooked. However, the consumers can then move to any of the other sellers, which means sellers who give away samples ("hits") can't capture many of the future sales. Also, in highly competitive markets, sellers can't raise their future price sufficiently to recover the costs of the free samples.

So, the more monopolized the market, the more likely free samples can be expected. In fact, we should expect sellers of highly addictive goods to work hard at eliminating the relevant competition through, in the case of cigarettes, extensive focus on branding their products and through, in the case of heroin, expunging the competition from given territories (markets) through threats of violence, including murders. The more addictive the good, the greater the effort to monopolize the market. And violence and threat of violence is common in street-level heroin markets because the good is so addictive, giving the dealers strong incentive to protect their markets from intruders.

With all the problems people have with addictive goods (health problems, early death, and a miserable life before death), we might rightfully wonder why people—"addicts"—would take free samples in the first place. A prominent, often-heard reason is that some people are stupid or, less coarsely, irrational. Many people just don't properly consider the future consequences (either costs and/or benefits) of their current actions.

People also vary in their inclination to become addicted to a good. Some people can smoke and never develop a compulsion to smoke one cigarette after the other. Some people can't take a sip of wine without downing the whole bottle. The problem is that people often don't know before they take those first free samples how addictive they are to various goods. That is, some people who take the free "hits" are quite rationally gambling that they are not among the easily addicted class of consumers. Some first-time users win the gamble; others don't and pay handsomely for their taking freebies, not so much for the effects of the free sample as for the gamble they took, willingly and, maybe, rationally.

RATIONAL ADDICTION

Discussions of addictive goods are generally narrowly focused on the behavior of users who have already become addicted. Addicts have, within that narrow framework of the analysis, lost a degree of their ability to make rational choices about their future consumption of the addicted good. The addicted users are thus, in that constrained reality, subject to exploitation by the sellers because their chemical dependency and the absence of competitive sellers don't allow them to respond with ease to price hikes.

However, University of Chicago economists Gary Becker and Kevin Murphy have argued that there is another perspective on addiction that can give addiction a rational interpretation.[36] Their perspective is the choice framework for *potential* users *before* they take the first free samples. Before they become addicted, Becker and Murphy suggest, future addiction can be a choice that consumers can make quite rationally by considering and discounting the stream of future benefits and costs from consumption of the addictive good. If the expected, discounted future benefits exceed the expected, discounted costs, then the first free samples are taken. If the reverse is the case, then the free samples won't be taken. This is not to say that all users make rational choices. It is, however, to postulate that *some* (maybe many, if not most) users might *become* addicted quite rationally, with a reasonably complete consideration of the consequences.

Who might such rationally addicted people be? The group could include people who do not become highly addicted to the good, such as moderate drinkers who are able to maintain some control over their future consumption and can contain the future costs. The group might also include users who already have poor life prospects. For example, they are terminally ill, depressed, or suicidal. This group might reason that the addiction can't do much to shorten their lives or make living significantly worse. And, the group might also include people with substantial resources who rightfully calculate that if they get into trouble with an addiction, they can buy their way out of the addiction through expensive and exclusive rehabilitation programs before their lives are destroyed. Any number of celebrities (for example, Britney Spears, Mel Gibson, and Lindsay Lohan) have fallen prey to one or more addictions, only to go into "rehab." Indeed, given all the free publicity given to stars who have gone to the brink of ruining their careers, only to find the fortitude to correct their ways, an addiction could be a valuable career move. Any number of celebrities' careers seemed to have been revived by their downfalls and recoveries.

This is not to say that addiction is a rational move for everyone, but only for some under some circumstances. From the Becker/Murphy perspective, what we can say is that to the extent we heap praise and valuable air time on recovered stars, we might reduce the addiction among people in the general population who see the problems the stars face, but we could also increase the tendency of stars to take their chances with addictive goods.

The Becker/Murphy perspective also allows us to argue that the more addictive the good, the more responsive consumption can be to price changes *over the long run*, which can be orchestrated through excise tax policies. If the market for an addictive good is assumed to be limited only to the currently addicted users, a higher tax on the good—for example, cigarettes—might elevate the price of the good but also can be seen as not having much effect on consumption. After all, addicts have to have their "fixes."

However, if the market for the good is expanded to cover *prospective* users—those who are not yet addicted but who are rationally considering consumption of the addicted good—then a higher tax and a concomitant current and future price increase can be viewed as having a much greater effect on curbing consumption over time. This is because prospective users will include the higher price (caused by the higher tax) they will have to pay for the addicted good for some time into the future as a part of their cost calculations.

For some prospective users, the expected stream of future costs from the addiction can rise above the expected stream of the good's benefits. The cut in consumption for these prospective users in response to the projected price increase will be sizable, since they will be cutting their consumption not only in the current time period, but also for all future time periods. If taxes on the addictive good have been raised in the past, and prospective users begin to anticipate further tax increases into the future, then we can anticipate that even more prospective users will not take those first attractively priced samples.

Interestingly, Becker and Murphy joined with Michael Grossman to show empirically that the price responsiveness of consumption of cigarettes is substantially greater in the long run than in the short run than has been traditionally assumed.[37] Their line of argument suggests that the anti-smoking lobby has had a serious effect on cigarette sales because of the lobby's work to increase the future costs of smoking through the *prospects* of increasing cigarette taxes and tightening controls on where people can smoke. Many young *prospective* buyers must now be thinking that they will eventually have to pay through their noses for smokes and then will only be able to smoke in designated areas of even their own homes, where second-hand smoke cannot be a health threat to others, especially children (even the smokers' own children).

CONCLUDING COMMENTS

In taking up the topic of why so many things are free, I've had to be selective as to explanations covered because of inevitable space constraints. I've tried to explore explanations many readers will view as unconventional, possibly eye-opening. I have intentionally paid little attention to the argument that free goods are devices that businesses use to snare and exploit hapless consumers, not so much because such an explanation has no validity (people's stupidity probably explains much of what people do and don't do, as I have conceded all along in this book), but because such a discussion of hapless, unthinking consumers would add little to what readers probably already know. However, there are additional, more or less transparent conventional explanations for free goods.

An implication of the analysis in this chapter is that buyers often get things "free" because producers can charge more for some other product that must be purchased in order to take advantage of whatever is free. Many universities allow parents of students to park on campus free of charge, but only because the free parking increases the value of the on-campus education to students and their paying parents—and increases the prices the university can charge. Bars offer cheese cubes and crackers at happy hours because they can charge more for drinks (or they can avoid lowering their happy-hour prices for drinks by more than they do).

One of the more common explanations for free goods, especially on the internet in the forms of information and other digital goods, is that the cost of producing more units of some goods for use by producing additional consumers is zero, or close to zero. Moreover, there is so much of some goods that the cost of additional units is close to zero. The collective value of all inframarginal units (all but the last unit) can be high, but in competitive markets, prices tend to be pressed down to the cost values of the marginal units. Information on almost any topic is so abundant on the internet that it is simply difficult, if not impossible, for many producers of the information to do anything other than give away the information they produce. Once produced, producers incur little-to-no additional costs from allowing free viewings and downloads to all web site visitors

True, the *Wall Street Journal* has been able to make money by charging subscriptions to its article database, but few other newspapers in the country—not even the *New York Times* (at this writing)—believe that they can charge for their articles. So, almost all newspapers give away their articles with the hope that their web advertisements will cover their production costs (as well as cover their reductions in ad revenue from the decline in the subscriptions to their printed papers, caused in part by online articles). And even the *Wall Street Journal* might start allowing free access to its web site, not that it isn't making money from the sale of subscriptions, but because it might make its new owner, Rupert Murdoch, more money by increasing its daily hits with free access and downloads of articles, and be able to collect more from advertisers on its web site than it collects from its subscribers.

Many goods are given away by producers because the producers themselves get personal benefits from seeing their products used more widely than would be the case were a positive price charged. Many musicians allow free downloads of their web-based music because they value knowing that more people than otherwise will be listening to their music or because the download will lead to "market buzz" and greater future sales of other songs and tickets to their concerts. As noted, Dwight Lee and I allow free downloads of our text-based video modules because we receive pleasure from the thought that the downloads allow us to extend our "classroom" to all points on the globe. Our textbook publisher has agreed to the free downloads because the downloads might stimulate market interest and sales in the textbook at no added cost. Besides, when we tried to sell the modules, we were unable to sell enough copies to cover the transaction costs of managing the web-based sales. Then, why not "sell" them for nothing? There really was no better price available to us.

NOTES

[1] For details on the structure of firm costs both with a constant scale and increasing scale of operations, see McKenzie and Lee (2006 Chaps. 8 and 9) and review video modules 8.1 and 9.1–9.3 at http://media.merage.uci.edu/McKenzie/Modules.html.

[2] See McKenzie and Lee (2006) and go to http://media.merage.uci.edu/McKenzie/Modules.html.

[3] Starbucks has a menu of plans, which, at the beginning of 2004 included a $6.00 login fee good for 1 hour and $.10 a minute for every minute over an hour, a day pass for $9.99, a yearly pass for $29.99 per month with a penalty if terminated within the year, and a month-to-month plan for $39.00 per month.

[4] Sometimes there is a price for each unit, and sometimes the charge is the same up to a certain number of units, and then increases if additional units are taken.

[5] Alcoholic beverages are seldom covered in the price of a cruise because they are more costly than food and there is a greater variance in alcohol consumption than food consumption.

[6] Nelson (1970). Nelson's work is an extension of the seminal work on the economics of information by economist George Stigler (1961) discussed in Chap. 4. In the development of his theory of "experience goods," which he saw as a fairly broad category, Nelson included jewelry, typewriters, radios, televisions, tires, batteries, aircraft, boats, motorcycles, heating and plumbing systems, bicycles, automobiles, music instruments, and appliances (Nelson 1970, p. 319).

[7] Nelson (1970) found that an overwhelming majority of product reviews in *Consumer Reports* were for experience goods, especially durable goods.

[8] Indeed, I have to expect that by the time this book is released, the publisher will have to allow prospective buyers to sample it in both print and audio forms.

[9] Fitzgerald (1996).

[10] Lindstedt (1999).

[11] Parmar (2003). See also Lammers (1991) who found that chocolate samples given out in specialty stores increased the sales on the day of the sampling, but the purchases were relatively inexpensive, generally under $5. As reported in Heilman et al. 2005, other researchers found that samples had a relatively greater impact on sales than a temporary price reduction. Steinberg and Yalch (1978) found that food sampling increased the sales to obese people more than to non-obese people.

[12] Heilman et al. (2005). The array of marketing studies on the impact of free, in-store samples is very limited to date. However, if an array of new studies appears, it should not be surprising that the exact short run and long run impact of samples on sales will vary greatly. This is because the exact conditions under which free samples are distributed can affect sales, and these conditions are likely to vary greatly from study to study.

[13] The interviews were done at Costcos in Orange County, California on June 9, 2007.

[14] See Arthur (1989, 1990, and 1996).

[15] Wright (1995).

[16] From a memorandum from Bill Gates and Jeff Raikes to John Sculley and Jean Louis Gassee on "Apple Licensing of Mac Technology," June 25, 1985 (from the author's personal files).

[17] And it needs to be recognized that in the late 1980s, many industry experts were betting on IBM to dominate the operating system market. In *PC Magazine*, William Zachmann wrote in 1992, "I expect that OS/2 will not only succeed but will take a lot of wind from Windows' sails in the process. I think OS/2 is the odds-on favorite to replace DOS as the dominant desktop operating system... I see a big change toward OS/2 and away from Windows over the next year" (Zachmann 1992).

[18] Klein et al., (1998), p. 19.

[19] The Justice Department used expert witness Frederick Warren-Boulton to explain how computer "users become 'locked in' to a particular operating systems [sic]," adding, "The software 'lock-in' phenomenon creates barriers to entry for new PC operating systems to the extent that consumers' estimate of the switching costs are large relative to the perceived incremental value of the new operating system. Often, switching operating systems also means replacing or modifying hardware. Businesses can face even greater switching costs, as they must integrate PCs using the new operating systems and application software within their PC networks and train their employees to use the new software. Accordingly, both personal and corporate consumers are extremely reluctant to change PC operating systems" (Warren–Bolton 1998, pp. 21–22).

[20] Judge Thomas Penfield Jackson found in the Microsoft antitrust case that "it is a commercial necessity to preinstall Windows on nearly all of their PCs. Both OEMs [original equipment manufacturers] and Microsoft recognize that they have no

commercially viable substitutes for Windows, and they cannot preinstall Windows on their PCs without a license from Microsoft" (Jackson 1999, p. 21).

[21] Jackson (1999, p. 12).

[22] Fisher (1998, p. 15–16).

[23] Fisher (1998, p. 27).

[24] Klein (1998, p. 6).

[25] The Justice Department used Franklin Fisher to explain to the court how Microsoft's pricing strategy was "predatory": "A predatory anti-competitive act is one that is deliberately not profit maximizing, save for supra-normal profits to be earned because of the effects on competition" (Fisher 1998, p. 19), a definition Fisher used to charge that, "Microsoft's [predatory] actions as to price are not profit-maximizing in themselves but are profitable only because of their adverse effects on competition," which caused Fisher to assert that any price below the short-run profit maximizing price is necessarily "predatory" (Fisher 1998, p. 7).

[26] U.S. Department of Justice (1999, p. 8).

[27] McKenzie (2000).

[28] Business Software Alliance (2006, p. 1).

[29] Business Software Alliance (2006, p. 2).

[30] Business Software Alliance (2006, p. 4).

[31] Leeds (2001).

[32] Givon, Mahajan, and Muller (1995).

[33] Maltz and Chiappetta (2002).

[34] See Besen (1986), Besen and Kirby (1989), and Johnson (1984).

[35] Becker and Murphy (1988) have developed this "Theory of Rational Addiction" on which this section is based.

[36] Becker and Murphy (1988).

[37] Becker, Grossman and Murphy (1994) found that in the short run, a 10% increase in the price of cigarettes will lead to a 4% reduction in cigarettes sold. However, if the time period is extended, a 10% increase in the price of cigarettes will lead to a 7 to 8% reduction in sales.

Chapter 7

FREE PRINTERS AND PRICEY INK CARTRIDGES

When I started working on this book, I bought a new desktop from Dell Computers. As I completed the phone order with the Dell representative, he told me, "And, we have a special promotion underway this week. With your computer we can send you a printer free of charge, and with free shipping." Why not? How could I say no? I had no reason to ask about the quality of the printer, but I had no reason to assume that it was a high-quality printer either. I was not surprised when the printer arrived that it was more or less minimal on all quality counts. It was about the size of an oversized loaf of bread but did have surprising speed and print quality for being what must have been Dell's bottom-of-the-line printer. But then, why would I care? It was free.

After about a hundred pages of printing, the black ink cartridge (about the size of a small bottle of aspirin) went dry. When I checked on a replacement cartridge from Dell, my free printer all of a sudden became pricey; the cost of the cartridge was $74.95! Most readers have probably had much the same experience on buying a new printer for little or nothing, only to learn later that replacement cartridges cost as much or more than the printer. According to one technology analyst, when an ink jet cartridge costs only $30 and holds about an ounce of ink, a gallon of ink costs $3,840, or over 1,200 times the cost of a gallon of gasoline in Southern California (at this writing).[1]

Why do printer companies employ a pricing strategy of, supposedly, lowballing the price of the printer and highballing the price of the cartridges? In this chapter, I can review several commonly cited explanations, all of which have serious weaknesses, in an effort to learn how printer/cartridge pricing strategies can be devices for segmenting the buyers of printers according to their price sensitivity, which is affected by how much they expect to print over time, and by their time discount rates.

RELATIVE PRODUCTION COSTS AND BUYER ENTRAPMENT

The solution to the puzzle of why ink cartridges often cost as much or more than printers could be simple as many economists might think: The printers and ink cartridges could be priced in line with their relative production costs, but such an answer hardly seems satisfying by simple assessment of the relative value of the materials and skills required to make printers and cartridges. Printers look for all the world to be substantially more costly to produce than ink cartridges. Printers are bigger and technically more sophisticated, or so it seems. Clearly, the $30 price of a black ink cartridge, which holds only an ounce of ink, is way above (twenty times) the wholesale price of ink, which was (at the time of this writing) $195 a gallon, or about $1.52 an ounce.[2]

An explanation for the pricing strategy might also be that by dropping the price of printers, printer companies entrap mindless and unknowing, not-fully-rational buyers. Once printers have been purchased (or received free of charge), buyers must buy their cartridges from their printer's manufacturer that hold patents on their printers and cartridges.

While this explanation is widely believed to be the "full story" (and likely has a measure of truth in it), I don't find it anywhere close to being totally satisfying. The explanation relies too heavily on the proposition that printer producers are "smart," while their buyers are "dumb." Granted, the printer market likely has some buyers in it who recklessly, or irrationally, buy printers without a thought to the cost of cartridges. Granted, also, printer companies would be remiss in their duty to maximize profits if they did not take advantage of such mindless buyers by "sucking them in" with low-priced (or free) printers and then "sticking it to them" with high cartridge prices.

With those concessions made, I note that a major problem with the argument is that many printer buyers are smart and attentive, fairly rational consumers. Many (if not the overwhelming majority) of the buyers in that group are themselves people in businesses who buy lots of printers for their businesses and who no doubt have as strong a profit motive to find the best printer deals they can as printer producers have the incentive to find the best pricing strategy they can.

But what about printers bought for home use? Perhaps home printer users don't have the same level of incentive as business people to buy smart. However, many home printers are bought by people who work in business during the day. Should we really expect business people who buy smart when they are at work to morph, once they leave their workplaces, into consumers who buy dumb outside of their workplace? Might the difference in buyer behavior for work and home purposes be a matter of a change in operative incentives to buy smart and dumb as people change their economic roles? You might think many people would buy "smarter" as consumers than as business managers, especially those who work at large companies. This is because people at work are "agents" the costs of whose mistakes can be spread over a large number of coworkers and owners. At home, they and their few family members can share a higher proportion of the costs of mistakes in purchasing. Perhaps a better explanation for any morphing going on is not that people's relative dumbness, and sensitivity to *given* costs of mistakes, declines as they move from roles of workers to consumers, but rather that the sizes of their purchases go down as they move into their roles as consumers because they buy in far smaller quantities at home than at work.

Sure, printer companies might be able to entrap some buyers—but only for a time, or so it would seem. Should we not expect buyers—individual consumers as well as business managers and entrepreneurs—to learn from sad experience that printer companies have a record of entrapping them with below-cost-printer prices and then charging exorbitant, monopoly prices for cartridges? Might they all not learn that a cheap or free printer is not the "deal" printer companies make it out to be?

Frankly, when the Dell rep offered to send me a free printer, I had no doubt that the cartridge price would be exorbitant. I was in no way fooled, or felt entrapped to buy the Dell cartridges for my free printer. I knew the pricing game Dell was likely playing from the start from the fact that over the last 25 years I have surely been through a dozen or more printers, with differing qualities and prices. I threw out the printer Dell sent once the cartridge that came with the printer (which, by the way, could not have been more than one-third full when it was new) was spent and I saw the cost of a replacement cartridge. I am guessing that many readers of this chapter are exactly like me, reasonably informed on the relative prices of printers and their cartridges.

Should not many printer buyers eventually begin to ask about the prices of cartridges when they are considering their printer purchases and to compute (in a rough way, at least) the full cost of use of various printers? If printer manufacturers are willing to compete as aggressively as they do on speed and print quality margins for buyers' business (and printer speeds and print quality are remarkably higher than they were a decade ago) and if printer buyers learn from experience, then should printer manufacturers not be expected to compete, eventually, on their printers' *full cost of use*? If so, the lowballed printer/highballed cartridge pricing strategy would be expected to break down, *eventually at least*, with the price of cartridges falling and the price of printers rising to more properly reflect their relative costs of production.

LOW- AND HIGH-VOLUME PRINTER USERS

The printer/cartridge pricing strategy might be nothing more than an attempt by printer companies to charge different consumers as much as they are willing to pay for their "bundled good," the printer combined with a series of cartridges used over the life of the printer. As economists have conventionally argued, people who use printers a lot can generally be expected to get more value from their printers than people who use them sparingly. High-volume users should have a higher demand and might be expected to be willing to pay more for the bundled good than low-volume users, *if they have to do so*. One way to charge high-volume users more than low-volume users over time is to drop the price of printers and hike the prices of cartridges, or so the argument has been made.

THE RELEVANCE OF SEARCH COSTS

Again, while helpful, this explanation does not satisfactorily explain the long-term persistence of relative prices of printers that often appear to be below competitive levels and cartridges that often seem to be above what seem to be competitive levels. I see two problems with it. First, ink may indeed be "lucrative" or have a "high margin," as a *New York Times* reported in its assessment of Hewlett-Packard's rosy financials for third quarter 2007 (at the time this chapter was being finalized).[3] However, what is often overlooked, and not said in the H-P news report, is that the ink business at a printer company like H-P may only be "lucrative" or have a "high margin" because of concessions on price and the profitability of printer sales.

Second, the basic problem with the argument is that high-volume printer users, by virtue of their high use and expenditure levels, have more reason than low-volume users to remain alert to the total cost of using various printer/cartridge bundles from different producers. High-volume users can justify searching carefully for the best printer/cartridge deals in terms of total cost of use. By definition, they can spread their search costs over more uses than can low-volume users and therefore should be able to obtain lower use prices. Knowing that high volume users are more alert to total-use prices across various printers, printer manufacturers should be more inclined to give them relatively lower use prices measured, say, by the total cost of printing per page.

This last point leads to the explanation for the printer/cartridge pricing scheme that I find most satisfying: Low-volume printer users have little incentive to carefully weigh the total-use cost of various printers. Hence, it doesn't matter very much to them if their printer cartridges are "overpriced." They aren't necessarily irrational, or unwilling to make rational, cost/benefit calculations on various printer options. On the contrary, their failure to carefully weigh the full cost of printing can be grounded in a not-so-transparent cost/benefit calculation: The cost of a search for the best printer (in terms of total cost of use) can be greater than the cost savings from their getting the best printer deal that could include lower prices on cartridges, but only if they incurred the necessary search costs.

Low-volume users, in other words, can remain rationally ignorant of comparative use costs of various printers. Their search costs for the best printer/cartridge deal can easily exceed any savings they might realize from a printer with a lower use cost.

Interestingly, the printer Dell sent me was clearly (given its size, technical sophistication, and speed) intended for low-volume users. I threw it out when it ran out of ink because, given that I am a moderately high-volume printer user, I was fully aware that there were other, more durable printers on the market intended for high-volume users, with lower per-page use cost.

No doubt, Dell sent the printer to many other computer buyers, who probably were rational by remaining ignorant (correctly so) of the comparative total-use cost of various printer models, and who gladly were willing to pay the "exorbitant" prices for their cartridges after receiving their free printers. I put *exorbitant* in quotes because the total-use cost that low-volume users incurred from the free printer/high cartridge price could be "high," but lower than the total-use cost they would have incurred had they undertaken additional search costs and bought a higher quality, more expensive printer with a lower total-use cost—when costs are restricted only to the cost of cartridges, printer, and paper and when their search costs are ignored.

Hence, I should expect printer manufacturers to seek to segment their markets according to buyers' price sensitivity with various printer models, differentiated by, say, durability and speed (with durability and speed being key determinates of different groups of buyers' responsiveness to the total cost of use). I would expect that the printers targeted for low-volume users will be less durable and slower than printers targeted for high-volume users. I would also expect that, for printers targeted for low-volume users, the printer price would have a lower ratio-to-the-ink-cartridge price and would have higher total-use (per page) costs than would be the case for printers targeted at high-volume users. If both parties to the trades—printer companies and buyers—are gaining by the trades, who's to say who is exploiting whom? Put differently, both could be exploiting the other—and each loving being exploited.

CHAPTER 7

DIFFERENCES IN DISCOUNT RATES

No doubt, the present value of money received in various future years is not likely to be the same for printer manufacturers as they are for buyers, and the present value of money over time is likely to vary among manufacturers and among buyers. Printer manufacturers might have lower discount rates than many printer buyers because the printer manufacturers have a longer planning horizon and greater access to financial capital and lower interest rates on borrowed funds. Printer manufacturers may, however, receive lower interest rates because they have more established reputations for their financial integrity (as well as have more useful collateral) than do many buyers.

Many printer buyers may be new businesses without solid records of financial success. Such new business buyers might have little financial equity in their firms, other than their personal savings, which many new entrepreneurs drain to zero in order to get their ventures up and running even at a modest pace. Accordingly, they may very well have to pay a risk premium for any funds they borrow and invest in information technology of any kind, not just printers. They may have invested a lot of "sweat equity" in their new ventures and have borrowed all they can from all loan sources, including credit cards (at relatively high interest rates) just to make a go of their new ventures.

In addition, these businesses/printer buyers may not be confident that they will be around in the intermediate-term future, much less the long term. They may want to invest as little as they can in equipment such as printers upfront and string out purchases of their consumables like ink for printing. And then there are no doubt many buyers—perhaps young or very old people—who have high discount rates because they get great value from consuming today at the expense of greater consumption in the near and long term. They want to buy cheap printers so that they can enjoy other things now, not in the future. The young may simply have an urgency to get things now. The old might not want to delay consumption because they might not be around to enjoy future consumerables.

And make no mistake about it, buyers *do* vary in their discount rates, with some buyers having extraordinarily high discount rates and with their discount rate often going down as the payoff is extended into the future, at least according to research. Richard Thaler asked subjects how much in the way of a payment delayed by 1 month, 1 year, and 10 years would make them indifferent to their receiving an immediate payment of $15. The median response implied a discount rate of 345% for a one-month delay in payment, 120% for a one-year delay in payment, and 19% for a ten-year delay in payment.[4] No doubt, different subjects gave responses that were both higher and lower than the median responses.

To the extent that cost considerations allow, we should expect printer manufacturers to develop not only different classes of printers (for examples, dot matrix, ink jet, and laser jet) with different capabilities, but also various models of the same class of printers with different durability and speed—and give them different ratios for printer price to cartridge price, which can lead, understandably, to dramatically different prices per page printed. Obviously, that is what printer companies do. Depending on the model, the cost per printed page for an inkjet printer is between 10 and 15 cents; for a laser jet, between 2 and 5 cents.[5]

Those printer buyers with high discount rates (relative to the discount rates of the printer producers) will tend to be attracted to the printers with relatively low printer prices and high cartridge prices. For example, for buyers who, for whatever reason (no established credit record, youth, newness to business, impatience, lots of alternative investment opportunities with limited equity and loan sources, and/or bad credit records), have to pay high annual interest rates of, say, 18% for borrowed funds to buy a printer, you can understand why they might be inclined to buy low-priced printers and high-price cartridges. Such a deal allows them to avoid seeking a loan from banks or credit card companies. They might pay a premium price for cartridges, but the price of the printer use with a high cartridge price can be lower than it would be if they had to buy a high-priced printer with borrowed money and then bought lower-priced cartridges.

Those printer buyers with relatively low discount rates will tend to be attracted to the printers with relatively high printer prices and low cartridge prices. For example, buyers who can borrow money for only 6% would hardly be attracted to the printer/cartridge deals found attractive by people who have to pay interest rates of 18%. Why pay the premium on the cartridge price, designed primarily for people with high interest/discount rates, when you can borrow money cheaply to buy the higher priced but ultimately cheaper printer in use?

The different pricing schemes should *not* be interpreted as printer manufacturers necessarily "exploiting" either the low- or high-discount buyers. Through their menu of printer/cartridge options, the printer manufacturers are simply providing mutually beneficial deals to the different groups of buyers. Indeed, with the menu of printer models, as noted, buyers can self-select, choosing the pricing schemes that best fit their time preferences and credit worthiness, as well as their needs for printer durability and speed.

GAMING PRINTER/CARTRIDGE DEALS AND TECHNICAL AND CONTRACT SOLUTIONS

Of course, it needs to be stressed that not all buyers of cheap printers that come with expensive cartridges need have high discount rates. People who print in low volumes might buy a cheap printer because they print so little that they might not have to replace the cartridge for several years, if ever (given that they might buy an upgraded printer model by the time the cartridge runs dry).

Admittedly, the cost of producing printers and cartridges can impose a check on the variety of printer classes and models within classes. Also, if different printer/cartridge models could use the same cartridges, many savvy buyers would buy the cheap printer models and use the lower-priced cartridges meant for the expensive printers. However, the cost of making sure that cartridges are not interchangeable might be minimal, involving, perhaps, nothing more than a slight change in connectors on the cartridges or an adjustment in the printer software.

In addition, barring technological fixes for cartridge interchangeability, printer manufacturers can negotiate printer contracts with different (if not individual) business-buyers, offering the same class and model of a printer (with different ratios of printer price to cartridge price) to different buyers. The manufacturers can solve the prospect of buyers purchasing the printer cheaply and using cheap cartridges intended for use with the expensive printer with a contract provision that requires that all cartridges be bought from the manufacturer directly, or indirectly from designated resellers, at specified prices, with the possible addition of a contract clause that requires a minimum expenditure on ink cartridges within the term of the contract. Alternatively, the negotiated price can be stipulated in terms of so much paid per page printed (or so much per imprint, if the printer is also a scanner, copier, and fax machine), with the printer manufacturer supplying the printer, plus all the required cartridges needed. This is exactly how the University of California system finalized its printer deals with various manufacturers for all ten universities in the UC system. At my business school, we chose to lease several high-tech and very fast multifunction (scanning, copying, printing, and emailing) machines from Xerox that charges us so much per "image." ("Page" is no longer the relevant metric for measuring machine use since multifunction machines can rarely be used for printing.) My school is required to buy Xerox cartridges and is obligated to make a minimum payment per year.

However, I hasten to add that printer producers will be checked in the interest rates they can charge implicit in the high cartridge prices by competition from other more widely recognized sources of loanable funds, say, banks and credit card companies. If the implied interest rates embedded in ink cartridges for low-priced or free printers were higher than the interest rates other lenders charge, printer buyers would, naturally, borrow from banks and credit card companies to purchase higher priced printers with lower-priced ink cartridges.

Nevertheless, we should expect that ink cartridges for low-price or free printers would likely carry a "premium" price on their cartridges. This is because when printer producers lower their prices on their printers, or give them away free, in anticipation of charging premium prices on cartridges, they will necessarily be incurring the risk that many printer buyers will never use their printers enough for the printer producers to recover the suppressed price of their printers. Hence, ink cartridges would be expected to sell for prices that cover such risk costs, plus interest. In short, H-P's "lucrative" ink business may not be as lucrative as advertised because no accounting is made for the risk cost assumed in that portion of H-P's business.

When Dell sent me the "free" printer with my Dell computer I bought (mentioned at the start of the chapter), Dell surely recognized that many of the free printers would go to computer users like me, who will take anything when it is offered at no additional charge, but who will not buy the print cartridges. Dell's cartridge price had to be high enough to enable Dell to recover the suppressed price on my printer, plus the costs of sending out free printers to all other computer buyers who never follow-up with purchases of the pricey cartridges.

So, yes, buyers might understandably see ink cartridges as pricey, but that does not mean that they are not participating in mutually beneficial trades. Buyers might be expected to be better off with the pricey cartridges than without them, but not as well off as they would like, of course. Similarly, printer producers might also be participating in mutually beneficial trades, but their gains are not likely to be as great as buyers might surmise from comparing the ink cartridge prices with their manufacturing costs. Anytime producers *bank* (in a relevant manner of speaking) on making their money on delayed but hard-to-predict purchases into the future, their prices must account for the time and risk costs.

THE EVIDENCE ON THE RELATIVE PRICES OF PRINTERS AND THEIR INK CARTRIDGE

The line of argument developed in the previous section leads to two important conclusions. First, the total cost of using a printer will tend to fall with the durability and price of the printer. Available studies tend to confirm that conclusion. In one such comparative study, *PC World* magazine found that the total cost of use for 3,000 printed pages for a $50 Lexmark Z605 color printer was $517. A much more durable $180 Hewlett-Packard 6122 color printer had, for the same number of printed pages, a total-use cost of $393, nearly a quarter less than the Lexmark printer.[6]

Second, the ratio of the printer price to the ink cartridge price can be expected to rise with the durability and price of the printer. Tables 1 and 2 show the prices for a selection of Dell and H-P printers and ink cartridges, as reported on the companies' respective web sites in March 2007, and provide some (but hardly perfect) confirmation of expectations developed in the preceding section. The durability of the printers in each table designed for businesses is generally greater than the durability of the printers designed for home offices. The durability of the printers listed for each category (home and business) rises with their model numbers. The printer prices also rise with the model number and durability of the printers for both companies.

Table 1 Ratios of printer price to ink cartridge price for Dell Monochrome Laser Printers, ranked by durability, March 2007

	Printer price (1)	**Ink cartridge price (2)**	**Pages printed per cartridge (3)**	**Cartridge price per printed page (2)/(3)**	**Ratio (1)/(2)**
	(1)	(2)	(3)	(4)	(5)
Home office					
1110 (17ppm)	$99	$65	2,000	$.0325	1.52
1710 (27ppm)	$179	$100	3,000	$.0333	1.79
1710 (high capacity)	$179	$130	6,000	$.0212	1.37
Medium and large businesses					
5210n (40ppm; 200,000 pages/ month duty cycle)	$999	$170	10,000	$.0170	5.88
5310n (50ppm; 250,000 pages/ month duty cycle)	$1,199	$170	10,000	$.0170	7.025

The prices for the printers and their cartridges for the two companies are reported in columns (1) and (2) of each table. The ratio of the printer price to cartridge price *generally* rises, as can be seen in column (5). The cheapest Dell home printer, model 1110, which sold for $99, had a ratio of 1.52, which means that the printer price was about one and a half times the price of each replacement car-

Table 2 Ratios of printer price to ink cartridge price for Hewlett-Packard Monochrome Laser Printers, ranked by durability, March 2007

	Printer price (1)	**Ink cartridge price (2)**	**Pages printed per cartridge (3)**	**Cartridge price per printed page (2)/(3)**	**Ratio (1)/(2)**
	(1)	(2)	(3)	(4)	(5)
Home office					
1018 (12ppm; 3000 pages/month duty cycle)	$130	$70	1,000	$.0700	1.857
1020 (15ppm; 5,000 pages/month duty cycle)	$180	$70	2,000	$.0350	2.571
1022 (19ppm; 8,000 pages/month duty cycle)	$200	$70	2,000	$.0350	2.857
Small and medium businesses					
P3005 series[a] (35ppm; 100,000 pages/month duty cycle)	$549	$129	6,500	$.0198	4.256
4259 series[a] (45ppm; 200,000 pages/month duty cycle)	$899	$150	10,000	$.0150	5.99
5200 series[a] (35ppm; 65,000 pages/month duty cycle)	$1,439	$168	12,000	$.014	8.57

[a] HP has several printers in a series, generally rising in capability with the printer price.

tridge. The next cheapest model, which sold for $179, had a ratio of 1.79, an increase of 18%. However, the cartridge for the second more expensive model can print 50% more pages (3,000 instead of 2,000 before running out of ink), which suggests that for those who print a lot (and don't have high discount rates) the second printer is more attractively priced. The cartridge for the third, most expensive home printer, carries a price that is 30% above the cartridge price for the second most expensive printer, but it can print 6,000 pages, which means the ratio of printer price to cartridge price drops 24% to 1.37 per page, which, again, makes the third printer in the list a more attractive buy for those who print a lot (and don't have high discount rates). The per-page cost of the cartridge for the much higher-priced and much faster and more durable printers designed for small and medium-size businesses drops another 27% to 1.7 cents from 2.1 cents per page.

Casual observation of Table 2, which covers a selection of H-P printers, shows much the same pattern; the ratio of the printer price to the cartridge price rises with the durability and prices of the HP printers. While more sophisticated statistical techniques could be applied to adjust for differences in printer features, it seems clear that the tables provide good, albeit preliminary, evidence that printer companies provide buyers with a menu of printer/cartridge pricing options, allowing buyers to self-select according to the intensity of their printer use (and, hence, upfront price sensitivity) and manufacturers' and buyers' discount rates. Accordingly, the H-P cartridge cost per printed page drops by half when going from the first home printer to the second and holds steady when going to the third home printer.

In the bottom half of Table 2, concerned with H-P printers for small and medium-size businesses, the printer price-to-cartridge price ratio jumps dramatically from what those ratios were for H-P home printers (see column 5). However, the cartridge capacities of the business printers jump by even more, by three to six times the capacities of the home printers, which means, as expected from our foregoing discussion, the cost per page for the first H-P business printer is 43% less than the cost per page of the highest volume H-P home printer. Take another step up in H-P business printers and the cost per page drops another 21%, and then another 7% when moving from the second to the third listed business printer.

CONCLUDING COMMENTS

Consumer entrapment and stupidity are commonly heard explanations for a variety of firms' pricing strategies. I give them at least some credence, if for no other reason than I have repeatedly acknowledged: printer companies will, when they can, charge what the market will bear. The prevalence of consumer switching costs for any number of products leads me to believe that firms will hike their prices whenever the present discounted value of the price premium consumers have to pay over time is just below the costs they would have to incur to move to lower-priced products. I also understand that I regularly make wrongheaded purchases, and I observe others doing so and confessing to doing much the same. Then, of course, in any time period, there will be new, inexperienced buyers coming into the market, and producers can be expected to exploit such buyers.

Having made those concessions, I have to return to points pressed in the chapter:

- First, inexperienced buyers regularly become experienced buyers. That is, they learn from their own exploitation and from becoming more aware of alternative, better-price products.

- Second, the most economical strategy for consumers is often to remain "exploited" by higher prices than to incur the search and transaction costs to avoid being "exploited." Making "stupid" or "wrongheaded" purchases often constitute the most economical way to shop.

- Third, if many consumers can learn about their mistakes and can correct their decision rules and their decisions, we have to wonder how pricing strategies that play to consumer entrapment and stupidity can persist. The problem is that smart producers should be expected to rush to markets in which consumers can be entrapped and/or are inclined to make stupid buying decisions, and continue to do so without correction. The growth in competition among producers can be expected to correct much market exploitation. When pricing strategies persist in reasonably competitive markets, we should feel an obligation to look for other explanations for why prices are what they are, which is what I have done in this chapter.

- Fourth, many buyers, especially of printers, are just as sophisticated in their purchases as producers are in their pricing.

- Fifth, if printer companies adopt pricing strategies to exploit the unwary printer buyers, the printer companies have to be concerned that their much larger group of wary printer buyers will be driven to other printer companies, because they will also see the printer/cartridge price scams and will assess them as bad deals. Perhaps, printer companies might find ways of isolating the unwary buyers. Dell may have thought it had such a way when it sent me the free printer. I have to wonder how many printers it had to give away to wary and unwary computer buyers to increase its ink cartridge sales. Perhaps Dell was the unwary party in that give-away program.

My favorite explanation for the persistence of relatively low printer prices and high ink cartridge prices is that, once again, producers and buyers are, for the most part, making mutually beneficial trades. Producers offer a variety of combinations of printers and ink cartridges with different combined prices that enable producers to segment their markets. However, the variety of options can mean that buyers who can self-select according to their time value of money (whether measured by the interest rates they would have to pay on loans to buy equipment or by their own internal, subjective evaluations of money through time). Those buyers who have high interest or discount rates can be expected to buy the low-price printer/high-priced cartridge combinations. As buyers' interest or discount rates go down, the ratio of the printer price to cartridge price that buyers choose can be expected to fall.

In no small way, printer companies have used their pricing strategies to become, effectively, loan companies for their customers with the loans provided implicit in the upfront reduction in the printer price and the interest payments implicit in the premium charge on the ink cartridges.

The argument developed in this chapter does not lead to the conclusion that printer/cartridge combinations will never be "overpriced" for different groups of buyers. Of course they will. Again, producers can be expected to charge as much as they can, but in doing so, they encourage other producers to move into the printer market, causing over time a decline in the printer/cartridge prices taken together, as well as a gradual escalation in the quality of the printer/cartridge product combination—all to the benefit of consumers.

Cartridge World has opened a chain of stores that sells "remanufactured" cartridges for 50% less than the cost of new cartridges. In early 2007, Kodak announced that it would be the first company to offer a low-priced ($150–$300) printer series (EasyShare 5100 All-in-One) that uses replacement cartridges that cost just under $10, half of what other printer companies charge for the same level printer but also at half the volume of ink that their other printer cartridges contain (for example, H-P Photosmart c4280 All-in-One). However, Kodak priced its low-price printers at about $50 more than comparable models from other manufactures, as the *New York Times* technology columnist reported.[7]

Later in the spring of 2007, printer companies began announcing 50% reductions in the price of their cartridges. However, because companies dropped the amount of ink in each cartridge by two-thirds, the cost per printed page actually rose, according to Lyra Research, Inc.[8] The explanation given to journalists at both the *Los Angeles Times* and *New York Times* for the new cheaper line of ink cartridges with less ink is consistent with the theme of this chapter, that the new cheaper and smaller cartridge line gives consumers who print very little an opportunity to tie up less of their funds in cartridges.[9]

The more general point of the chapter is that when a product (printer) requires an upfront investment of some significant amount and also payments for some consumerable (ink), producers should be expected to offer their buyers a menu of product options. That's what printer companies do.

However, I note in drawing this chapter to a close that cell phone companies have upfront equipment, phones, and consumerables, minutes, to sell. And cell companies offer bundles of phones and minutes with much the same pricing strategy adopted by printer companies for printers and cartridges. Go check out their rate plans, carefully considering (in light of points made in this chapter) how in their plans, the price of the phone often varies inversely with the prices for the minutes used each month and with the length of the contracts. That is, the higher the price of the phone, the lower the cost of the monthly minutes. The longer the contracts, the lower the cost of the phone and minutes. Those different pricing combinations are understandable within the analytical framework developed in this chapter. The phone companies are simply playing to the different discount rates of different consumers.

The lesson to be learned from our discussion of the pricing of printers and their ink cartridges is that there is money to be made not only through the sale of products, but also through implicit loans to buyers by offering a menu of products with different relative prices on the main product and the consumerable that is used with the main product.

NOTES

1. Fasoldt (2004).
2. Fasoldt (2004).
3. "Sales of ink and laptops push H.P. past forecast" (2007).

Chapter 8

~

WHY MOVIE TICKET PRICES ARE ALL THE SAME

In almost any introductory economics textbook and course, students are regularly taught that in competitive markets, prices are determined by the laws of supply and demand. That is, prices will settle where markets clear, or at the price at which the quantity consumers are willing and able to buy exactly matches the quantity suppliers are willing and able to offer. If demand rises, the market price will rise, and vice versa. That kind of analysis doesn't explain the constancy of movie ticket prices across movies, no matter how successful they are.

DIFFERENT PRICE FOR DIFFERENT FOLKS

Supply-and-demand theory implies that goods with greater demands that reflect, say, greater quality or popularity can be expected to have higher prices, all other considerations (especially cost) held constant. So it is that rock bands who break into the public spotlight with a "hot" selling album generally charge higher ticket prices for their concerts than their lesser-known counterparts, with long-established stars—the Rolling Stones, The Grateful Dead, and Paul McCartney—being able to charge more than almost all newcomers.

Goods with lower costs are expected to sell for less in competitive markets than higher cost goods, everything else held constant, again. Furniture made in China with cheap labor sells for less in the USA, in spite of the heavy transportation costs, than does furniture made in the USA (which is a major reason domestic furniture companies have been terminating production in the USA one after the other over the past decade or more.)

If the cost of producing a good falls (which translates into an increase in supply), its supply increases, and the price of the good can be expected to fall. Accordingly, the prices of computers have tumbled over the past three decades with the commoditization of many computer parts. The prices of MP3 players, even the market-category dominating iPods, can be expected to fall with time (at least after adjusting for inflation and quality improvements), as more producers achieve lower cost structures, inspired by technological improvements; as the current profitability on MP3 and MP4 players cause more producers to enter the market; and as producers find fruitful ways of effectively challenging Apple's market dominance (with a market share of above 75% at this writing).

Even standard monopoly theory teaches that firms with at least some market power—that is, firms that have limited competition within their markets—should be able to charge different prices as demand rises and falls with the popularity of the products they sell, with the popularity varying with the seasons of the year (automobiles) and sometimes with the days of the weeks (restaurant meals) and time of day (bar drinks). Monopolies can also be expected to price discriminate, charging buyers with high demands more than they charge buyers with low demands. Of course, this is the case only so long as the low-price buyers cannot profitably resell their purchases to the high-price buyers, a point stressed in more than one of the earlier chapters.[1]

Accordingly, many monopoly toll-road companies and state transportation agencies (in this case virtually sole-source suppliers) charge higher prices for toll road (and toll lane) usage during rush hours than at other times of the day. The same can be said of many subway systems. Utility companies often engage in peak-load pricing, lowering their rates in the middle of the night, when few people are awake to use electrical appliances, and hiking their rates during late afternoon, when industrial and home uses of electricity are typically at their daily peaks. Practically all concert halls and stage theaters charge different prices for seats in different sections because they know that demand depends on the location of the seats. Even movie theaters engage in peak-load pricing (as we saw in Chap. 4), they just don't do it in response to popular and not-so-popular movies.

THE PUZZLE OF UNIFORM TICKET PRICES AT THE MOVIES

The immense variation of pricing strategies in so many markets make movie ticket prices look like one big pricing puzzle mainly for the virtual absence of variation in ticket prices for different first-run movies in the same theaters and even across theaters in the same geographical markets. In the movie industry, there is also no price adjustment for the opening success of different movies. The ticket price for a movie that is a blockbuster the first weekend of its run will typically carry the same ticket price the following week as a total box-office dud. And ticket prices, especially in the USA, don't typically vary for different days of the week and different seasons of the year. No doubt, the question "Why?" has crossed many readers' minds.

When *Spider-Man 3* opened in May 2007, it set a record with total world ticket sales of $151 million for its first three-day weekend. *Pirates of the Caribbean 3* followed later in May with a three-day opening total of $142 million in global ticket sales. Both movies carried adult ticket prices on their first and following weekends of $10.50 at Regal Theaters (in Southern California).[2] *Miss Potter*, which had sold only $30 million in global tickets sales through its first six months of screenings between its opening in December 2006 and May 2007, carried the same ticket prices.[3] Why?

The puzzlement over the uniformity of movie ticket prices is especially acute, given that

(1) The production costs of major motion pictures can run from a few million to several hundred million dollars (*Spider-Man 3* and *Pirates 3*, released in 2007, each had production costs of about $300 million[4]), with the average production cost of a major film in 2006 at $100 million (a third of which is in marketing costs),[5]

(2) movie going varies substantially by the day of the week (with the peak days being Fridays and Saturdays[6]) and by the season of the year (with the summer months and the Christmas school break being the annual peaks[7]), and

(3) some first-run movies are evident blockbusters on their opening weekends, which means they draw overflowing crowds and, in a matter of a few weeks, gross tens, if not hundreds, of millions of dollars in revenues, while other first-run movies are evident duds practically at their openings and have losses that run into the millions, or practically to the limit of their production costs.

The ticket pricing puzzle is further compounded by the fact that movies differ dramatically in terms of their star power, special effects, advertising budget, media reports before (and after) openings, pre-opening reviews, length, and the number of screens where different films can be seen. Academy-Award-nominated movies carry the same ticket prices as films that never had a chance at a nomination, and even the ticket prices of Oscar-winning movies don't budge a dime in the days and weeks following the Academy's globally televised awards ceremony. Indeed, art films, which, from their inception, are not expected to have big box-office draws, will often have the same ticket prices as expected blockbusters, for example, the next *Harry Potter* movie, the demand for which might be expected to be heightened by the enormous success of all previous books and movies in the series.

Clearly, a commonly expressed concern that differentiated movie ticket prices for different movies will unsettle moviegoers, reducing their demand for all movies, must be looked upon with some skepticism since so many moviegoers are used to price variations, as noted, for other forms of entertainment and even for different classes of moviegoers.

Theaters' fear of violating social norms of pricing "fairness" *might* be an explanation for the uniformity in ticket prices. That is, movie theaters might figure that differentiated prices would be seen by moviegoers as attempts by greedy studios and theaters to take advantage of the success of movies, but the fairness concerns should dissipate when moviegoers learn that higher ticket prices can be associated with higher studio production costs and higher rental rates theaters must pay distributors for successful movies, and when moviegoers realize that higher prices for successful movies can mean that they will not (as frequently) be turned away at the box office.

Indeed, it is odd that movie DVDs range widely in price from, say, less than $5 to more than $25, depending on the success of the movies in theaters and the time since the movies ended their run in theaters. Their rental rates of movies at Blockbuster can also differ, depending on their theater success and time since they were released—over which there appears to be no fairness concern voiced by consumers or consumer groups.

Why would moviegoers be any more concerned about differentiated box-office ticket prices than over differentiated DVD prices?

Maybe moviegoers are especially sensitive on fairness grounds to theater thicket prices than to DVD prices, but without more evidence on the issue, the fairness justification for uniform ticket prices appears highly suspect, and probably should be relegated to the bin of nice sounding but vacuous explanations when considered carefully. This is especially true since there is a case to be made on fairness grounds that the producers, distributors, and exhibitors should be able to garner some of the gains from a particularly valuable film since they (not moviegoers) were crucially involved in the making of the film.

Granted, an argument might be made that under a pricing regime of constantly changing movie prices, people might head for theaters unaware that the ticket price of a movie they want to see has spiked upward because of its recent success, only to realize on arrival that the ticket price is more than they want to pay. However, it would appear that such frustrations could be managed through ticket prices being widely posted in advertisements and on the internet. The solution for potential moviegoer frustration is simple admonition: "Check before going off to the theater." And should we not expect people to learn to check ticket prices if price variation were the norm? Broadway theatergoers, air travelers, and vacationers have surely learned that prices for theater and airline seats and hotel rooms vary markedly—by location, season of the year, and day of the week.

As research has shown, consumers can be disgruntled by price increases unassociated with cost changes, but it also appears from research that consumers can welcome price reductions.[8] If such is the case, moviegoer disgruntlement with price variation can be assuaged by starting all movies with prices that include premiums to account for discounts that are expected to be granted on films that prove to be non-hits, if not total flops. With ticket discounts, the end result can be the same, non-uniform ticket prices, with ticket prices reflecting variation in market demand and production costs, as supply-and-demand-curve theory suggests should be the case. Films with limited draws and publicity in their opening weeks will have lower ticket prices for the duration of their theater runs than films that have sellouts and attract a great deal of publicity. Some moviegoers might shift from popular to unpopular films, but that always is the case with price differentiation and can still be a profit-maximizing strategy for studios, distributors, and exhibitors— *if* non-uniform ticket prices were truly a better pricing strategy for movies as a product group over largely uniform prices. But then we have to wonder if even discounts on ticket prices after the opening weekend could be counterproductive, given that the discounts can effectively announce that discounted movies are not worth seeing, which can send the "marketing buzz" into reverse.

Another explanation for ticket price uniformity could be that theaters believe that moviegoers are insensitive to price changes. That is, they are unlikely to respond sufficiently to price reductions for total box-office receipts to go up with ticket-price cuts. That is, the demands for different movies, no matter how popular, are inelastic. But such an argument does not preclude non-uniform ticket prices.

Supply-and-demand theory predicts that movies with higher demands (all other factors equal) should be expected to have higher prices. In addition, if movie demands are truly inelastic, then consumers should be insensitive (relatively unresponsive) to price increases, as well as price decreases. Firms should be expected to hike their prices—and, hence, gate receipts—until consumers are forced to pay attention to prices, and respond sufficiently that box-office receipts fall with further price increases. It would certainly be a weird world in which all films had the same price sensitivity at every price and, hence, every movie had the same revenue (and profit-maximizing) ticket price.

PAST PRICE VARIATIONS

What is particularly mysterious about the persistence of uniform movie ticket prices in modern times is that in the first half dozen years of this century, movie theater chains have been filing for bankruptcy one after the other as attendance has eroded because of the expansion of DVD sales and DVD piracy, with the movies purchased being shown on large (to enormous) home plasma television systems whose sound and picture quality can easily supersede theater sight and sound quality at its best. If there were a true moral issue at all in the sameness in movie ticket pricing, it would appear to be that movie theaters *should* be able to differentiate movie prices to pull themselves out of the red, as well as capture some of the economic gains that, given theater losses, appear now to be captured by moviegoers who have nothing to do with getting the films on the truly big theater screens.

Still, in spite of what seems to be unrelenting economic pressures for theaters to move away from uniform prices, uniform prices remain entrenched as the *modus operandi* for the theater industry. Researchers have concluded that the only tenable explanation for the persistence of uniform prices, which is necessarily "inferior when compared to alternative regimes, is that prices remain uniform simply because that is a part of the industry's history and structure."[9] The problem with such a line of argument is that if non-uniform pricing were indeed superior to uniform pricing, and researchers apart from the industry can easily ascertain such is the case, it would appear that the industry is unnecessarily suffering massive losses and is leaving substantial profits on the table. This means that the stocks of theater chains should be seriously suppressed, which is obviously an open invitation for (greedy) corporate raiders to do their thing, that is, buy up enough shares in theater chains to gain control over management and install a new and improved non-uniform ticket pricing regime.

Takeovers of theater chains would seem to be so patently obvious, if they were consistently and significantly mispricing their films that, given takeovers are rare, uniform ticket prices might not be the inferior pricing strategy that elementary, supply-and-demand analytics would suggest it is. If uniform prices were demonstrably inferior to variable pricing, with lots of profits to be had from the switch, uniform ticket prices would not be observed for so long because of the competition in the market for corporate control.

Granted, movie ticket prices do vary in some ways, for example, from one area of the country to another. That is, ticket prices are typically higher in New York City than in Greenwood, South Carolina, presumably reflecting the difference in the cost of theater space and labor, if not demand for theater seats, in the two different venues.[10] As noted in Chap. 4, movie theaters also engage in some price discrimination, charging seniors and children less than non-senior adults and charging lower prices for matinee screenings than for evening screenings (although the price difference may be no more than 50 cents or a dollar). Moreover, some movie houses scattered thinly around the country charge no more than a dollar or two for second and third-run movies (mainly movies that have screened in first-run theaters, have been released on DVD, and may have even aired on television).

These forms of ticket variation make the uniformity in ticket prices for all (or practically all) major first-run movies in given theaters and across theaters within given markets all the more puzzling. As noted, some first-run movies— *Titanic, Star Wars, Spider-Man, Pirates* and *Harry Potter,* for example— do exceptionally well, grossing hundreds of millions of dollars and making ten of millions for their supporting producers. Others (the names of which are hard to remember) are losers from the start. The extant variations in box-office tickets noted above certainly support the presumption that movie theaters are actively seeking to maximize gate receipts by adjusting ticket prices to at least some market conditions. Why not other market conditions, not the least of which is the initial popularity of the film itself?

The movie theater industry emerged in the late nineteenth and early twentieth centuries with a great deal of price variation, with ticket prices at one time set (would you believe?) by the length of the film, as well as the season of the year and day of the week. In more modern times, movie theaters have occasionally experimented with variable pricing. In 1970, theaters in the District of Columbia that cut ticket prices for Monday through Thursday showings by two-thirds reported an increase in box-office receipts and a more than doubling of popcorn sales.[11] Other theater chains in the 1980s and the early 1990s tried discount days on ticket prices.[12] The interesting outcome from theaters' experimentation with price cuts is that the cuts did not survive, which is good reason to suspect that uniform pricing in theaters could be an optimum pricing strategy, given the market and legal environment in which the industry must operate today. This means that movie theaters might not be leaving money on the table, or not enough to cause theaters to shift *en mass* toward non-uniform ticket prices, at least not yet. Otherwise, as noted, savvy investors could be expected to take over the theater chains and change their pricing strategies.

WHY UNIFORM TICKET PRICES

Economists Barak Orbach and Liran Einav attribute uniform ticket pricing to such factors as the history, conservatism, and regulation of the industry, mainly because they seem to believe that the success of movies is reasonably predictable.[13] After all, to the extent that the success or failure of movies is predictable, then differentiated ticket prices should be expected. The greater the predicted success in terms of attendance, the greater the demand, and the greater the price that can be charged. If the market success of movies is easily predicted, then price uniformity doesn't make economic sense (or so it would seem). This is because investors are leaving profits on the table with a virtually costless way of pocketing the greater profits: vary movie ticket prices.

A partial explanation for the persistence of largely uniform ticket prices today can be found in the observation that before the early 1970s, non-uniform ticket prices across films and even theater seats were not uncommon, or so Orbach and Einav stress. In the 1940s, ticket prices differed markedly—across several grades of movies, A, B, and C. Ticket prices for low-budget B-westerns were lower than for big-budget A-musicals. With the advent of television in the late 1940s and its spread in the 1950s and 1960s, providing a constant flow of B- and C-level visual entertainment, studios understandably began to concentrate their attention on A-level movies, or forms of entertainment television couldn't match or wasn't inclined to match. Some growing uniformity in ticket prices should have been expected with the narrowing in movie variation and targeted audiences.

But still, such an explanation leaves unexplained why ticket prices for highly differentiated A-level movies are the same. Economist Arthur De Vany challenges the view that the success of movies is predictable in a series of studies on the movie industry he undertook with several co-authors over the past two or more decades, which he has assembled, with revisions and updating, into a book titled *Hollywood Economics*. His conclusion about the movie industry is stark: "Nobody knows anything" (a quip first made by screenwriter William Goldman but widely validated by entertainment industry researchers).[14] Not only that, industry operatives can't know much of anything about the potential success of particular movies until they are released because of the nature of the movie business, beset with "extreme uncertainty" aggravated by key characteristics of moviemaking:

- First, all films are different, often radically different, products. The success of one product says little about the likely success of other products.

- Second, a movie is a packaged deal, with practically all expenditures on it made upfront, before the first screening. This means that costs must be sunk before anyone knows anything about what attributes of the film contribute to its success.

- Third, movies are works of art often made by combining the talents of hundreds of people—from cameramen to film editors to costume designers to composers to actors and actresses—whose efforts must be carefully coordinated. Slight changes in the talent and/or in the case of coordination can give rise to nontrivial changes in the final product. Oscar nominations can be left on the hard drives of the film editors or in the translation of composers' scores into music heard on the screen.

- Fourth, producers of movies really do not know the full nature of their product until moviegoers see the movie and experience it. Perhaps one of the attractions of going to movies is that the process is one of discovery for moviegoers, which also makes movies openings the start of a discovery process for producers. De Vany observes that the basic problem of any entertainment industry, and especially true of movies, is "that audiences don't know what they like until they see it; every film is a discovery and audiences transmit their discoveries to others in a dynamic cascade of information. The process of many individuals choosing among movies and transmitting their knowledge to others amid a changing slate of competing movies induces a very complex dynamical behavior that leads to wildly diverging outcomes."[15] That is to say, the success, or lack thereof, of a movie depends greatly on "marketing buzz" or on an "information cascade," which can be the boon and bane of movie producers.

- Fifth, fifty to a hundred movies are in theaters at any point in time, which is to say that every movie faces substantial competition for the screens of theater owners and the eyeballs of moviegoers. Moreover, the competition is ever changing as new movies are released and others end their runs. Plus, newly released movies have the ever-growing competition from movies released on DVD or on the internet. Movie producers have a tough time estimating the demand for their films because they have little idea of what their products really are and what the competition will be.

- Sixth, as Orbach and Einav have found, relatively high correlations exist between, say, movies' budgets and their box-office gross,[17] but such correlations are more or less meaningless because of tremendously high variance of the impact of an individual movie's budget and box-office gross.[18] That is to say, while movies' successes tend to rise with their budgets, many movies with big budgets have been major flops. Many movies with small budgets have been blockbusters. Hence, budgets are very poor predictors of box-office success. The same can be said for the star power of the directors and actors and actresses in films.

Star power on the production team or in the cast can offer movie producers some modest downside financial protection on the opening of a film and a slightly improved odds of the firm making money, but star power is also a poor predictor of the financial success of a movie's full theater run.[19] Consider Oscar-winning star Jack Nicholson who had a string of successful movies over his career (*One Flew Over the Cuckoo's Nest, As Good as It Gets, About Schmidt* and *Something's Gotta Give*, to name just several of his hit movies) before he starred in *Anger Management*, a total artistic and financial flop. Stars in successful movies can follow with appearances in non-hits and outright duds. De Vany argues that perhaps the best advantage stars offer a film is a signal of quality to moviegoers. This is the case because stars are often offered the best roles in the best movies, and stars (at least those who survive for long) may be stars not only because of their acting skills, but also because they have a demonstrated record of selecting quality roles and movies.[20]

Regardless of how they achieve their success, stars that are able to offer producers even a modicum of predictability of success are stars whose required payments can drain the movie of any predicted profitability attributable to the stars themselves. De Vany has found that having stars in movies does not increase their chances of success: "Star movies cost a lot of money and have a tendency to run over budget. A single person just can't put that much protection value on the screen in a story involving many characters and events. It is really a strong performance in an outstanding film that makes a star. Movies make stars, not the other way around."[21] Besides, if stars in films could make the movie, which could give rise to above-normal ticket prices, then it would appear that the stars should be able to soak up a major portion of the revenue gains from higher ticket prices.

Ditto for sequels and prequels. Many sequels and prequels that follow successful movies have often done well but then others have also bombed, which means, again, they do little to ease the difficulties studio executives face in selecting movies and feeling confident about the outcomes of their selections.[22]

Ditto for the size of the openings of movies. The size of openings in terms of attendance and gross receipts can't say very much about the gross receipts for the movie's entire theater run because of the unknown nature of the word-of-mouth recommendations. All that can be said from statistical analysis is that hit movies tend to have slightly higher grosses on the opening week or two than non-hit movies. The box-office receipts from hit and non-hit movies tend to build for the first 4 weeks, after which the hit movies' receipts continue to build while the non-hit movies take a plunge, but it is hard to say beforehand what causes the break in the two paths or which movies will take the two paths.[23]

Ditto for the impact of movies' marketing budget on their success. As De Vany concludes, "There is no evidence to show that marketing has much to do with a film's success. Marketing is mostly defensive anyway; a studio has to market its films just to draw attention in a field where everyone is shouting. If you don't shout too, you will be drowned out and may not be noticed."[24]

Through years of turning and twisting movie industry data, De Vany concludes that people in the movie business can't really know much about the business simply because no movie is typical. De Vany asks, "Why are executives paid so much, far beyond what the studio moguls of the past received? Maybe it is because they are good at deciding which movies to greenlight. But, it isn't true. Nobody has a formula, not even top executives"[26]

Another reason that predicting the success of movies is so tough is that the movie business is subject to the "winner's curse," which means that the winning bidder for a resource or product tends to pay too much. Movie producers face the winner's curse in several markets: the (auction-based) market for the rights to successful books that have the potential for being translated into movies, the market for noted screenwriters and their scripts, the market for producers and directors, and the market for actresses and actors.

The winning bids for these various forms of talent will reflect the most optimistic assessments of a movie's potential. Many of these winning bids will prove to be overly optimistic, which means that movies can be expected to have a high failure rate in terms of the producers' bottom lines. In a survey of over 2,000 movies, De Vany found that 78% lost money and only 22% were profitable, which means that the industry taken as a whole suffers a net average loss, in spite of the tremendous financial successes of a handful of movies (only 3% or so movies are runaway blockbusters).[27]

The uncertainty surrounding the financial prospects for new movies has been greatly aggravated by piracy that can begin on a global scale weeks before a movie is released in theaters. The industry estimated in 2007 that piracy reduced the movie industry's annual global revenues by about a tenth.[28] Indeed, the number of internet venues where new movies can be downloaded can reach into the thousands, and can easily be greater than the number of theaters where the movie is shown.

According the De Vany and David Walls, piracy is particularly problematic for studios since few movies turn out to be successes and the successful movies are the ones most likely to be pirated on a grand scale. Moreover, the pirated copies can have a snowballing effect since any reduction in ticket sales due to downloads can lead to a more rapid collapse in the number of theaters showing the pirated movies. Many potential moviegoers may not go to heavily pirated movies because they can lose their standing in the widely reported rankings of movies by ticket sales (not by the count of people who see the movies).[29]

As noted at the start, conventional supply-and-demand-curve analysis predicts different prices for goods with different demands. On economists' blackboards, they can show different prices for differentiated demands because they know, by *assumption*, what the demands for the differentiated goods are. In the movie industry, no one can predict the same degree of price differentiation because so little is known, and can be known, about the actual market conditions faced by movies at the time of their release. Under conditions of the kind of extreme uncertainty that exists in the movie industry, a rule of charging the same for all new releases is hardly nonsensical. This is especially true when studios and theaters can adjust the count of screens for new releases that show great promise on opening (or are not subject to the usual "extreme uncertainty"). And when the *Spider-Man* and *Harry Potter* sequels were released in the spring and summer of 2007, those films opened on thousands of screens, suggesting that the executives might have known more than the concept of "extreme uncertainty" suggests.

Moreover, in conventional supply-and-demand-theory, one often explicitly stated assumption is that a change in the price will not affect demand (meaning the full price-quantity relationship)—not now or in the future. A price change affects only the quantity demanded in the specified time period. Hence, if current revenues can be raised by a price hike, then the price should be hiked. Nothing in the way of future revenues is lost, because the future demand for the product is unaffected by current consumption. This is not the case in the movie industry, beset, as it is, with the potential for huge information cascades. In such a market environment, short-run revenue gains can be more than offset by future revenue losses as the cut in the initial attendance due to a price hike can reduce the potential size of the information cascade. De Vany argues that a ticket price hike at a movie's opening could scare off moviegoers who could, if they saw the movie, help jumpstart the information cascade: "The lower rate of information transfer would lead to a shorter run and a lower total level of demand. The ability to extend the run makes an almost perfectly elastic supply response possible, so that price need not rise to choke off excess demand."[30]

And it needs to be stressed that movies have a very limited window of opportunity for proving their worth, with the mean run-life in theaters at under 6 weeks and with the median run-life of 4 weeks—and many films have only a week or two of run-life.[31] According to De Vany, "A film has less than a 25% chance of lasting 7 or more weeks and less than a 15% chance of lasting more than 10 weeks."[32] By the fourth week after their openings, as noted above, movies tend to diverge with the gate receipts of the few profitable hit movies continuing to grow and the gate receipts of the many non-hits typically taking a radical dive.[33]

As noted movie producer Robert Evans has remarked, "A film is like no other product...it only goes around once. It is like a parachute jump. If it doesn't open, you are dead."[34] Under such market conditions, playing with price differentiation across films and theaters can be tantamount to cutting your ripcord just to see what you can get away with on the jump.

Conventional market analysis is founded on the presumption of increasing costs of production. In face of a greater demand, the price needs to rise in part because of supply constraints—or more accurately, increasing (marginal) cost of production—and because of the need to curb any excess quantity demanded. That is to say, the supply response is limited, which elevates the usefulness of a price hike. In the movie industry, the supply constraints are far less binding, especially in recent decades of gradually falling movie attendance and the expansion (or rather overexpansion) of theater screens and seats, a major source of the theater industry's financial distress over the past decade.[35] If a movie proves to be a hit early in its run and more seats are needed, the hit movie can simply be shown in larger theaters, on more screens, and with greater frequency during each day. In other words, the supply of theater seats is highly elastic, if not often perfectly elastic, at a close-to zero marginal cost of seats.[36]

Under market conditions of risk and uncertainty, producers mitigate their product development problems by producing a number of products and then selling them (all at once or over time) in bundles. The expectation can be that most of the products developed for the bundles will be failures. The profits from the few hits can more than offset losses from the many flops, but no one can know in advance which is which. Studios are able to adopt such a strategy in the production of a portfolio of movies during the course of, say, a year.

However, the movie industry has been forced to operate under serious antitrust constraints imposed by the Paramount decision in 1948, cited in Chap. 4. Under that court decision, studios were forced to divest themselves of their theaters and to offer individual movies to individual theaters on a movie-by-movie bid basis. While the Paramount ruling is now widely believed to be misguided, it nevertheless still means that studios can't be vertically integrated with theaters, with vertical integration being one means by which the industry could, if it were free of the Paramount decision, dampen the risk and uncertainty inherent in movie making. Studios also can't distribute their films in bundles, which is another means of reducing risk and uncertainty in the industry. This is to say that because of the Paramount decision, there is more risk and uncertainty in the movie industry than there needs to be. With that legally added risk and uncertainty, we might reasonably expect ticket prices to be more uniform than they would otherwise be. If the financial outcomes for bundles of movies are more predictable than for individual movies (which should be the case because the variation in receipts across several movies can be expected to be less than for individual movies), then we would expect to observe more variation in ticket prices across bundles than exists when movies are sold separately.

DVD RELEASES

If extreme uncertainty is an important explanation for uniform ticket prices for first-run, untested movies, then we shouldn't be surprised that when movies are released on DVDs or are available for downloads over the internet, prices for different movies begin to diverge. When DVDs are released, studios have a great deal of information on the theater success of movies and can, as a consequence, better estimate the demand for DVDs than when the movies were first released in theaters. Some price divergence should be expected on the "new release" racks in the aisles of Best Buy, Costco, Wal-Mart, Target, and Circuit City, and all other major DVD retailers. When first released on DVDs, movies that were blockbusters when they made their run in theaters generally sell for more than movies that were non-hits, which sell for more than total disasters. Obviously, the demand for many (not all) movies predictably fades with time. As a consequence, their DVD prices can fall to under $5 within a matter of a few years.

However, we should not expect the divergence on DVD prices of movies to be as great as the difference in the theater success of movies. This is because highly successful movies tend to have longer theater runs than less successful movies. That is, many potential DVDs for highly successful movies are soaked up in greater ticket sales over their longer runs, potentially dampening their DVD demands, and the price that they can charge. Also, when movies reach the DVD stage, studios must understand that many DVD sales will depend on how many people who saw the movies in the theater will want to see the movies a second and third (or more) times on their home theater systems. Any number of movies that are highly successful in their theater runs—for example, *Spider-Man 3*—may have been successful because of the information cascade that got going when they were first released in theaters. People went to see *Spider-Man 3* because others were going to see it, and the movie became a social event, a value that is lost in the DVD market where people buy movies to watch them alone or with a few friends and family members. Other movies that are not so successful—say, *About Schmidt* or *Something's Gotta Give* could inspire a relatively greater demand in the DVD market because Jack Nicholson's performance was so superb that people want to see the movie over and over again to catch the nuances in Nicholson's acting.

Finally, as a reminder, the demand for DVDs released by studios can be dramatically undercut by movie pirates who, understandably, will focus their sales efforts on the successful films that moviegoers want to see time and again.

CONCLUDING COMMENTS

Given that movies are made at dramatically different levels of technical sophistication and costs and the attendance and financial outcomes vary so dramatically across movies, the observed sameness in ticket prices appears, at first glance, very odd—and at odds, supposedly, with much conventional market theory. However, on close inspection, uniform ticket prices should be expected—because of the variations in movies as works of entertainment and business ventures, with each movie having a uniqueness and newness of its own. The variation in movies on almost all margins is so great that "typical movie" is a rough concept at best. If there were something useful to the concept of "typical movie" (and outcomes of individual movies could be predicted with some degree of confidence and the industry were not subject to extreme uncertainty), then surely the demands for individual movies over their theater runs could be distinguished and ticket prices would unlikely be expected to be so uniform. As it is, in the absence of an ability to know much about the demand for any given movie (as De Vany and others have demonstrated, at least to my satisfaction), the best price for any particular movie is, more or less, a total guess, and the ticket price charge on movies in the immediate past on other films is as good a guess as any to what should be charged to maximize studio, distributor, and theater profits. The price charged for other films has the added advantage of not rocking the industry boat. In short, industry operators might understandably have an aversion to having venturesome pricing strategies that can add to the substantial risk and uncertainty they already face.

Granted, an argument can be made that a uniform ticket price for all movies is not necessarily a stable price. From time to time, industry operatives might try pricing strategies other than doing what everyone else is doing.[37] After all, trying other non-uniform prices is tempting, given the extent of producers' sunk costs and the very low (if not zero) marginal cost of adding seats in response to lower prices. However, industry operatives might have found what game theorists have discovered: cooperation on price among competitors, without explicit collusion, can best be achieved by following the rule of "tit for tat." That is to say, theaters may have decided to continue to do what others have been and are doing until someone does otherwise and cuts the price, at which time, all can respond in kind. The outbreak of a ticket price war can add to the degree of risk and uncertainty in the industry. Studios and distributors may be helping theaters maintain the price status quo somewhat by including in their movie licenses with theaters a minimum per viewer charge on the screening of films.

When some of the "extreme uncertainty" in the movie industry has been overcome through the record of ticket sales at theaters, movie prices can be expected to diverge somewhat. Although much research is needed on the statistical tie between movies' theater success and DVD prices and sales, it does appear that prices for movie DVDs diverge far more than theater ticket prices. DVD prices obviously do vary with the age of the movies: The older the movie, the lower its DVD price (assuming the same theater success). There are also differences in prices of newly released movie DVDs.

NOTES

1 If you skipped Chap. 3, a point of explanation might be useful: If resale is relatively costless and the high and low price gap is significant, those consumers charged low prices can be expected to sell to consumers being charged high prices. Alternatively, middlemen can be expected to arise, buying from low-price consumers and selling to high-price buyers. Either way, the amount purchased by low-price buyers can be expected to rise while the amount purchased by high-price buyers can be expected to fall. The results should force an equalization of prices charged to both groups of consumers by the producer.

2 Waxman (2007, p. B1).

3 According to DragonDynasty, as accessed May 29, 2007 from http://www.boxofficemojo.com/movies/?id=misspotter.htm.

[4] Waxman (2007, p. B1).

[5] As reported by the Motion Picture Association of America at http://mpaa.org/researchStatistics.asp, accessed on June 15, 2007.

[6] According to one report, 70% of average box-office revenues for all movies are generated on the weekends (Orbach and Einav 2006, p. 16)

[7] The average share of the American population that attended a movie during the 1985–1999 period has an annual peak of about 14–15% at Christmas, only to gradually fall through the beginning of the year (with temporary attendance peaks on President's Day and Easter) until Memorial Day, after which average weekly attendance as a percentage of the American population jumps through July 4 and continues on a slightly declining plateau through late summer, or until attendance plunges through Memorial Day. Thanksgiving weekend offers another temporary peak before attendance begins to climb back toward the Christmas annual peak. (See Obach and Einav 2006, figure 4 on p. 16.)

[8] Thaler (1980).

[9] Orbach and Einav (2006, p. 1).

[10] Orbach and Einav (2006, p. 5) found that ticket prices in some cities can be three times what they are in other cities.

[11] Headley (1999).

[12] King (1992).

[13] Orbach and Einav (p. 13) observe, "For example, production costs and gross box-office revenues have been found strongly correlated, with a simple correlation coefficient of .5 to .7 for each year between 1985 and 1999 (Prag and Casavant, 1994; Einav, 2004). Sequels perform quite similarly compared to the originals, at least in terms of the order of magnitude (Ravid 1999). Furthermore, much of the uncertainty regarding a movie's success is revealed after its first weekend on the screens (Einav 2004), so at least in principle admission prices can be adjusted on the first Monday after the release date."

[14] De Vany (2004, p. 275 and Chap. 2.). Harvard economist Richard Caves summarizes the so-called "nobody knows principle" for new movie releases this way: "[P]roducers and executives know a great deal about what has succeeded commercially in the past and constantly seek to extrapolate that knowledge to new projects. But their ability to predict at an early stage the commercial success of a new film project is almost nonexistent" (2000, p. 371).

[15] (2004, p. 7).

[16] De Vany (2004, Chap. 1).

[17] Orbach and Einav (2005).

[18] De Vany (2004, Chap. 6).

[19] De Vany (2004, Chap. 6).

[20] De Vany (2004, p. 136).

[21] De Vany (2004, p. 5).

[22] De Vany (2004, Chap. 6).

[23] De Vany (2004, p. 5 and Chap. 3).

[24] De Vany (2004, p. 4).

[25] De Vany (2004, p. 65).

[26] De Vany (2004, p. 244).

[27] De Vany (2004, pp. 214 and 234). Vogel came to much the same conclusion on the ratio of losing to winning movies (Vogel 2004).

[28] In 2006, the movie industry global revenues were estimated to be $26 billion (Motion Picture Association, http://mpaa.org/researchStatistics.asp, accessed on June 15, 2007). The MPAA also estimated that its member studios lost $2.3 billion in box-office receipts in 2005, the latest year for which data was posted at http://mpaa.org/piracy_internet.asp, again accessed on June 15, 2007).

[29] De Vany and Walls (2007).

[30] De Vany (2004, p. 45).

[31] De Vany (2004, pp. 15 and 73).

[32] De Vany (2004, p. 7 and Chap. 1).

[33] De Vany (2004, Chap. 2).

[34] As quoted in Litwak (1986, p. 84).

[35] Filson (2004).

[36] De Vany (2004, p. 45).

[37] Just as this chapter was being finalized, Regal Theaters in Southern California began experimenting with a general admission/adult ticket price for its "arts theaters" (theaters that run mainly low-budget, documentary, and foreign films) of $9.50 for Sunday through Thursday evenings. The ticket prices at the arts theaters went up to $10 for Friday and Saturday nights. By way of comparison, the general admission/adult ticket prices at all other Regal Theaters in the area was $10.50 for all evenings of the week.

Chapter 9

WHY SO MANY PRICES END WITH "9"

If you go to Apple's online iTunes music store, one fact stands out: All individual copy-protected songs sell for $.99.[1] Why $.99 and not $1? Does Apple really believe it is fooling online music shoppers into thinking that $.99 is a far lower price than $1, or is Apple taking advantage of shoppers who read $.99 as, say, $.90 (or even $.09)? If you buy an artist's album at iTunes, the price varies with the popularity of the artist, the number of songs on the *album*, as well as the album's age, but always has $.99 tacked onto the end of the price—say, $9.99, $12.99, and $14.99.

If prices don't end with $.99, and instead use whole-dollar numbers, many of them still end with 9, as in $99 or $999. As do other major daily newspapers around the country, the *Los Angeles Times* frequently carries a number of advertising inserts. On Fridays, the *Times* always has a four-page advertising insert from Fry's, a chain of big-box electronics stores scattered across the country. Each insert carries the prices of 150 or so items—from computer parts to televisions to kitchen appliances—virtually all with prices that end with 9. For example, the insert advertises $14.99 for a PQI 512 MB thumb drive, $79 (or sometimes $69) for an iPod Shuffle, $799.99 for a JVC camcorder, and $1,249.99 for an HP notebook computer. Indeed, of the 48 items listed on the front page of Fry's color insert on the day I wrote these words, only two items (just 4%) did *not* have prices ending with 9: $10.50 for a package of 50 recordable DVDs and $0 for an aluminum external hard drive case (an $11 posted price minus an $11 rebate).

Fry's is hardly unusual in choosing prices that end with 9, as is fully evident to anyone who casually flips through newspapers and magazines, walks down the aisles of grocery stores or other retail outlets, or takes note of gasoline prices posted on all service station marquees. Many of the prices not ending with 9 ended with close-by numbers: 8, 7, or 5, as in $599.97, the price of a mattress/box spring set advertised in the *Los Angeles Times* the day I took note of Fry's prices.[2]

If buyers are fooled by these "odd prices,"[3] "just-below (even) prices,"[4] or "psychological prices,"[5] as marketing researchers dub them, buyers' folly is truly remarkable because odd-numbered prices are so common and have long been embedded in posted prices.[6] On iTunes, the list of songs by any artist is accompanied by an entire column of nothing but $.99 prices. How can consumers, who are smart enough to find their ways to the iTunes web site and figure out how to download songs to their iPods, miss the pricing sleight-of-hand, if that is what is afoot? Don't consumers learn from the bombardment of odd prices that sellers are out to trick them? Teenage buyers certainly seem to get it. On my asking a sizable group of high school students who had downloaded individual songs from iTunes for the prices of the songs they had bought, all correctly answered $.99, or rounded up to $1. No one said "90 cents."

If buyers are fooled by the just-below prices, then they are fooled systematically, given the pervasiveness of just-below prices. You have to wonder how sellers, who, like buyers, are drawn from the human race, have the smarts to fool so many buyers. Sellers are also buyers. Are sellers fooled when they are buyers? How can they be so smart in one role, selling, but so deficient in another, buying? If buyers aren't fooled by odd prices, then sellers must be fooled into believing that buyers are fooled by prices ending in 9 (or other non-even-dollar endings). Otherwise, why don't sellers round up their prices and garner the extra cent or dollar in profits? For economists, such questions form the foundation of a real mystery, the solving of which might require more than a modicum of detective work.

Perhaps consumers of practically all goods are *stupid* (as distinguished from *rationally ignorant*), as seems to be suggested by the common presumption that odd prices can be explained by sellers taking advantage of their buyers' *price illusion*, and doing so time and again, as might be the case at grocery stores and filling stations. As conceded in past chapters, people's stupidity might explain a lot of things, but not why people are stupid in the first place about something so simple as the last digit of a price when they are presumed to be reasonably capable of determining which sophisticated goods they should buy. In addition, the people-are-stupid explanation for 9-ending prices is not much of a value-added insight, or advancement in argument, beyond what can be heard in casual conversation by people who are untrained in economics—and the stupid-people explanation doesn't appear to be supported by evidence.[7] Anyone who is too stupid to *know* that all gallons of gasoline are sold with 9/10 cent on the end, even when that fraction is displaced in a very small font, is certainly too stupid to be pumping gas, much less driving.

If the people-are-stupid explanation can be set aside, then why do so many prices end in 9, or have other odd-ending digits, or fall just below some even dollar amount? More than likely, since so many merchants charge odd prices, no one should be surprised if the explanation for odd or just-below prices is multifaceted, given that individual merchants sell different products and face widely varying marketing conditions. Market researchers and economists have several ways and levels of explaining odd or just-below prices, as we will see in this chapter. There certainly is reason to believe that 9-ending prices have a rational foundation either on the part of the sellers, the buyers, or both.

JUST-BELOW PRICES AS HISTORICAL ARTIFACT

One line of explanation for odd pricing posits that odd prices are an artifact of history (which suggests that people don't only price with the purpose of maximizing profits; they price because they are locked into a historical rut). In the late 1800s, when British goods came into the country, their prices were rarely even dollars because of the required currency conversion. Supposedly, because British goods were believed by American consumers to be, generally speaking, of higher quality than domestic goods, American retailers started attaching odd prices to their American-made goods they sold in the hope that such prices would signal consumers that American-made goods were of equal quality to foreign-made goods or that such prices would mislead consumers into believing that American-made goods were foreign made.[8]

However, posted odd pricing could not have been widespread until the advent and spread of fixed pricing among American merchants after the Civil War. Before then, prices were largely determined by haggling between buyers and sellers.[9]

According to one line of historical argument (which is almost certainly apocryphal), odd pricing's roots can be traced to the competitive efforts of a single businessman in Chicago in the mid-1870s. In 1876, newspaper publisher/journalist Melville Stone wanted to set up a new newspaper with the intent of outcompeting established newspapers that sold for a nickel with a not-so-novel marketing strategy: He would sell his paper for far less—indeed, for a penny! But there were few pennies in circulation at the time, which limited his street sales. As the story is told, Stone convinced a number of Chicago retailers that a just-below price of, say, \$1.99 would be viewed by many customers as much cheaper than \$2—maybe as low as \$1—and such just-below prices would be a boon to their sales. Such pricing would result in more pennies in circulation—and more newspaper sales for Stone, especially since, at the time, so few other goods sold for a penny. To overcome the penny shortage that resulted at the time, "Stone journeyed to Philadelphia, bought several barrels of pennies from the mint, and brought them back to the Windy City," all to encourage just-below pricing and sales of his newspaper.[10]

Nice story, but this explanation for the source of just-below pricing must remain suspect, since Stone sold his newspaper business shortly after establishing it to become the first director of the Associated Press. Moreover, there surely had to be less costly ways of getting pennies into circulation (for example, leaving them on table tops in bars!). To have just-below pricing spread throughout Chicago, much less the country, there probably had to be advantages for such pricing to merchants in general, not just for a lone publisher of a penny newspaper. Moreover, the Stone story must remain a dubious explanation for just-below pricing because such pricing strategies didn't become pervasive in the country until the 1920s.[11]

Another explanation posits that odd pricing took hold from the promotional efforts of Macy's in the early 1900s, which introduced $.99 sales that were, reportedly, very successful. Presumably, such just-below pricing spread not only because of Macy's success with its sales, but also because merchants found just-below prices to be an effective means of curbing employee theft.[12] If prices were set in round numbers—say, $1 or $10—customers could all too often give clerks the required payment without having to wait on change, and without the clerk having to record the transaction on then newfangled cash registers, or so the argument is made. Prices ending with 9 would require clerks to give back change, which would require them to record each purchase, not several purchases at once. Moreover, old cash registers had different rings when the recorded price was $1 and when the recorded price was $.99, with, supposedly, giving storeowners a chance to monitor their clerks cash register use.[13]

This explanation has a ring of truth, given that store clerks are hardly the most trustworthy of employee groups. Pilferage by store clerks represents a larger portion of store merchandise losses than does shoplifting by customers.[14] No doubt, cash registers, especially when paper tapes were added, helped owners monitor their clerks. However, again, the ring of truth must be considered limited in terms of the *persistence* of just-below prices, given the advent of sales taxes in the early 1930s and their spread in the 1940s and 1950s, with 45 of the 50 states now having sales taxes.[15] Few cash sales in the country can now be made without change being required. The growing use of checks and credit cards over the past half century has surely further eroded the ability of clerks to siphon off cash from store sales, which means that the persistence of odd pricing must be both an artifact and a clear indication that sellers are now incurring employee monitoring costs that, consequently, curb stores' profits and undermine their stock prices.

If just-below prices are simply an artifact of the country's commercial history, with no lingering economic foundation, then money is being left on the table. As argued before, shrewd investors should be expected to buy up retail stores' shares at prices deflated by pervasive just-below pricing policies. Once they have control, they can change stores' pricing policies and then sell their shares after the stores' profits and stock prices take a one-time upward jump. However, as opposed to going away, just-below prices remain prevalent. Again, consider the pricing of music on iTunes or the pricing of electronics at Fry's that are left totally unexplained by commercial history.

Alternative explanations for just-below pricing, which we can now consider, can be grouped into three categories:

- Those explanations that reflect the economics of gathering information by consumers,
- Those explanations that rely on emotional or psychological reactions by consumers to various prices, and
- Those explanations that are founded on the proposition that just-below prices serve "coding" functions for sellers.

JUST-BELOW PRICING AND INFORMATION ECONOMICS

Economic explanations for just-below prices start with a recognition of the multitude of purchases buyers have to make and the considerable complexity of individual purchases, all requiring consumers to encounter myriad information flows bombarding them from all directions in markets. This information includes the many attributes of products (color, size, technical sophistication, etc.), store amenities, product reviews, promotional efforts from producers, contract terms of the sales, and prices charged. Buyers can be expected to try to deal with the onslaught of information—all of which cannot be absorbed except without considerable time spent and intensity of attention paid to details—by making decision rules to ignore some sources and bits of information in order to release mental capacity to cope with other, more important sources of information (although the analysis in these sources is not always based on strictly rational, economic calculations).[16]

Under time constraints and mental limitations to process information within those time constraints, there is a "cost to thinking."[17] Buyers can decide, very rationally, to ignore rightward, least important digits of posted prices. This is to say, buyers can just pay progressively less attention, and incur fewer attention and decision costs, to the digits the further they are to the right. After all, the mental processing requirements for any given digit might very well be the same. However, in a price of, say, $.99 for a song on iTunes, the cost of misreading (or ignoring) the digit in the hundredth digit placement of the price is a fraction of the cost of misreading (or ignoring) the digit in the tenth digit placement. Similarly, in a price of a music album on iTunes, say, $12.99, the cost of misreading (or ignoring) the hundredth digit is an even smaller fraction of the cost of misreading (or ignoring) the first digit to the right of the dollar sign.[18]

Hence, given buyers' rational inattention, it should not be surprising that buyers pay less and less attention to the digits to the right of the decimal place as the price of the product escalates. That is, the $.99 for the price of a song on iTunes should be expected to receive greater attention by buyers than the $.99 portion of the price of an album on iTunes, $12.99. This might be expected to be the case because of buyers' efforts to devote their scarce mental capacity to the more important and costly digits. As buyers' incomes rise, they might be expected to buy more albums than individual songs and more of everything else, which can mean that they have progressively less time and mental computing capacity to deal with all of the $.99s tacked on to their greater array and number of purchases. Hence, as people's incomes rise, sellers can be expected to use prices that play to buyers' decision rules that take rightward digits off their buying radar screens, which can further encourage buyers to ignore the rightward digits, because they can anticipate what the rightward digits will be.

BUYERS' DECISION RULES

By not considering the most rightward and even the second most rightward digits, buyers might, from time to time, pay a higher price (or 8% more, on the sale of albums at the iTunes site, for example) than is expected. However, the buyers' rule of considering *only* $12 in the price $12.99 can still be rational, because the costs incurred from considering all digits of all prices can mean the incurrence of more costs of time and attention than can be recouped from carefully considering all digits of all prices and never paying more than is expected.

Understandably, when consumers ignore the rightward-most digit, then it makes little economic sense for sellers to charge, say, in the case of the iTunes album, $12.90 or $12.91. An extra 8 or 9 cents in price and profit can be picked up by making that most rightward digit a 9, which implies that no one should be surprised when 9 is found with far greater frequency than 8 or 7, or any other number down to 0.

By devoting mental processing power to all digits equally, buyers can, with greater frequency, make mistakes in reading and understanding the more costly leftward digits. But this only means that buyers can be expected to apply attention to rightward digits only so long as the accuracy gains from doing so do not exceed the accuracy losses from paying less attention to leftward digits. From a strictly economic perspective, if there were no cost to buyers considering the rightward digits, and there were only gains from allaying the unexpected expense of paying the rightward digits, then there would be no reason for buyers to not consider all digits equally, no matter how high the price. There would then be no reason for the just-below prices—except where cost and competitive market conditions demanded such prices for purposes of clearing markets (making sure that the quantity of the good available exactly equals the quantity demanded) and maximizing profits, which would likely make just-below prices as rare as (if not rarer than) even-dollar prices are now.

RATIONAL IGNORANCE AND PRICE

In short, from the perspective of information economics, buyers can be expected to optimize the allocation of their scarce time and mental resources, which is to say that they can be expected to optimize the absorption of information in the same way and to the same degree that they optimize all other bundles of goods they buy. This also means that buyers will be *rationally informed* about prices of what they buy—and, at the same time, *rationally ignorant* or *rationally inattentive*—and better off for being so.[19]

Buyers might indeed seem to be stupid, as some casual and academic explanations for just-below prices seem to suggest, but if they are, their stupidity can be, in a meaningful economic sense, the smart way to go.

Following economic logic, as buyers' opportunity costs rise, we might anticipate that they would pay less and less attention to the rightward digits. It simply makes progressively less economic sense to consider the rightward digits as the value of buyers' opportunities rise, especially since higher opportunity costs can imply greater income and more purchases that, as noted above, can constrict the time and mental capacity that buyers willingly devote to the rightward digits on particular purchases. Naturally, the perspective of economics also suggests that as products become more complex and sellers make greater promotional efforts that add to information flows in markets, buyers might be expected to pay even less attention to the rightward digits on more goods.

The economic perspective on information flows and prices offers a potential explanation not only for just-below prices, but also for why just-below prices did not become prevalent until the 1920s. In decades before the 1920s, people had less income and could make fewer purchases of products that were less complex, and they had lower opportunity costs, making detailed consideration of rightward digits more economical. Also, it must also be remembered that penny differences eighty or more years ago were then valuable differences. Since the 1920s, no one should be surprised that just-below prices have spread with the growth in people's opportunity costs and incomes, the sophistication and complexity of products, and promotional campaigns. This means that if advertising and other promotional efforts have become relatively cheaper over time, then a spread of just-below prices should be expected as buyers try to conserve their time and mental resources for the leftward parts of prices.

If buyers ignore rightward-most digits only because of cost/benefit considerations, then we should expect that, in instances where buyers can process price information with greater facility, or where prices are readily extracted from the clutter of market information, we should expect fewer 9-ending prices. Researchers have indeed found that the incidence of 9-ending prices is significantly lower for goods and services sold over the internet than in brick-and-mortar retail stores. The explanation given is that with search engines, prices of goods and services on the internet are more easily obtained and compared, leaving buyers with more mental processing capacity to absorb the rightward-most digit(s).[20] (However, it needs to be noted that to the extent that buyers can quickly look at more prices, one could argue that the cost of attention to numbers to the right of the decimal goes up with the ease of internet searches.)

BUYERS' DECISION RULES AND THE LOGIC OF 9-ENDING PRICES, AGAIN

When just-below prices become dominant, buyers might not be fooled as much as might be supposed by sellers believing that buyers are not considering the rightward digits. This is because with the prevalence of just-below pricing, buyers can be expected to develop decision assumptions and rules, one of which can be that the $.99 part of any price will be in place even when buyers do not notice it. Buyers can anticipate that sellers understand that buyers are rationally ignoring those digits, which means buyers might rarely be fooled and exploited, or so any number of marketing scholars have argued.[21] Indeed, with the assumed pricing rule in place—"always expect an unnoticed $.99 at the end of prices"—buyers might be fooled only when the $.99 is not attached to even-dollar prices.

Of course, the prevalence of just-below pricing can fortify sellers' use of just-below pricing. If buyers never consider the rightward digits but assume, out of decision rules they devise from assessed probabilities from past experience, that the $.99 portion of the price will be there, then sellers lose little to nothing in sales by tacking on the $.99 portion of the price. By adding the $.99, sellers have not raised, effectively, the price that buyers really assume is being charged. Purchases are not then materially affected by the $.99 addition.

From this perspective of buyers adopting rational decision rules on price endings, it is understandable that iTunes charges $.99 for each song. If it charged $1, consumers might rightfully expect the $.99 to be tacked on, jumping the buyers' *assumed* price to $1.99 in their minds. The 1 cent added to the $.99 price is not, effectively, a 1% increase in the buyers' *assumed* price. Rather, it gives rise to 101% rise in the *assumed* price.

Apple and Fry's, in other words, may not be charging a just-below price to take advantage of music buyers' foolishness, or capacity to be duped, by a slight shaving of the price. Instead, the companies are working with the information gathering and decision rules that buyers have established. In effect, Apple and Fry's are forced to end many of their prices with $.99 by buyers' rationality in devising cost-saving decision rules. Nine-ending price can be path dependent, which means once it is adopted many other sellers must adopt 9-ending pricing. Just as we argued with regard to network goods, 9-ending pricing can reach something of a tipping point, meaning sellers start using the pricing strategy because they expect other sellers to start using it.

Apple and Fry's are caught in something of a prisoner's dilemma. The companies might very well prefer to simplify purchases by charging even-number prices. However, with just-below pricing so prevalent among so many sellers, Apple and Fry's have to recognize the rationality and, therefore, prevalence of buyers' decision rules. Apple and Fry's can't get all other sellers to give up just-below pricing. So, the two companies (and so many others) do what they must: adopt just-below prices even when they know all sellers—and buyers—would be better off, in the long run at least, if everyone were to charge even prices (or prices that did a slightly better job of clearing the market).

To the extent that buyers do develop decision rules about rightward-most digits in prices, the 1 cent drop in the price of a good from $1 to $.99 might suggest that consumers are highly responsive to a very "small" price reduction, making their demand for the good appear to be price sensitive or highly elastic at that price point, which researchers have found to be the case.[22] However, the consumer responsiveness to the price change (or their elasticity of demand) can be exaggerated. This is because for buyers, the relevant price in terms of their responsiveness is the *perceived* price, and that price does not drop by only 1 cent (in the case of a price change from $1 to $.99). Rather, it drops by $1.00, or from $1.99 to $.99, or by more than 50%.

In effect, just-below pricing can be a form of "pollution" for both buyers and sellers that has no necessary market solution. In the case of actual pollution of atmosphere, for example, each polluter can reason that her pollution is not a consequential source of the environmental degradation. Each may only pollute because to abate the pollution would force the pollution abater to suffer a cost other producers do not incur and, hence, a competitive disadvantage. So, all polluters continue with their pollution, in spite of not liking the aggregate environmental consequences.

In the case of just-below pricing, sellers might want to get rid of just-below pricing, but they all continue to price just-below even-dollar prices because buyers expect just-below pricing, which is aggravated by so many sellers charging just-below prices. Sellers that try to change just-below pricing can suffer, as do polluters, a loss of competitive position for buyers' dollars.

PSYCHOLOGICAL PRICING

Many marketing researchers' work is founded more on the psychology of pricing than on the economics of pricing (sometimes because of professional hostility to economic theorizing).[23] That is, buyers respond to particular prices, especially odd prices, because of "price illusion," "habit and inertia," the way the mind reacts to numbers as stimuli, given cultural and social constraints, often with no consideration of any cost/benefit calculus.[24] For example, researchers posit that just-under prices are prevalent because consumers are inclined to "round down."[25]

Why? They just do it or they don't, and not necessarily for any rational reason. The important issue for researchers steeped in psychology is what the data show. For example, market researchers and practitioners may posit that buyers can see a $3.99 price as $3 (and maybe some change), because of "price illusion" or mental tendencies to "round down," and then devise experiments or surveys to determine the extent to which buyers round down. Consider this non-exhaustive array of research studies, the first several of which had "disappointing" (mixed) findings:

- One researcher designed an experiment in which he gave respondents cards with four different products with different just-below and even-dollar prices on them. The researcher told the subjects that they had a 50–50 chance of winning the amount on the card, and then asked them how much they would have to be paid before they would be indifferent between the sure payment and the gamble. If the subjects were inclined to round down, the researcher expected the respondents to accept a lower sure payment for the odd-priced products than for the products that were priced in even dollars. Over experiments with five groups of subjects, the reported findings were inconclusive: Only two of the five groups responded as expected.[26]

- Other researchers repeated the above experiment or developed other price-recall experiments with much the same inconclusive findings on whether respondents systematically round down prices.[27]

- A large mail-order catalog-based retailer tried an experiment in the 1930s in which six items were given prices of $.50, $.80, $1, $1.50, and $2 in one part of the catalog. In another part of the catalog, the same items were given "customary prices" of $.49, $.79, $.98, $1.49, and $1.99, respectively. The catalog was mailed to 6 million people. The rounded prices had mixed results. For one or more products, they cut sales in half; on others they had no "appreciable effect," and on others "sales were disproportionately large." This outcome caused the mail-order company to abandon its testing, especially since the 1-cent increase on one item resulted in a loss of $50,000. The economist reporting the findings noted, "One thing is clear: competition was itself a custom limited by the history of institutions, and by the psychology of the competitors. The searcher after profits would continue to pay his respects to both."[28]

- In real-estate markets, the research findings about the impact of "just-off" (or just-below) some round number are mixed, with one study showing that just-off prices extend the time houses are on the market[29] and another study showing that such prices increase the actual selling prices of houses.[30] Yet another study found that actual selling prices of houses are unaffected by just-off prices, while the amount of time houses are on the market is shortened by just-off prices.[31] (However, it needs to be noted that these findings could be influenced by what exactly the researchers consider to be just-off [or odd] prices).[32]

Other researchers have been more successful in finding evidence of respondents tending to round down:

- One researcher asked respondents to determine which of two price reductions suggested the better deal: When the price was reduced from $.93 to $.79 or when the price was reduced from $.89 to $.75.[33] While the price reduction is the same, $.14, in both cases, the second price cut is actually a greater percentage reduction. However, most respondents chose the first price cut. A likely explanation is rounding down. The first price reduction appears as a $.20 price cut if only the first digits are considered ($.90 and $.70), whereas the second price cut represents a $.10 price cut ($.80 and $.70).

- Other researchers found from experiments that subjects are more likely to underestimate odd prices than even-dollar prices,[34] with the left-hand digits in prices more frequently recalled than right-hand digits.[35]

- When a store discounted the price of one branded product from $.83 to $.63, sales rose by 194%, but when the price was lowered further to $.59, sales jumped by 406%. On another branded product, when its price was lowered from $.89 to $.71, sales increased by 65%. When the price was dropped to $.69, sales expanded by 222%. Unfortunately, the store made no attempt to account for changes in the store's promotional efforts as the price was lowered in the two steps, leaving questions about how much of the sales jumps could be separately attributed to the odd pricing.[36]

- New Zealand researchers posited that while the price and quantity sold of a good will likely always be inversely related, price cuts at certain pricing points—namely, $10, $20, $50, and $100—should lead to a greater-than-expected increase in the quantity demanded.[37] They tested different prices for six products—a block of cheese, a frozen chicken, a box of chocolate, a hair dryer, an electric kettle, and a blender. They did indeed find what they expected, a greater-than-expected demand at odd prices for the three products that had $.99 attached to some dollar price ($4.99, $5.99, and $9.99), but they did not find an odd-pricing effect for those goods when the price was lowered by 4 cents ($4.95, $5.95, and $9.95). The researchers also found the expected odd-price effect when the price of the $20 good was lowered to $19.95, when the $50 good was lowered to $49.95, and when the $100 good was lowered to $99.[38]

- French researchers arranged for a grocery store to change the price of a package of cheese every 2 hours, going back and forth between €1.99 and €2.00.[39] They found the incident of sales went up, but not to a statistically meaningful extent, when the price was €1.99 However, the 99-ending Euro price did increase total consumer expenditures on cheese by almost 29%.

- When the lead researcher on the just-covered study joined with other colleagues in another attempt to estimate the demands of another set of three products (fly spray, cheese, and an electric kettle) with another set of subjects interviewed at a New Zealand shopping mall, much the same conclusion was reached; the probability of the good being purchased jumped when the price was lowered from an even-dollar price to one ending in $.99. Again, there was no jump in expected sales when the price was reduced by another 4 cents.[40]

- Other psychology-based explanations for just-below pricing include the prospect that just-below ($.99) prices indicate that the price of the product has not been recently raised,[41] that the product is on sale,[42] and/or that such prices are the lowest prices around.[43] One such study involved the distribution of 90,000 mail-order catalogs with identical items sent to people on different distribution lists with price endings of $.88, $.99, and $.00. The items with $.99 price endings outsold the identical items with price endings of $.88 and $.00 by 8%.[44] Still other researchers have found that using prices ending in 9 resulted in an average expansion in sales of 10% over other prices ending in any other digit over twenty product categories.[45] In another report on separate studies involving different prices for the same items in catalogs distributed to different mailing lists, researchers found that 9-ending prices resulted in a 35% increase in sales over prices ending in 0, 4, and 8. When they separated out the 9-ending on established and new prices, they found that the 9-ending prices increased sales on both categories of products—new and established—by the exact same percent, 22%.[46] But when still other researchers investigated the relative impact of a 9-ending price for syrup, the findings were, again, inconclusive.[47]

PRICES AS CODE

One reason for just-below prices never mentioned in the marketing literature is that prices are sometimes used by sellers as code. A price ending in $.99 can serve as the *standard* or *regular* price, if not the *lowest* price ever on the good.[48] A price ending in $.97 can be used to identify those products on sale for, say, 40% off, whereas a price ending in $.93 might identify items on sale for 25% off. The price codes can be very useful to stores, such as Limited Too, that accept returns without receipts. The clerks giving the refund can easily figure how much the person paid for the item. If the price ends in $.99, the clerk knows to give the customer the full printed price on the price tag. Customers returning articles with a price ending in $.93 get the price on the tag minus 25%, following our example above.

CONCLUDING COMMENTS

Given the pervasiveness of just-below pricing, it would probably be surprising if such pricing strategies did not have a rationale and a payoff to firms that use them. While the available findings on the effect of just-below pricing are mixed, certainly the weight of the evidence warrants this generalization: *just-below prices work in terms of hiking sales for many products sold by many sellers.* Given the variety of circumstances in which goods are sold, it should not be inexplicable that just-below prices are not always effective.

No doubt, some consumers are not fully rational in their purchases. That's hardly a novel claim about much consumer behavior. However, it is not at all clear from evidence that just-below prices necessarily reflect the exploitive, consumer-welfare-reducing powers of sellers, as many marketing scholars and practitioners seem convinced is the case. As noted, consumers can be better off even when they at times pay slightly higher prices than they think they are paying (or would be willing to pay if they always knew exactly what prices are being charged). Economizing on information gathering and processing clearly makes sense for consumers in today's complex market environments. By not considering the rightward-most digits of prices with great care, consumers can consider more prices and can make fewer mistakes on the more costly leftward-most digits within prices.

Furthermore, as noted in this chapter, the prevalence of just-below prices might reflect decision rules consumers make. Since so many prices end with $.99, for example, buyers need not consider those digits at all. They can assume they are there. This suggests that sellers use just-below prices to avoid consumers assuming that even prices end in $.99 when, in fact, they don't.

NOTES

1 In mid-2007, Apple started selling individual songs without copy protection for $1.29.

2 On the same day the Fry's insert appeared, the *Los Angeles Times* had another two-page insert from Pepboys Auto with 56 out of the 76 prices advertised (or 74%) ending with $.99. Another 16 prices (21%) ended with $.98, leaving only four prices with other truly odd endings ($11.88, $1.79, and $21.48).

3 See Hawkins (1954).

4 See Gabor and Granger (1964).

5 See Mason and Mayer (1990),

6 Originally, in the marketing literature, "odd prices" referred to prices that appeared to end disproportionately in odd numbers—3, 5, and 9—as opposed to even numbers—2, 6, 8, and 0—with odd-number-ending prices dominating even-number-ending prices. One early survey of several thousand prices found that 64% of the prices ended in 9, 19% ended in 5, and 9% ended in 3. The even-number prices—8, 6, 4, 2, and 0—accounted for 2%, 0%, 0%, and 1%, respectively (Twedt 1965). Now, the term "odd prices" generally refers to prices other than prices ending in zero. For more recent surveys of the distribution of the last digit of prices, with price endings in 5 and 9 accounting for up to 80% of all prices, see Friedman (1967), Monroe (1973), Stiving and Winer(1997), Schindler and Kirby (1997), *Daily Mail* (2000), and Lee et al. (2006). A widespread pattern of just-below prices has also been found in Germany (Hogl 1988), in New Zealand (Holdershaw 1995) and around the globe (Gilmour 1985).

7 Georgoff (1971) found that "price illusion" might explain consumer behavior for some consumer groups buying some products but not other groups and other products, with the net effect being weak evidence for the proposition that price illusion explains the prevalence of odd pricing. Although Lambert (1975) found some evidence that odd pricing was associated with price illusion under some circumstances, other researchers (Dodds and Monroe 1985) found no evidence to support the view that buyers' perception of the value and quality, and their willingness to buy was affected by odd and even prices.

8 Georgoff (1971).

9 Georgoff (1971).

10 Morris (1979, p. 44).

[11] Adams (n.d).

[12] This historical explanation for the spread of odd pricing to all corners of the globe has been repeated by a number of scholars: Rudolph (1954), Twedt (1965), Harper (1966), Sturdivant (1970), Kreul (1982), and Gilmour (1985).

[13] According to one account, "Department stores first started the (odd) pricing method to eliminate shrinkage at the cash register caused by the pocket-bookkeeping among many sales personnel. Even-priced merchandise often would be paid an exact amount direct from shoppers' purses. The clerk would then serve another customer or two before ringing up a sale. When prices were changed to off amounts, making change from one's pocket became obvious and correct change usually required a trip back to the register" (as quoted in Twedt (1965, p. 55), citing Garvey Corporation (1964, p. 6).

[14] Bolger (2002).

[15] Fox (2002).

[16] For samples of discussions about how information flows affect consumers' consideration of different digits within prices, see Brenner and Brenner (1982), Schindler and Wiman (1989), Schindler and Kirby (1997) and Sims (1998). For more general discussions about the theory of consumers devising simplified decision rules, see Dawes (1964), Tversky (1972), and Shugan (1980).

[17] Shugan (1980).

[18] Sims (2003), and Lee et al. (2006).

[19] Sims (2003), and Lee et al. (2006).

[20] Lee et al. (2006, pp. 13–16).

[21] Marketing researcher Anil Kashyap has made the argument this way: "Buyers may use rules of thumb when searching for items and comparing prices...If firms are aware of this tendency by consumers, they may set prices so as to exploit the use of the rules" (1995, p. 266).

[22] Schindler (2006).

[23] Skouras, Avalonitis, and Indounas (2005, p. 363) write, "The weakest part of [microeconomic theory] is surely the notion of utility-maximizing by rational consumers. This is not only implausible as a general description of buyers' behavior but there are many instances in the everyday experience of most people that seem to contradict it. Moreover, the work of psychologists and several psychological experiments have shown beyond any reasonable doubt that rationality and utility-maximisation [sic] can hardly be considered as universal and ever-present traits of consumer behavior" (Kahneman, 1994; Kahnehan and Tversky, 2000; Thaler, 2001).

[24] Ginsberg (1936); Boyd and Massey (1972). Researchers have found that while the internet prices of American internet-based products predominantly end in 9 and few end in 8, the opposite is the case in the five Asian countries studied (Hong

Kong, Singapore, Malaysia, China, and Japan). They suggest that Asian firms use 8 more frequently than American firms because the number 8 is a symbol of luck in Asian societies (Heeler and Nguyen n.d.)

[25] Whalen (1980) and Schindler and Warren (1988).

[26] Lambert (1975).

[27] Alpert, McGrath, and Alpert (1984); and Schindler and Kibarian (1996).

[28] Ginsberg (1936).

[29] Palmon, Smith, and Sopranzetti (2004)

[30] Allen and Dare (2004)

[31] Salter, Johnson, and Spurlin (n.d.).

[32] House prices of $99,950; $99,800; $99,750; and $99,250 are considered just-off prices, whereas "round prices" include $100,000; $99,900; $99,500; and $99,000 (Salter, Johnson, and Spurlin n.d). To my way of thinking, $100,000 and $90,000 are clearly round prices, but it is not at all clear to me why $99,800 would be considered a just-off price but $99,900 would be considered to be a round price.

[33] Monroe (1979).

[34] Schindler (1984); Schindler and Wiman (1989).

[35] Schindler and Kibarian (1993).

[36] Nagle (1987).

[37] Gendall, Holdershaw, and Garland (1997).

[38] One finding with the research is problematic: When the prices of all six goods were raised to levels above the even-number prices (that is, to $5.10, $6.10, $10.10, $20.10, $50.10, and $110), the probability of the goods being purchased was curbed, but only slightly, suggesting that the elasticity of the different demands for hikes in price above even numbers was very low, which could make such price increases profitable because the price hikes could lead to revenue increases while the curbed sales would lead to a reduction in production costs.

[39] Gueguen and Jacob (2005).

[40] Gendall, Fox, Wilton (1998).

[41] Schindler (1984).

[42] Quigley and Notarantonio (1992); Schindler and Kibarian (1996).

[43] Harper (1966), Schindler and Kibarian (1996); Schindler (2006).

[44] Schindler and Kibarian (1996).

[45] Blattberg and Wisniewki (1987).

[46] Anderson and Simester (2003).

[47] Little and Ginese (1987).

[48] Schindler (2006).

Chapter 10

THE ECONOMICS OF MANUFACTURERS' REBATES

In Chap. 9, we noted how many product prices advertised in Fry's Electronics full-page multicolor newspaper ads ended with 9, such as in \$2.99 or \$299. What is also notable about Fry's ads is how many of the products carry manufacturers' mail-in rebate offers. On the day I began writing this chapter, Fry's included 53 products in its full-page ad, a third of which carried manufacturers' mail-in rebates. Two of the advertised products had two rebate offers, with one declared "Free" after the rebates were deducted from the posted price.

But then, Fry's ad mirrors a national apparent affection among manufacturers for rebates. About a third of all personal computers and their peripherals and over a fifth of all digital cameras, camcorders, and LCD TVs are sold with rebate offers. Overall, according to one long-time rebate researcher, there are about 400 million rebate offers annually in the USA alone, all worth about \$6 billion.[1]

With so many rebate offers floating through the economy, you have to wonder about the sanity of manufacturers offering rebates. Why don't they just lower their prices and be done with it? Rebates have management costs of their own. Obviously, the added revenues from rebates have to be greater than the added costs, but how can that be? Granted some shoppers can be sucked in by rebates, but you might think they would learn from experience, at least eventually, what the rest of us know, that rebate offers must be approached with care.

THE NATURE OF REBATES

What's a "rebate"? Simply put, a rebate means that customers can get a price break from the normal or posted price of the products of so many cents or dollars by exerting some future effort, which generally means mailing the rebated products' manufacturers (or their out-sourced fulfillment centers) completed forms, along with proof of purchase and possibly other documentation. Usually, the rebate specifies the dollar payout, the redemption period (or the timeframe within which the rebate must be sought in order to receive the payout).[2]

Customers can be given two, four, or eight weeks (or even as long as a year) from the date of purchase (or delivery) during which the rebate must be sought.[3] If the rebate is not sought within the redemption period, the rebate can be denied, which means that customers end up paying the full posted price and, possibly, getting a product that they would not have bought were the rebate offer not made.

Rebates have become such a prominent promotional tool that students of rebates have their own lingo:

- **Lift** is the increase in sales from a rebate promotion.

- **Breakage** is the percent of customers who never seek redemption of their coupons, or fail to obtain a rebate because they didn't meet the requirements.

- **Slippage** is the percent of redeemers who never cash their rebate checks.

Of course, as many readers know from personal experience, manufacturers can make the rebate redemption process more or less onerous for customers with, naturally, manufacturers' goals of affecting the redemption rates and profit maximization in mind. For example, to receive rebates, customers might be directed to a website where the redemption process can be completed or where customers can be given additional instructions and contact information for mailing completed redemption applications, coupons, box tops, product bar codes, serial numbers, and/or one or more other proofs of purchase.

Transparently, manufacturers' ability to "manage" redemptions and payouts no doubt opens opportunities for rebate abuse, especially among manufacturers who do not count on repeat business and/or who face serious prospects of going out of business. A manufacturer might set the redemption time frame to run for only a week beginning exactly three weeks after the purchase date, with the manufacturer counting on customers not reading the fine print on the rebate offer at the time of purchase and/or forgetting about seeking redemption within the limited redemption window. Also, rebate checks can be mailed in envelopes purposefully designed to look like junk mail, with the hope that the checks will be discarded, and/or the checks might have a short deadline for cashing—all with the intent of increasing the slippage rate.[4]

With the millions of redemptions sought, many dishonest attempts to manipulate the breakage rate can get mixed up with honest mistakes on the parts of manufacturers tendering rebate offers or with manufacturers' efforts to contain buyers' rebate fraud. *Business Week* reports that Samsung Electronics of America settled a complaint from the New York Attorney General in late 2005, agreeing to pay $200,000 to 4,100 rebate redeemers who were denied rebate payments only because they lived in apartment buildings and Samsung's rebate program allowed only one rebate redemption per street address (a restriction Samsung could have imposed to suppress rebate scammers who might attempt to buy multiple units of the rebated items, only to resell them after receiving the rebates).[5]

One self-described "rebate junkie" sent in the required paperwork for a $100 rebate on a TiVo video recorder that cost $300, expecting to receive his rebate check within the promised six-to-eight-week period after mail in. Eight weeks later, the customer called TiVo to find out where his check was, only to be told that the matter had to be "researched." He received his check 20 weeks after purchase, or three months after the company's self-imposed payout deadline.[6]

Perhaps TiVo's delay can be chalked up to inevitable worker mistakes, oversight, and ineptness when so many rebates must be handled. However, we can only note that a greater number of mistakes can and should be expected than otherwise when companies do not penalize themselves for not meeting their payout deadlines. The "rebate junkie" got the same check after twenty weeks that he should have received after eight weeks. A "self-enforcing" rebate offer would be one under which the manufacturer commits to escalating the rebate payment after the due date for the rebate (but, of course, such rebate contract provisions can feed back into a lower initial rebate and, possibly, more onerous rebate terms).

With the growth in rebate offers, there has been a more-than-proportional growth in customer complaints. The Better Business Bureau found that rebate complaints received nearly quadrupled between 2001 and 2005, or from 964 to 3,641, with many of the complaints dealing with rebate fraud.[7] Tim Silk and Cornelia Pechmann found that at the end of 2006 there were ten appellate law suits underway at the time relating to rebate complaints, with 19 consent decrees already handed down by various courts, 11 laws regulating rebates enacted at the state level, and another 44 proposed laws under review at the state level (with several proposed bans on rebates in several states, including Delaware, Massachusetts, and New Hampshire).[8]

One study found from a survey of 35 managers of rebate fulfillment centers that the redemption rate could range from a low of 1% of total unit sales to 41%, depending on the product category.[9] Earlier studies found that seven out of ten buyers whose purchases were influenced by rebate offers (not including buyers who did not make their purchases because of the rebate offers) did not redeem their rebates.[10] Yet another study revealed an average breakage rate of 40% across a wide range of products with the total breakage equaling about $2 billion in 2005. A *Business Week* reporter equated the $2 billion in breakage with an increase in producer profits that, as we will see, is a misguided deduction (even if the $2 billion estimate on total breakage is correct).[11]

But then, breakage does have an upside for consumers. There no doubt would be far fewer rebate offers if the breakage rates were far lower than they are. A number of digital scanner manufacturers went bankrupt because their rebate programs had a close-to-100-percent redemption rate.[12] Low-end personal computer manufacturer eMachines and Microsoft terminated their high rebate offers when low breakage rates made them too costly, which suggests a link between the breakage rate and the value of the rebates: the greater the expected breakage rate, the higher the dollar value of the rebate offer can be.[13] However, a corollary rule within the rebate industry works to cap the dollar value of rebate offers: the higher the rebate value both in absolute dollar value and as a percentage of the product list price, the greater the redemption rate.[14] According to one rebate fulfillment record, the redemption rate on a $10 rebate for a $100 products is about 10%. A $50 rebate on a $200 product yields a typical redemption rate of 35%.[15] Another study found that given the price of the product and other rebate requirements, increasing the rebate value from $1 to $5 to $10 to $20 would lead to redemption rates, respectively, of 7%, 17%, 27%, and 50%.[16]

THE REASONS FOR REBATES

One of the obvious reasons for rebates is the stimulation of sales of rebated products. According to the survey work of Silk and Pechmann, given the price of the product and rebate requirements, increasing the rebate from $1 to $5 to $10 to $20 can increase sales, respectively, by 31%, 64%, 90%, and 135%.[17]

Why don't manufacturers simply lower their prices to retailers instead of offering rebates? If rebates stimulate sales, would not lower wholesale prices that can cause retail prices to drop do for sales what rebates do? The easy answer to the question of why rebates (and not other promotional forms) is that, as noted, not all rebates are redeemed. The unredeemed rebates *could* add to the profits of the manufacturers offering the rebates. I have italicized *could* because rebates are typically offered to all customers, not all of whom were attracted to the product by the rebate offer. Of course, many customers can be expected to be attracted to the store and the rebated product by the rebate (or else why would Fry's continue to offer them?). Accordingly, the rebates can add to sales.

One study found that rebates could expand sales by up to 150%, depending on the absolute dollar value of the rebate and relative value to the purchase price.[18] This means that such additional sales *can* add marginally to profits. The net profitability of the rebates depends, obviously, on exactly how many additional sales the rebate offers stimulate and how many price cuts the manufacturers suffer on sales that would have occurred without the rebate offers, as well as the added production costs.

We now have one tentative reason why Fry's ads carry rebates for many, but not all, advertised products: Different products appeal to different groups of buyers with different responsiveness to rebate offers and with different inclinations to redeem the rebates. Rebate offers on some products are profitable for manufactures, while others are not—because different buyer groups respond differently to rebate offers. This means we need to explore further why some people redeem their coupons and others don't, but we must first recognize the impact of rebates on product demand.

REBATES AND PRODUCT DEMAND

Because rebates give retailers something to advertise, they can often increase retailer traffic as well stir interest in—and marketing buzz around—the rebated products and can be expected to boost sales at the retail level above what they would otherwise be. This is especially the case when many buyers who are attracted to the product because of the rebate do not seek redemption of their rebates. That is, retailers' demand for the rebated products can rise. The greater the product demand at the retail level, the higher the price the manufacturer can charge retailers and the higher the price (without rebate) retailers can post.[19] The *increase* in the posted price will not likely equal the rebate, but still it should not be forgotten that the rebate can be partially offset by some increase in posted price.[20] How much a manufacturer can raise its wholesale price because of a rebate promotion depends on exactly how sensitive retailers and their customers—both those who are affected and unaffected by the rebate offer—are to a price change. Of course, how high the posted price can be hiked also depends on the size of the rebate offer and the redemption requirements.

CHECKS ON REBATE DIFFICULTY

As noted, manufacturers can make the rebate process more or less difficult and risky for customers by manipulating the redemption period and the requirements that must be satisfied for receiving rebates checks, with research showing that the complexity and difficulty of the redemption process do affect the breakage rate.[21] However, manufacturers and retailers must surely realize that customers can translate manufacturers' rebate requirements into transaction costs that, to the extent that customers notice and weigh such costs at the time of purchase, can reduce the assessed net value of any given size rebate. This means that manufacturers can be checked in terms of how onerous or easy they make their redemption processes by the impact of redemption requirements on customer demand. The more onerous the redemption requirements, the lower the increase in product demand—and the lower increase in the posted, before-rebate price that can be charged.[22]

Granted, many consumers do not read the fine print on rebate offers, and can be largely ignorant of the exact requirements for any particular rebate offer that catches their attention (all very rationally). However, that doesn't mean that consumers cannot impute some cost (albeit lacking in accuracy) for the requirements from casually scanning the content of the large print in the offers and the length of the fine print section of the offers, from hearing the rebate experience of others, and from considering their own experience with past rebate offers and redemption experiences. The more complex and taxing the rebate requirements appear from casual inspection (drawn from, for example, the length of the fine print section of the offer), the greater customers' imputed cost—and the lower the increase in their demand, the lower the increase in the posted product price manufacturers and retailers can charge.

Of course, manufacturers must be concerned about the redemption rate, which can be affected by how onerous the redemptions rules are. The less onerous the rules, the greater the redemption rate and the lower the rebate needs to be to achieve any given profitability goal from a rebate promotion.

OPTIMUM REBATES

What are manufacturers to do? The only thing they can and must do is optimize as best they can across several relevant variables: the size of the rebate (in absolute dollar terms and relative to the product price), the payout period and deadline, and any other redemption requirements that affect customers' costs. There is no point in our trying to specify exactly what manufacturers should do to achieve optimum profitability from rebates simply because the sensitivity of consumers to price changes, rebates, and redemption rules can vary considerably across products. Hence, many rebate promotional efforts are likely to be explorations, with the goal of finding the optimum rebate strategy. Many producers can be expected to grope their way toward rebate strategies that work well (profitably) for their products. Others will find that rebates are not worth their costs, possibly because they don't have customer groups who differ sufficiently in their price and rebate sensitivity or because too many customers, if not virtually all, redeem their rebates. TiVo terminated its $100 rebate promotion on a $300 video recorder (the rebate offer our "rebate junkie" had trouble with) because the breakage was close to zero. When producers have that kind of rebate response, they might as well lower their posted prices and save the expense of the rebate promotions and redemptions.[23] The inclination of firms to terminate rebate offers that have high breakage rates helps to explain why the *average* breakage rates are relatively low.

FIRM REPUTATIONS

The effectiveness of rebate promotions depends upon the manufacturers' and retailers' reputations for reasonable rebate terms and honest dealing. A reputation for reasonable terms and honest dealing can save customers the cost of reading the details of the rebate offer (which can be extensive) and the risk cost that consumers incur when an upfront payment must be made and a future cost must be incurred to obtain a payout that will be received even further in the future. Stores' and manufacturers' reputations for reasonable terms and honest dealing can increase customers' estimate of the value of the rebate and, accordingly, can increase sales and the posted price by more than otherwise. Such reputations can also reduce the required size of the rebate to produce any given sales expansion and posted price increase.

Few customers of sidewalk vendors in New York City or Washington, D.C. would put much stock in rebates the vendors might offer, mainly because there are few ways the vendors can make credible commitments that their wares are not knockoffs, much less that the rebate offers are worth more than the paper they are written on. On the other hand, Dell Computers can make rebate offers credible simply because any degradation of the company's reputation through misleading and deceptive rebate offers can translate into lower future sales and profits. Thereby, Dell's rebate offers can be viewed as self-enforcing contracts, meaning that the company can suffer a nontrivial cost as a result of any misbehavior (or what economists call "opportunistic behavior") on their rebates.[24]

Still, the reputation of all rebate offers as a promotional instrument can influence people's assessments of particular rebate offers, especially from producers with whom customers do not deal frequently. To the extent that consumers' bad experiences reduce the credibility they ascribe to all rebate offers as a promotional advice, the less effective rebates can be in stimulating demand and the less frequently they will be used.

Moreover, the sizes of the rebate and the price-after-rebate can influence the attraction of the rebate offer in what might seem an odd way: The larger the rebate relative to the price—and hence the closer the after-rebate-price is to zero—the more guarded many buyers can be toward accepting a rebate offer.

During the time this chapter was being developed, a friend sent me an email, noting that a 2-gig USB thumb drive could be purchased on the web site SlickDeals.com for "free." On going to the web site, I learned the USB drive was "free" only after rebate. The purchase price was $50, but the purchase was accompanied by a $50-rebate offer—*provided that the rebate application, receipt, and other proof-of-purchase documentation was received by the rebate fulfillment center no later than 30 days after the purchase date, not the delivery date.* I was effectively presented with a choice between a sure thing—I could buy a 2-gig thumb drive at Fry's for just under $20 (with no accompanying rebate offer)—or I could take what was an offer of a rebate gamble from SlickDeals, the prospect of getting the after-rebate-price of $0.00 (ignoring postage and the time cost of seeking the rebate) or ending up paying $50 for the 2-gig drive because my rebate redemption was denied, perhaps because I missed the redemption deadline and/or I didn't have the right documentation for the rebate.

The critical problem for me was that the thumb-drive manufacturer could delay shipping the thumb drive until I didn't have time to meet the redemption deadline, set, as noted, at thirty days *after* purchase. Delivery could have been important for obtaining redemption because it might be only then that I would have all of the required documentation.

Given the nontrivial risks involved, I did not buy the "free" drive. My decision not to buy the free drive was fortified by the fact that I had little basis for assessing SlickDeals' and the USB manufacturer's trustworthiness. I also harbored the reasonable hunch that the manufacturer had to be counting on a fairly high breakage rate to make the $50 rebate offer work. I concluded that the manufacturer could, and just might, manage the breakage rate via the timing of its shipments. My natural risk aversion—fortified by my working heuristic that "any deal that appears to be too good to be true very likely is too good to be true"—

led me to conclude that I would have to be offered a negative after-rebate-price for me to be attracted by the rebate (and even then a negative after-rebate-price would make me all the more suspicious of the manufacturer's intentions on managing the redemption rate).[25]

The point of my story is crucial: many rebate offers are made in the context of gambles, given that rebated products are positioned next to competitors without rebate offers but with posted prices lower than the posted price of the rebated products.

Firms do have ways of compensating for their own inability to make their rebate offers credible and effective. Akai is an electronics producer with a brand that is not widely known. Many readers of this book may never have heard the name, much less know that Akai sells, among other products, fifty-inch plasma-screen televisions that are excellent in terms of the picture quality (I have one). Such unfamiliar manufacturers might have a tough time using rebate promotions and selling their products through unknown retailers. Put another way, Akai would have to hike its rebate offer to have the same impact on sales that a better-known manufacturer—say, Sony—would have. However, Akai retailers' reputations for making credible commitments can substitute for Akai's lack of reputation. I bought my Akai television from Costco, a company I trust (almost totally) because of past experience with buying and then returning products and with seeking rebates on products Costco sells. Of course, Akai will need to "pay" Costco for its endorsement of Akai's brand and rebate offers by lowering the price it charges Costco.

BREAKAGE ECONOMICS

Why breakage? For economists, the breakage rate is grounded in fairly straightforward cost/benefit calculations, which can enable firms offering rebates to use them to segment their markets and to engage in price discrimination.

THE COST/BENEFIT CALCULUS

A strictly *economic* answer to the question of why breakage depends on presumed differences in potential buyers in terms of their ability and interest in noticing rebate offers. Customers also differ in their ability and willingness to process and to think about the information on the offers and to devise an appropriate and reasonably accurate cost/benefit analysis of the observed offers.[26] And many customers buy products with no intention of considering rebates if or when they are present. They are oblivious to the rebate offers (intentionally) and find the posted price more than reasonable. At purchase time, they judge the money reward from rebates to not be worth the redemption costs. They might understand that a few rebates are good deals, but they choose to ignore all rebate offers because it is hard (costly) for them to identify what they believe are the relatively few good deals among the many not-so-good and bad deals. That is, the expected value (discounted for time and risk) of the opportunity costs they would have to incur to go through the redemption process for a range of offers exceeds the expected value of the realized payouts.

One explanation for why some customers develop heuristics to seek redemption on all (or almost all) rebates and others do not is that customers differ in their opportunity costs. Hence, breakage rates between 0% and 100% on different products can, in part, mirror the spread of buyers' opportunity costs.[27]

Moreover, incurrence of the cost of seeking a rebate is a virtual certainty, while the probability that payout will be forthcoming if the rebate materials are mailed in can be significantly less than 100%. This is because, first, rebate offers can be fraudulent and, second, at the time of purchase, no customer can be sure that all of the required documents for a rebate payout will be known and, if known, will be kept in order for submission when the rebate redemption window is open.[28] Then, all customers will recognize the prospect that the manufacturer offering the rebate will make mistakes, requiring the customer to incur additional costs to make sure the manufacturer makes good on its rebate commitment (as was needed when the "rebate junkie," mentioned above, sought his $100 rebate from TiVo).

From the perspective of conventional microeconomics, the calculus for rebate redemption is simple: buyers determine rebates' costs and benefits, appropriately discounted, and seek redemption when the discounted expected redemption benefits exceed the expected costs. The issue is not so simple to psychologists and behavioral economists, as will be seen in Chap. 11, after we consider in the remainder of this chapter probably the most robust explanation for rebates, price discrimination.

REBATES AND PRICE DISCRIMINATION

We have argued in this chapter that, like coupons and so many other promotional devices, a prime reason for rebates is not simply to be nice to some consumers by giving them a break in the price. Rather, rebates allow manufacturers and retailers to hike their product demands and posted prices and then to segment their markets and to price discriminate. That is to say, rebates allow manufacturers to charge price-insensitive buyers a higher price than they charge price-sensitive buyers.[29] Those customers who have the time to search out rebates and to redeem their rebates are likely to be customers who have time to know better the prices and features of a wide range of competitive products. Hence, they are in a position to be relatively responsive to the price reduction implied by rebate offers. They can be expected to switch purchases from an array of products to the products carrying the rebates and relatively lower prices. Lowering the price to price-sensitive buyers can be revenue enhancing when the percentage jump in sales from price-sensitive buyers is greater than the percentage reduction in price.[30] Given the greater sales, manufacturers will incur greater production costs. So long as the increase in revenues is greater than the increase in costs, greater profits can be garnered from the price-sensitive segment of the market.

One often unrecognized consequence of rebates noted in this chapter is that consumers who are not swayed by rebate offers can end up paying a higher price than they would have without the rebate promotion. This is the case because, again, rebate offers allow manufacturers to get the price-sensitive consumers to identify themselves by responding to the rebate offers. Having done that, they have isolated the price-insensitive buyers for a price hike. Such a price hike for the price-insensitive buyers can be profit enhancing simply because the percentage reduction in the quantity they buy can be lower than the percentage increase in the price they pay, the result of which is that the manufacturers' total revenues from price-insensitive consumers can go up. With fewer sales to price-insensitive buyers, manufacturers' costs associated with serving the price-insensitive buyers can go down. With revenues going up and costs going down, profits from the price-insensitive buyers go up.[31]

CONCLUDING COMMENTS

One retailer probably spoke for many rebate observers when he quipped, "Manufacturers love rebates because redemption rates are close to none ... they get people into stores, but when it comes time to collect, few people follow through. And this is just what the manufacturer has in mind."[32] As we have seen, the retailer is wrong on both the facts and the presumption of many manufacturers' intent. The rebate redemption rate varies across products and, as we have seen, is often far above "none," perhaps, on average, close to 40 or so percent. And we noted that manufacturers have gone under because redemption rates have been excessively high, if not close to 100%, when the rebate in absolute dollars and relative to the posted price, has been high.

Moreover, few manufacturers should be expected to seek a zero, or anything close to zero, redemption rate. A zero, or close to zero redemption rate could imply that the attractiveness of the rebate has been limited, which suggests that the rebated products' demands and prices have been left largely unaffected by the rebate offers. Hence, we should not be surprised that when rebates are offered, redemption rates are somewhere in the middle of the 0%–100% range.

Our answers to the question "Why rebates?" have hardly been exhaustive.[33] However, at the same time, the arguments presented clearly indicate that rebates are motivated by multiple economic considerations, as should be expected of all pricing strategies. The more important economic arguments often missed by analysts and buyers is that rebates can induce higher manufacturers' profits through the impact of rebates on elevating market demand, which can push posted prices upward and through the extent to which rebates can permit price discrimination. By segmenting markets, rebates enable firms to raise their posted prices by more than would be expected from an increase in product demand alone, mainly because the posted price would be paid mainly by price-insensitive buyers (and buyers who either consciously decide after purchase not to incur the redemption costs or who forget about redeeming their rebates altogether). The lower, after-rebate price will go to buyers who, by their willingness to incur the redemption costs, are likely to be price sensitive. Both the price increase for the price-insensitive buyers and the price reduction to the price-sensitive buyers can generate greater revenues. They can also provide the economic (above-competitive) profits for producers to innovate and to improve established products. However, the economics of rebates hardly makes for a full explanation for the prevalence of rebate offers. As we will see in the next chapter, manufacturers can count on rebates having a variety of psychological effects, or so many academics and marketers stress.

NOTES

1. As reported in *Business Week* (Grow 2005).
2. The most notable difference between coupons, considered earlier in the book, and rebates is that coupons require buyers to expend effort before purchase while rebates require buyers to expend effort after purchase.

[3] Silk and Janiszewski (2006, p. 15) drew a random sample of 315 rebate offers at wheresmyrebate.com and found that 60% of the offers had redemption periods of 14 days and 38% had redemption periods of 30 days.

[4] Silk and Pechmann (2007).

[5] Grow (2005).

[6] Grow (2005).

[7] As reported by Grow (2005) and Odell (2006).

[8] Silk and Pechmann (2007). Rebate complaints, for example, have dealt with such matters as delays in payouts and advertisements not clarifying that to get the advertised price, buyers had to redeem a rebate and delays in rebate payouts. In court cases, the complaints have involved allegations that the rebate window had expired, that sales people had claimed a rebate was available when such was not the case, that a buyer had not been told in a telephone sale that the redemption window lasted for only 30 days; that the rebate offer could not be combined with other promotional advantages, that the rebate offer did not disclose that one or more other products had to be bought to obtain the rebate, and that proper notice was not given that all or a portion of the rebate had to be returned if service was cancelled before the end of the contract period (Silk and Pechmann 2007).

[9] Silk and Janiszewski (2003, p. 8). Another (1987) study found an average redemption rate of 70% of unit sales (Jolson, Wiener, and Rosecky 1987).

[10] Jolson, Weiner, and Rosecky (1987) and Hoch, Dreze, and Purk (1995).

[11] Grow (2005).

[12] McLaughlin (2002).

[13] Silk (2003, p. 9).

[14] Dhar, Morrison, and Raju (1996).

[15] Grow (2005). An increase in the rebate from $20 to $40 on a $120 item will nearly double the redemption rate, according to the survey of managers of rebate redemptions centers already mentioned (Silk and Janiszewski 2003, p. 9).

[16] Silk and Pechmann (2007, table 1).

[17] Silk and Pechmann (2007, table 1).

[18] Silk (2003, p. 7).

[19] One caveat needs to be introduced. When the discussion concerns how rebates can increase product demand and posted price, we mean that rebates can increase demand and price *above what demand and price would otherwise (without the rebate) be.* Rebate promotions can be used by manufacturers to prevent the demand and posted price for their products from declining. If rebates prevent declines in demand and price, then it is still the case that demand and price will be higher than they would otherwise be.

[20] Of the eighteen products with rebate offers on the Fry's ad that ran the day this chapter was started, all carried prices higher than alternative products that were displayed at Fry's or that could be found on the web through Newegg.com

[21] Jolson, Weiner, and Rosecky (1987); Soman (1998); Norr (2000); McLaughlin (2002), Shim (2002), and Spencer (2002). Silk and Janiszewki (2003); Tat and Lee (2001); and Tat, Cunningham, and Babakus (1988) report the comments of a Federal Trade Commission director who observed, "Some companies are quick to offer attractive rebates, but often make them so difficult to redeem that consumers give up" (as quoted from Shim (2002), and a marketing consultant who noted, "If you have to take a knife and cut through heavy cardboard to get a bar code, the [redemption] rates drop precipitously" (as quoted from Norr 2000).

[22] As a consequence, we might reasonably conclude that the greater customers' assessed cost of the redemption requirements, the smaller the impact of any given size rebate on market demand, which can imply a smaller increase in sales and the posted price of the product. A corollary deduction is that the greater customers' assessed cost of the redemption requirements, the greater the rebate's dollar value must be to have the same impact on product demand and the posted price. Likewise, the shorter the redemption period, the smaller customers' assessed probability that they will redeem the rebate and the lower the rebate value must be—which implies a more restricted increase in product demand and posted price due to the rebate. This also means that the shorter the redemption period, the greater the rebate must be to have the same effect on sales and posted price. Similarly, the further into the future the manufacturers' self-imposed payout deadline is, the lower the (present discounted) value of the rebate—which implies, again, a more restricted increase in product demand and posted price.

[23] Even with virtually every buyer redeeming their rebates, it doesn't follow that the firm made an unprofitable promotional move, simply because the greater volume of sales price could have led to revenues rising by more than costs. It simply means that a rebate promotion was not needed to achieve the greater sales and greater profits and that additional profits could not be garnered through price discrimination. Sales could have been increased at lower cost with across-the-board price reductions.

[24] Telser (1980).

[25] My concern over the manufacturer of the USB drive failing to provide the required rebate information in a timely manner has been realized in at least one case filed with the Federal Trade Commission. According to Silk and Pechmann (2007), the FTC ruled that one company provided in many cases the required rebate information only after the redemption window had closed.

[26] Shugan (1980).

[27] Even if a customer's expected discounted opportunity cost (measured by, say, foregone earnings) incurred in getting the rebate were $90 to obtain a $100 rebate, the customer might buy the product (and a host of all other rebated products) without noticing the rebate and, if the rebate offer is noticed, without any intention of redeeming the rebate and without actually redeeming it, all very rationally. The economics of such a non-redemption position can be straightforward. While the redemption effort can be postponed, with its discounted value reduced appropriately to the current dollar equivalent, the payout will necessarily be even further into the future (perhaps 2 or 3 months later), possibly making each rebated dollar worth less than each dollar of cost incurred (in the mind of the buyer).

[28] Consumers can easily make understandable redemption mistakes. For example, redemption might require submission of a "bar code" on the package, but the package might have several bar codes with no clear indication of which bar code is required for the rebate payoff (Silk and Pechmann 2007). The redemption window may be no more than 7 days, and the details required for a complete redemption application might not be known until after purchase (Silk and Janiszewski 2006; and Silk and Pechmann 2007)

[29] Chen, Moorthy, and Zhang (2005).

[30] For more technical presentations of price discrimination between identified segments of the market, see McKenzie and Lee (2006, pp. 443–445), a discussion that is covered in video form, video module 11.5 at www\home\mckenzie\public_html\ModulesaftePublication101206.html.

[31] Of course, some manufacturers will not offer rebates because they do not have a sizable segment of their buyers who are sufficiently insensitive for a price hike to cause revenues from them to rise. Such manufacturers might as well lower their posted prices rather than go to the expense and trouble of offering rebates.

[32] As quoted in Greenman (1999).

[33] For example, very little has been said about how rebates can, and have been, used as a market research device that enables manufacturers to assess their products' elasticities of demand (or the price sensitivity of consumers). The feedback from rebates, revealed in increased sales and redemptions, can suggest to manufacturers whether and by how much they should raise or lower their prices to improve their profitability. Rebates may also be a means by which manufacturers can affect, or rather manage, the advertising and stocking decisions of their retailers. Retailers will likely hike their inventories of the rebated products, perhaps reducing the shelving space available for competitors' products.

Chapter 11

THE PSYCHOLOGY AND EVOLUTIONARY BIOLOGY OF MANUFACTURERS' REBATES

In the previous chapter, we took up the easy explanations for rebates, all grounded in economics. Rebate analysts steeped in psychology and evolutionary biology see the rebate issue as more complex than economists do. Their modes of analysis involve added concepts of subjective weighting of the costs and benefits of consumer decisions, endowment effects, the salience and vividness of product and rebate features at the time of purchase, and the time inconsistency of consumer choices, revealed in procrastination and forgetfulness. Evolutionary biology theory suggests that rebates play to certain mental tendencies that have been "hardwired" into our brains eons ago.

SUBJECTIVE WEIGHTING OF COSTS AND BENEFITS

Psychologists and behavioral economists insist that consumer decisions are not founded exclusively on economists' simple present-value cost/benefit calculations, discounted for only time and risk.[1] They suggest many consumers add an additional adjustment factor. That is, they apply *subjective weights* that stand apart from the subjective values embedded in costs and benefits (as economists normally compute them in expected, present discounted value terms). Following the lead of psychologists Daniel Kahneman (who won the Nobel Prize in Economics in part for work summarized here) and the late Amos Tversky, consumers can weigh costs (or losses) and benefits (or gains) differently, with prospective costs (and losses) looming more influential, and having a greater weight, in consumer decisions than prospective benefits (or gains).[2] Hence, people can, in general, be expected to be cost (and loss and risk) averse (which is totally understandable since, generally speaking, "cost," "loss," and "risk" carry negative subjective evaluations). One study fortified this position by finding that people will not take a 50% chance of losing $50 unless that prospective loss was set against a 50% chance of gaining more than twice $50, suggesting a $50 gain is valued less than a loss of an equal amount.[3]

Economists are inclined to discount future costs and benefits by some constant interest rate. One hundred dollars of benefits received a year from now is worth today a little more than $94, assuming a discount rate of 6% (and with no risk, an assumption made for purposes of simplifying the analysis). That same $100 received 2 years into the future is worth $89 today, again, assuming a 6% discount rate per year. Economists are inclined to make similar calculations for future costs.

Psychologists, by way of contrast, argue that consumer decisions over time are more complex. They suggest that costs incurred and benefits received at different points in time can be weighted differently by different consumers. If consumers do apply different weights to costs and benefit over time, the *gap* in the computed *subjective* values of the gains received 1 and 2 years from now can be greater than indicated by the above present value figures, $94 and $89, respectively. Consumers could see the gains as *subjectively* worth the equivalent of, say, $93 and $84, respectively. At the same time, the subjective evaluation of a short delay in a gain at some point in the distance, if not intermediate, future can be less than the same delay in the near term. This is because the delay can be viewed as more painful in the near term than at some distant point in the future.

One study found that when volunteers were asked if they prefer to take a $19 payment today or a $20 payment tomorrow, most subjects chose the $19 today. However, when they were asked whether they would prefer a $19 payment 365 days into the future or a $20 payment 366 days into the future, most chose the 1-day future delay.[4] The difference in the applied discount rates for the two choices is enormous. The apparent inconsistency in observed choices can be viewed as consistent only by understanding that choosers were applying different weights to the 1-day delays at different points in time.

The same kind of differential weighting can be applied to costs as they are incurred further and further into the future. However, there is no reason to believe, psychologists insist, that people's weights applied to costs over time need to move in lockstep with the weights applied to gains. Indeed, the weights used for adjusting downward future expected gains will likely rise slower than the weights used for adjusting downward expected future costs.[5] This can mean that the costs incurred immediately can be given a higher subjective value relative to gains than is the case for costs incurred in the future. In such cases, manufacturers should seek to postpone into the future any rebate costs consumers must incur.

For example, consumers might subjectively assess a $10 rebate redeemed immediately after purchase as worth $10, but might also subjectively assess the redemption costs at $12, causing the consumer to be unmoved by the rebate offers. However, when the rebate gains and costs are pushed off into the future, the subjective weights of the costs and rebate value can change, and can do so at different rates, so that the above rebate offer will, at some future point, be viewed favorably. For example, the consumer might view the $10 rebate as being worth $9.50 4 months from the date of purchase, after adjustments for time, risk, and subjective weight. The future redemption costs might then be assessed at $8.00, making the rebate offer at the time of purchase attractive. Hence, some rebates might not affect current sales, and might not be redeemed because the offers entail costs that are too close at hand for them to have a favorable subjective gain/cost comparison for buyers.[6] Put another way, manufacturers can make their rebates of a given size and with given redemption requirements more effective by having the redemption window begin and end at some point in the future, a conclusion that seems to be born out by the many rebates that are not redeemed, a deduction also supported by experimental evidence.

Psychologist Dilip Soman offered 60 university summer school students $1 to fill out a short survey on their attitudes, opinions, and interests.[7] The students were also offered the chance to take the dollar by filling out the first questionnaire and quitting the experiment, or they could, in one version of the experiment, earn a larger payout, $8, if they waited 4 weeks and filled out a second longer questionnaire.[8] While the length of the questionnaire did appear to affect the willingness of the students to take the delayed reward offer, 40–60% of the students were willing to accept the delay in the reward, supposedly in part because of the delay in the effort that would have to be made, or so Soman concluded.

Soman devised another experiment in which he offered 300 undergraduate and graduate students, put into a dozen groups, a reward of $20 to drive 10 miles to buy three different products.[9] The percent of students accepting the offer went from 28% when there was no delay in the reward and effort to 40% when there was a 3-week delay in both.[10] Accordingly, Soman is confident of the lesson to be drawn from his studies: "I experimentally demonstrate that an incentive involving a given face value and effort will affect choice to a greater extent when the incentive is provided with a temporal delay as compared with when it is provided immediately."[11]

Of course, manufacturers must keep in mind that real world and experimental evidence also reveals that while extending the redemption window into the future can make the purchase deal, with a rebate, more attractive today, beyond some extension into the future of the redemption window, the redemption rate can begin to diminish.[12] In one such study conducted by Amos Tversky and Eldar Shafir, students were offered $5 to fill out and return a questionnaire, with different groups of students being given different timeframes within which they could obtain the reward: 5 days, 3 weeks, and no deadline. Over the three reward periods, the return rates were 60%, 42%, and 25%, respectively. [13] The fading of the redemption rate can be a consequence of people's tendency to forget about the rebate,[14] but the redemption rate can fade with time because, beyond some point, the rebate's weighted value begins to fade while the weighted costs escalate as they become more immediately pressing.[15]

ENDOWMENT EFFECTS OF PURCHASES WITH REBATES

Psychologists and even economists have long recognized a phenomenon that helps explain the prevalence of rebates, while not squaring neatly with economists' concept of rational behavior (conventionally and narrowly conceived): people are not willing to pay as much for a good that they don't have as they would charge in selling the good if they have it.[16] For instance, from my experience teaching at a major sports university where I have raised the issue class after class, year after year, I am confident few students at major sports universities would pay $500 for a ticket to a sold-out football game with their arch rivals. If students refuse to pay $500 for tickets when they don't have tickets, they are effectively revealing that they have something better to do with $500 than go to the game. When they have a ticket (often given to them free or at cut-rate student prices by their universities) that is being scalped for $500, they should be willing to sell their tickets, because they, supposedly, still have something better to do with $500 than to go to the game. However, scalpers are often willing to pay students $500 for their tickets for sold-out games. Nonetheless, almost all students with tickets pass up the sale price and go to the game without ever thinking of the potential inconsistency in their behavior between their unwillingness to buy the ticket for $500 when they don't have a ticket and their unwillingness to sell at $500 when they have one.

Some economists might attribute the difference in students' purchase and sale prices to the risk of being caught doing something illegal, scalping. While the argument can't be totally dismissed, such an explanation is problematic because the threat of being caught scalping a ticket has to be close to nonexistent for most students.

Granted, my experience is anecdotal, or hardly scientific, but my classroom experience is still consistent with other more scientific studies. Behavioral economist Richard Thaler sees the "anomaly" in my sale/buy price anecdote as sufficiently prevalent in human experience to warrant a "parsimonious explanation" that he dubs the *endowment effect*.[17] The endowment effect is the inertia built into consumer choice processes due to the fact that consumers value goods that they hold more than the ones that they don't hold.

Kahneman, Thaler, and Jack Knetsch have found supporting experimental evidence of the endowment effect in a relatively simple experiment.[18] They gave mugs to one group of subjects who then set their sell prices at two to three times the prices of the subjects who did not have mugs but were willing to buy them. These researchers trace the difference in selling and buying prices back to people's natural proclivity to be risk averse, to weight losses more heavily than gains. This implies that out-of-pocket expenditures to buy a cup are subjectively weighted more heavily dollar for dollar than opportunity costs from not selling the cup. Hence, a $10 purchase price for a mug is subjectively weighted more heavily than a $10 payment received for the same mug that is possessed. To equate the subjective values of the two transactions, the sale price must be above the purchase price.[19]

The endowment effect might help explain some (but not all) rebate breakage in the same way that the endowment effect has been used to explain the popularity among retailers of money-back guarantees on merchandise returns made within a specified period of time, say, 60 days.[20] A customer not totally familiar with the benefits of a product might buy the product, reasoning that with the money-back guarantee, the most that can be lost is the transaction costs of returning the product when use reveals that the product isn't worth its purchase price. The customer will buy the good so long as the expected value of the product's use for the refund period is greater than the transaction costs associated with buying and returning it. Once the good has been used for a while, however, the good becomes a part of the customer's endowment, which means that its value jumps (according to theory and evidence from behavioral economics). In order to be enticed to return the good, many people would demand more in "money back" than they paid for the good. Hence, money-back guarantees can stimulate sales and enhance profits.

Much the same argument can be used to explain some rebate breakage. Some customers might reason at the time of purchase that the good is not worth the listed purchase price, but is worth the purchase price minus the (net discounted value of the) rebate. Once purchased, the endowment effect kicks in, elevating the value of the good, and the selling price imputed for the good. The redemption of the rebate for some consumers can no longer be needed to justify continued use of the good. Moreover, the value of the rebate after purchase declines because it becomes a foregone gain. At the same time, the effort that needs to be expended to redeem the rebate can be construed as a heightened cost because as time goes buy, its incurrence becomes more immediate and more heavily weighted in the rebate redemption decision.

SALIENCE AND PROCRASTINATION

Cognitive psychologists posit that consumers in making purchasing decisions give undue weight to salient or vivid features of a product and events close to the time of purchase.[21] Psychologists Richard Nisbett and Lee Ross pose a thought experiment that they use to clarify the meaning and importance of salience for everyday consumer choices.[22] Suppose a person has studied an array of automobiles for purchase, with two principle features in mind, the cars' expected longevity and safety record, eventually narrowing the final choice to two cars, the Volvo and Saab. From examination of automotive reviews in *Consumer Reports*, the buyer decides on the Volvo because of its reported superior safety record.

However, the night before the expected purchase day, a friend exclaims at a cocktail party on hearing the choice of a Volvo: "A Volvo! You've got to be kidding." The friend relates that his relatives had reported serious mechanical flaws with their Volvos, resulting in considerable repair bills.[23] Nisbett and Ross argue that most car buyers will give undue weight to the cocktail party comment, even though the story of a bad experience is only one more data point that, even if included in *Consumer Report's* sample, would have left the magazine's overall assessment of the Volvo's superior repair record undisturbed.

Experimental results seem to affirm the Nesbitt/Ross hypothesis, at least according to economist George Akerlof, among others.[24] One group of freshmen who had declared majors in psychology were given in one experiment the mean evaluations for psychology electives the prospective majors could take.[25] Another group of students were given the mean evaluation scores, plus were asked to listen to a panel discussion among upper-class psychology majors who provided evaluations of the courses. The students exposed to the panel discussion disproportionately chose courses that were the focus of the panel discussion and that were rated, during the discussion, as above average by the panel members.

In an analysis of 141 rebate advertisements in newspapers, slightly more than two-thirds of the ads gave emphasis to the after-rebate price by using a larger font than was used for the before-rebate price. Only a fifth of the advertised rebate offers used the same font size for the before-rebate price, the rebate, and after rebate price.[26] Soman, in a study considered earlier, believes his work on the timing of rebate redemption windows on the breakage corroborates the Nesbitt/Ross work, arguing that one reason a requirement for immediate redemption of a rebate can undercut purchases is that the redemption effort becomes a more salient feature of the rebate as the required redemption effort is moved closer to the purchase date.[27] Moorthy Chen and Soman found that emphasizing the after-rebate price, as opposed to the posted price in advertisements increased the purchase rate from 43% to 68%.[28]

Akerlof also posits a part economic/part psychological theory of procrastination, founded on the argument that "present costs are unduly salient in comparison with future costs, leading individuals to postpone tasks until tomorrow without foreseeing that when tomorrow comes, the required action will be delayed yet again."[29] By all accounts, procrastination is common, and rising (for reasons that are not yet clear). More than a quarter of surveyed Americans admit to being frequent, if not chronic, procrastinators, with men being slightly more inclined to procrastinate than women. In 1978, the percentage of Americans identifying themselves as chronic procrastinators was 5%. In 2006, the rate was up to 28%.[30]

Procrastination, of course, can have both good and bad effects. By repeatedly delaying the incurrence of costs that are immediate at each time period, people can quite rationally make the delays pay. At the time of purchase, buyers may only have a vague idea of the true costs and benefits of undertaking any particular action at various future points in time. As time goes by, consumers can often reasonably expect to be able to gather more and better information on what the relative costs and benefits are and, hence, what they should do. They can also ascertain with greater accuracy the best (cost-minimizing and gain-maximizing) time period for taking action. In short, not all procrastination is a bad behavioral strategy.[31] Repeated delays—procrastination—can be all the more welfare enhancing when buyers recognize their propensity to misjudge at purchase time the future costs.

Many students of procrastination, however, focus exclusively on the negative consequences of procrastination. Accordingly, researchers have found that much procrastination can make procrastinators poorer, more stress prone, less healthy, and more unhappy.[32] Procrastination in college can lower students' grade-point averages, which can diminish students' lifetime opportunities.[33]

Akerlof argues that individuals can, in misjudging costs over time, all too often exhibit preferences that are "time inconsistent," resulting in eventual outcomes that stand contrary to the procrastinator's own long-term interests and welfare.[34] He uses his proposed theory of procrastination to explain a range of behaviors, not the least of which include drug users who after every hit affirm that they will not take another, overweight people who gain weight while supposedly dedicated to diets, and savers who continually postpone saving time period by time period until they enter retirement with inadequate income streams. In each short-run time period, the salience of the immediate subjectively weighted cost looms larger than the subjectively weighted cost of undertaking restraint in the next time period. The unfavorable balance between the force of immediate and future weighted costs at the start of each time period can force an unwanted long-term outcome—unless a deadline is set. Akerlof argues that deadlines can be, and are, productive because they force an increase in people's assessments of future costs relative to immediate costs, truncating procrastination.

The relative salience of immediate over future redemption costs, and the resulting procrastination calculus, might also help explain low rebate redemption rates. At the time of purchase, the rebate value looms salient in the purchase decision, since it is so visible in the product's advertisement, while the costs of redemption can be obscure, hard to assess, and set back in time. When faced with the opportunity to seek the rebate, buyers will find the current redemption costs more salient, and will weight them more heavily, than redemption costs postponed until the next time period. By going through such calculations day by day, or week by week, the consumer can delay seeking redemption of rebates until just before the redemption deadline.

Of course, the consumer can, along the way, forget about the rebate. Using the Akerlof perspective on procrastination, it is understandable how drawing the end of the redemption period closer to the purchase day can, beyond some point, increase the redemption rate, as researchers have found to be the case.[35] By having a long redemption period, procrastinating can lead to forgetting or setting aside any intention to redeem the rebate (especially if the assessed costs of redemption change with time). And, as noted earlier, research has shown that extending the deadline into the future can lead to greater forgetting about rebate redemption.[36]

EXPLANATIONS FOR PEOPLE'S OBSERVED DECISION MAKING

Again, economists posit that people's behavior with regard to rebate offers or any other form of product promotion can be grounded in rational precepts. However, as also noted, much work done by psychologists and behavioral economists brings into question the extent to which people behave rationally, at least *as rationality* is posited in conventional microeconomic analysis.[37] There are myriad nontrivial anomalies to (perfectly) rational behavior, with Thaler making a substantive career of pointing out a growing list of "paradoxes and anomalies in economic life."[38]

Why aren't people more rational than they have been observed to be? There are at least two levels of explanations. One level of argument is grounded in people's limited mental capabilities. People just aren't able to make the kind of finely tuned cost/benefit calculations economists assume they are capable of making. The second level of argument draws on evolutionary biology (and evolutionary psychology) to explain why people face the mental limitations they do in assessing costs and benefits.

LIMITATIONS OF THE BRAIN

Economists' implicit, if not explicit, presumption that people's rational capacities are unbounded hardly squares with the transparent fact that the brain has physical limitations, the most obvious of which is size (the typical adult human brain weighs only 3 pounds) and storage and thinking capacities (an adult brain has only 100 billion neurons, critical to their memory and thinking capabilities).[39] Given the myriad of sensory data—sights, sounds, tastes, and touches—people encounter on a daily basis, the brain simply can't notice, register, absorb, evaluate, and store the full range of experiences in all of their details. The brain simply does not have the necessary neurons. The brain must select a portion—perhaps a minor portion—of all possible experiences that will be noticed, absorbed and evaluated, and then the brain can only store selected parts of experiences in compressed form.[40]

Our mental construction of experiences is thereby defective in two ways: First, many components/aspects/dimensions of everyday experiences are never considered. Second, those parts that are recorded are incomplete, making any stored perception of the world more like a "portrait" than a "photograph," "reflecting the artist's hand every bit as much as it reflects the things portrayed."[41]

Hence, it should not be surprising that purchases are often based on the more "salient" product features, meaning, for purposes of this chapter, that the details of rebate offers, especially the redemption requirements, are not always considered with care by many consumers. They simply don't have the mental wherewithal to notice, absorb, evaluate, and then register and recall accurately such numerous details. Of course, if buyers faced only one rebate offer, and no other promotional efforts, as they shop, they might then be able to know the fine details of the given rebate offer, but, again, a given rebate offer is almost always part of a great buzzing confusion of information buyers face everywhere as they shop.

When there are delays in buying decisions (including the receipt of the gains and incurrence of the pains from buying decisions), such delayed decisions must then be founded on isolated, highly selective, incomplete, and compressed-form mental images of the real experiences, with the brain necessarily having to fill in the many details that had to be left out when the experience was first stored in the brain. We should not be surprised if, on recall of experiences from memory, the brain often fills in details incorrectly or as it sees fit, for its own distorted purposes, perhaps to lend support to the correctness of past decisions.

When imagining the future, and assessing future costs and benefits, the brain must rely on what it has to work with—incomplete, compressed, imperfectly formed and evaluated experiences stored in memory. This means that imagined futures will be heavily influenced by imperfectly and incompletely stored present and past experiences and will also be made all the more imperfect by people's inclination to leave out details of imagined future experiences—since no one can know exactly what the future holds and since the future will unfold from a host of interdependent and interacting forces that are literally unknowable. And the further in the future the experience is, the smaller the variety and count of details of the experience people imagine.[42]

As Daniel Gilbert succinctly summarized a vast literature on our ability to deal with imagined futures: "When we imagine future circumstances, we fill in details that won't really come to pass and leave out details that will. When we imagine future feelings, we find it impossible to ignore what we are feeling now and impossible to recognize how we will think about things that happen later."[43]

The lesson to be drawn from this literature on people's mental/rational capacities for our study of rebates is clear: Manufacturers play to our mental/rational capacities and limitations when they offer rebates, counting on many buyers to mis-assess the costs and benefits of rebate redemptions at some point in the future. Rebate offers that seem attractive at the point of purchase for many buyers can simply be unattractive in cost/benefit terms when the time comes for those same buyers to expend the necessary effort to redeem the rebates. At the time of purchase, buyers might very well think of their opportunity cost of redeeming rebates as being approximately equal to the cost they would incur at the time of purchase, appropriately discounted, for the fact that the costs will not be incurred for some time. When the future roles around, buyers' then opportunity costs could have escalated for any number of reasons that could not have been imagined at the time of purchase (a change of jobs, an illness in the family, a reduction in the price of vacation trips, etc.).

Just because buyers who fully intend to redeem their rebates might not do so because they did not imagine their futures completely and correctly, it does not follow that buyers are worse off for having used their mental capacities as they have. After all, mental limitations necessarily restrict buyers' welfares, just as buyers' physical limitations do. In storing and recalling experiences and then projecting future experiences, all imperfectly, buyers might be doing nothing more than optimizing their employment of their scarce mental faculties. It might be nice to think that people could be better off *if* they didn't have to store incomplete, compressed images of experiences and to imagine imperfectly and incompletely future costs and benefits. It would also be nice *if* buyers could learn and earn more than they do. Buyers who do not redeem rebates because of the deficiencies in the way they store, retrieve, and imagine the future might still be better off than they would have otherwise been. Because buyers have to economize on their limited mental faculties in the cases of given (if not all) rebates, they can more efficiently use their limited mental capacities to avail themselves of a greater variety of opportunities that could yield more value than the value lost on, say, rebates that are not redeemed.

THE EVOLUTIONARY FOUNDATIONS OF REBATE BREAKAGE

An important theme of this chapter has been that many rebate offers affect some purchases and don't affect others because people's circumstances and preferences differ, leading even rational buyers to different purchase and rebate redemption decisions with the overall breakage rate often way below 100%. Another important theme has been that buyers exhibit what on the surface appear to be nonrational, if not irrational and time-inconsistent, behaviors. Some buyers allow salient product and rebate features to unduly control their purchase decisions, with the end result being that they are unable to maximize their utility very well over time, as Akerlof posits. That is, many people seem to make mistakes systematically and seem unable to calculate costs and benefits over time with anywhere near the precision economists assume for rational decision makers. As noted, some of these thought-to-be mistakes can be chalked up to differences in the applied subjective weightings of costs and benefits for different product features, including rebates, especially for different points in time into the future. Other presumed consumer mistakes can be attributed to the brain's efforts to economize on its storage and retrieval capacities that are bound to lead to behaviors not in accord with economists' conventional construction of rational behavior that, for all intents and purposes, assumes away mental limitations on cost/benefit calculations for current and future time periods.

If the brain has built-in limitations on accurately calculating costs and benefits, what explains those limitations that, in turn, explain why buyers might weight salient product and rebate features in their purchase decisions and why buyers might be poor judges of their future ability and willingness to redeem rebates? Evolutionary biology theory might provide a partial answer.

Charles Darwin, the much heralded father of modern evolutionary biology, argued that all attributes of existing living organisms today are adaptations to evolving local circumstances.[44] Those extant attributes fittest for given sets of circumstances were the most likely to be passed down to succeeding generations. The beaks of finches are effectively shaped over long stretches of time by the success that finches with the fittest beaks have in feeding and, hence, breeding. As evolutionary biologists and psychologists maintain, the human anatomy grew and adapted in response to evolving circumstances. Through natural and sexual selection, those ever-varying human attributes most fitted to circumstances were passed along, improved, and refined through succeeding generations, with human's physical and mental capacities more or less shaped as we find them today by the time humans started farming, ten or so thousand years ago.[45] Jerome Barkow, Leda Cosmides, and John Tooby, leaders of the burgeoning field of evolutionary psychology, have given a common explanation for why "psychological mechanisms" are, more or less, hardwired into the constructions of human brains, meaning they have not been subject to significant change by culture or the particular individual and group circumstances since humans became farmers. They write in introducing their edited volume on *The Adapted Mind*:

> What we think of as all human history—from, say, the rise of the Shang, Minoan, Egyptian, Indian, and Sumerian civilizations—and everything we take for granted as normal parts of life—agriculture, pastoralism, governments, police, sanitation, medical care, education, armies, transportation, and so

> on—are all the novel products of the last few thousand years. In contrast to this, our ancestors spent the last two million years as Pleistocene hunter-gatherers, and, of course, several hundred thousand years before that as one kind of forager or another. These relative spans are important because they establish which set of environments and conditions defined the adaptive problems the mind was shaped to cope with: Pleistocene conditions, rather than modern conditions. This conclusion stems from the fact that the evolution of complex design is a slow process when contrasted with historical time. Complex, functionally integrated designs like the vertebrate eye are built up slowly, change by change, subject to the constraint that each new design feature must solve a problem that affects reproduction better than the previous design. The few thousand years since the scattered appearance of agriculture is only a small stretch in evolutionary terms, less than 1% of the two million years our ancestors spent as Pleistocene hunter-gathers. For this reason, it is unlikely that new complex designs—ones requiring the coordinated assembly of many novel, functionally integrated features—could evolve in so few generations...Moreover, the available evidence strongly supports this view of a single, universal panhuman design, stemming from our long-enduring existence as hunter-gatherers.[46]

In short, according to many evolutionary theorists, many of our physical and mental faculties, predispositions, and frailties have remained virtually unchanged for thousands of years.[47] It was during the Pleistocene epoch that humans developed capacities for memory, sight, and sound, as well as sexual orientations and inclinations and gender-based behaviors (a subject to which we will return when considering modern gender-based earnings differences in Chap. 13). Moreover, evolutionary biologists and psychologists maintain that it takes at least 50,000 years for a new improved mental adaption to spread throughout the human population—the so-called "evolutionary lag"—that, again, implies that there have been little to no changes in people's mental hard-wiring since the Pleistocene epoch.[48]

There are reasons to harbor some skepticism over claims such as the one above about the end of human adaptation to the conditions of the times ten thousand years ago, because the claim is based on scant information about what conditions were actually like way back then and because the claim assumes something of a linear physical and mental adaptation rate. Human evolution development could have accelerated in our recent past, given the accelerating changes in conditions (including knowledge, education, wealth, technology, and size of markets), a point stressed by evolutionary biologist Michael Rose.[49]

Here, we can only draw out deductions that could be valid, assuming the Cosmides/Barkow/Tooby-type claim is reasonably sound. If so (or to the extent that the claim is valid), to understand the rational limitations of modern humans, we need only point out that when our rational capacities (and limitations) were being shaped, life was everywhere tough and short (with life expectancy less than 30 years). Subsistence living was pervasive and controlling. Social existence was largely relegated to small groups of fewer than 150 members, and more frequently, with 50 members.[50] The gains from trade were accordingly limited by the size of the highly constricted market, if gains from trade existed at all, due to the narrow confines of the relevant group (a point Adam Smith stressed).[51] The concept of progress in individual and group welfare was, very likely, largely a non-governing thought, since progress, such as there was, could only be witnessed at the slow pace of evolutionary time (with the progress hardly noticeable in any generation until after the first millennium A.D.). Recent research suggests that for typical workers (not landlords, aristocrats, and capitalists) the trend in real wages could have been flat until the early 1800s, with only temporary rises in real wages above the trend between 1000 A.D. and 1800. The rise in worker real wages began with the advent of the Industrial Revolution, only to accelerate during the 1800s.[52]

Without question, during the Pleistocene epoch, humans, and their hominid ancestors, experienced a dramatic enlargement of their brains, especially of the frontal lobe, that front part of the brain devoted largely to processing and storing experiences and to analytical thinking of the type that is at the heart of disciplines such as economics. As Gilbert has noted, it was in the latter Pleistocene epoch that humans developed the capacity to imagine their futures and to base behaviors on those imagined futures.[53] At the same time, the subsistence living must have been a harsh taskmaster, and a limiting constraint on people's ability to accurately imagine their futures and to accurately weigh future costs and benefits. Our ancestors must have had little margin for error in the use of their physical and mental abilities. Those humans who misused or didn't use cost effectively their physical and mental abilities are probably not among our ancestors. Those who did—that is, those who weighed costs and benefits with some relative care, precision, and ease—are heavily represented in our ancestors. They had a survival and reproductive advantage over those who were not so adept at assessing various opportunities' costs and benefits. Those who didn't try to achieve subsistence at the lowest cost for self, related kin, and in-group members did not see their genes passed down through the generations. They disproportionately succumbed to the inevitable outcome from sub-subsistence living. No wonder economists have, for more than two centuries, been able to spin an extensive microeconomic theory founded on the premise that modern humans seek to minimize costs and maximize gain. Cost/benefit analysis applied to immediate conditions could have become a part of successful earlier human's adapted mental hardwiring.

At the same time, it is also altogether reasonable that people's rational capacities were circumscribed during the Pleistocene epoch, especially when making cost/benefit calculations for good and bad things that could happen in the future. This is because people's futures were short under the best of circumstances. (Retirement planning was an option hardly worthy of consideration!) In the Pleistocene epoch and before, if people diverted much time and energy away from subsistence living and toward contemplating and accurately assessing the costs and benefits of future behaviors, they might not have the necessary subsistence to make it into their reproductive futures.

As any number of researchers have observed, humans are today just not very good at projecting themselves into the future and from those future points in time to very accurately assess what they should do today and what they will likely do in the future when the future arrives.[54] From an evolutionary perspective, there should be no wonder why people today might overweight what they have—their endowments—relative to the things they might obtain. Endowments increase people's confidence that they could subsist and have reproductive futures. The evolutionary perspective also suggests why people might be loss and risk averse, since current losses can raise the specter of living below the subsistence level. Risk can similarly imply incurrence of current, upfront costs that lower the standard of living to below-subsistence levels in the immediate and intermediate futures.[55]

Moreover, from an evolutionary perspective, people can be expected to unduly jump at opportunities that have current gains and future costs that understandably could be underweighted. Again, current gains harbor the prospects of people making it into the future.[56] The future costs may never have to be incurred because there could very well not be a future, especially if short-term gains are not sought with some undue aggressiveness. From an evolutionary perspective, we should not be surprised by research that has found that people exude an overconfidence in predicting performance of future tasks, including the redemption of rebates,[57] that shows people underestimate future time costs to accomplish future tasks, including rebate redemptions,[58] and that they are often inclined to procrastinate on the completion of tasks until they forget about them.[59]

Our genes may simply be well programmed to undertake cost/benefit analyses in the short run, and from short-run to short-run, but we modern humans may not be so well programmed to accurately take account of the long-run consequences of a sequence of short-run decisions. Hence, our minds may be well adapted to deal with salient product and rebate features, as well as to procrastinate on the incurrence of costs.[60]

Put simply, rebate (and other promotional) offers can be viewed as promotional efforts designed to play to human mental limitations that have their origins in the survival conditions of our distant past. Granted, as Rose has argued with force, his study of the evolution of fruit flies, whose lives are short and whose evolutionary changes can be easily observed through hundreds of generations that occur in a relatively short periods of time, significant evolutionary changes in human physical and mental faculties could have occurred in the several hundred generations since the end of the Pleistocene epoch, especially since the emergence of agriculture and trade alone could have progressively increased the time humans have had to adapt their skills at weighing the gains and pains of future activities. However, there is still no reason to believe that humans should now be able to engage in rational, future-oriented decision making with the precision that economists have traditionally assumed.

CONCLUDING COMMENTS

Several of the more important psychological arguments relating to rebates (and other promotional strategies) can often be explained by people trying to economize on their scarce mental faculties that were largely shaped long ago when survival and reproduction—not rebate redemption nor other pricing strategies—were the critical issues of the day.

Having conceded that point, caveats are in order. With all the frailties people possess that can be said to be hardwired in their brains to one extent or the other, it might appear buyers are at the mercy of forces beyond their control. That is hardly the lesson to be learned. Rather, people are obviously capable of learning and overriding their natural inclination to succumb to, say, the salience of promotional efforts and endowment effects. Knowledge of how rebates, or any other pricing strategy is used, can empower people to shop in ways they might not otherwise be so inclined. This book must have been written with that prospect in mind.

The future prospects for rebates should not be viewed as holding their current promise. The problem lies in the extent to which the credibility of all rebate offers depends on buyers' experience with particular rebate offers in the near and intermediate terms, which is necessarily related to how all firms conduct themselves in terms of honest dealing on rebates. To the extent that the costs of individual firms' misbehavior on rebates settles on other firms (or firms' misbehavior undermines the general credibility of rebates), we should expect excessive misbehavior on rebates by all firms collectively, which can be reflected in growing consumers' complaints and disregard of rebate offers. This means that with time, no one should be surprised if rebates become less and less effective in stimulating consumer demand for particular products, in hiking their prices, and in enabling firms to segment their markets for the purpose of engaging in price discrimination. Rebate misbehavior can cap manufacturers' use of rebates, if not cause manufacturers' and retailers' promotions of rebates to recede, at least over time. Interestingly, in Fry's newspaper ad mentioned at the start of Chap. 10, because it carried so many rebate offers, a quarter of the advertised products were prominently advertised as carrying attractive prices *free of the complication of rebate redemptions*. And, as this chapter was being finalized major stores—Best Buy and Office Max—announced that they were getting out of the rebate business because they and their customers had had so many problems with them.[61]

NOTES

1 Technically, economists have long known that for purposes of utility maximization, costs and benefits must adjust for the *variance* in potential outcomes, with the working rule being that the greater the variance, the greater the adjustments (Lee 1969).

2 Kahneman and Tversky (1972).

3 Researchers have found that in a 50/50 gamble, as prospective gains were elevated relative to the prospective loss, there was more electrical and blood-flow activity in the so-called "reward area" of the brain, as measured by CAT scans. As the prospective loss grew relative to the prospective reward in 50/50 gambles, the measured neural activity in the reward area of the brain declined (Tom, Fox, Trepel, and Poldrack n.d.).

4 Lowenstein (1987). Assuming an interest rate of 6%, the difference in the present value of \$19 today and \$20 tomorrow is \$.997. The difference in the present value of \$19 in exactly a year and \$20 in a year and 1 day is \$.95, indicating the economic/money value of the 1-day delay currently is nearly 5 cents greater than the economic/money value of the 1-day delay a year into the future.

5 Kahneman and Tversky (2000a and 2000b). As noted in the chapter on the prices of printers and printer cartridges, Thaler (1981) found that when his subjects were asked how much of a payment they would need to be indifferent between that future payment and an immediate \$15 payment, the subjects gave responses that resulted

in the median responses implying a 345% discount rate for a 1-month delay in payment, a 120% discount rate for a 1-year delay in payment, and a 19% discount rate for a 10-year delay in payment.

[6] This line of argument is an implication of the more general theoretical work of Thaler (1980), Mowen and Mowen (1991), Shelley (1994), and Silk (2003).

[7] Soman (1998).

[8] The subjects were put into different groups that were given a chance to fill out different length questionnaires (4 and 8 pages) with different delays (2 and 4 weeks) and different payouts ($4 and $8, depending on the length of the second questionnaires) (Soman 1998).

[9] Soman (1998).

[10] When the reward was $20 and the distance was 10 miles, the percent of the students accepting the delayed reward went from 40% when there was no delay to 64% when there was a three-week delay. When the reward was $40 and the driving distance was 20 miles, the percent of students accepting the deal rose from 16% with no delay in the effort and reward to 60% when the delay was three weeks (Soman 1998).

[11] Soman (1998, p. 427).

[12] Silk (2003).

[13] Tversky and Shafir (1992).

[14] According to one survey, 22% of buyers of products with rebate offers indicate that they intended to redeem the rebate but forgot to do so within the redemption window (Silk 2006).

[15] Silk and Janiszewski (2006), Shapiro and Krishman (1999), and Scher and Ferrari (2000).

[16] Thaler (1980).

[17] Thaler (1981).

[18] Kahneman, Thaler, and Knetsch (1990).

[19] Behavioral economists argue that the out-of-pocket expenditures are viewed by buyers as *losses*, while the opportunity costs are viewed as *foregone gains*. For more evidence offered in support of the endowment effect and the subjectively different weights given to out-of-pocket expenditures and opportunity costs, see Becker, Ronen, and Sorter (1974); Weiss, Hall, and Dong (1980).

[20] Thaler (1980).

[21] See, for example, Lowenstein (1996).

[22] Nisbett and Lee Ross (1980, as reported by Akerlof (1991).

[23] Nisbett and Ross (1980, p. 15); and Nisbett, et al. (1976, p. 25).

[24] Akerlof (1991).

[25] Borgida and Nisbett (1977).

[26] Kim (2006).

[27] Soman (1998). The salience of the rebate value, given that it might be vividly displayed in large colored type, with the terms of redemption often obscured, can help explain why A.C. Nielsen reported that nearly two-thirds of the 9,214 consumers of personal computers and peripherals surveyed said that the rebates were either very important or somewhat important in their purchase decisions, but yet so many of the surveyed buyers failed to redeem their rebates (Ricadela and Koenig 1998).

[28] Moorthy and Soman (2003).

[29] Akerlof (1991, p. 1).

[30] Steel (2007), as reported by Borenstein (2007).

[31] This is a theme developed by Abrahamson and Freedman (2007).

[32] Steel (2007).

[33] Education professor Bruce Tuckman found in the class performance of 116 Ohio State University students in his courses with a large number of activities and assignments with deadlines that student grade point averages were negatively correlated with their professed degree of procrastination. The "most severe" procrastinators had grade-point averages of 2.9. "Moderate procrastinators" had grade point averages of 3.4 and "low procrastinators" had grade-point averages of 3.6 (as reported by Grabmeier 2006).

[34] Akerlof (1991).

[35] Silk (2003) and Silk and Janiszewski (2006).

[36] Scher and Ferrari (2000) and Shapiro and Krishman (1990).

[37] This vast literature has been adequately reviewed by Rabin (1998).

[38] Thaler (1991).

[39] Gilbert (2006, Chap. 1).

[40] See Pinker (1997) for more details on "How the Mind Works," which is the title of Pinker's book.

[41] Gilbert (2006, p. 85).

[42] Loewenstein (1996); Loewenstein, O'Donoghue, and Rabin (2003); Loewenstein and Angner (2003, pp. 351–391); and Liberman, Sagristano, and Trope (2002).

[43] Gilbert (2006, p. 238).

[44] Darwin (1859).

[45] In his closing words to *The Origins of the Species:* Thus, from the war of nature, from famine and death, the most exalted object which we are capable of conceiving, namely, the production of the higher animals, directly follows. There is grandeur in this view of life, with its several powers, having been originally breathed into a few forms or into one; and that, whilst the planet has gone cycling on according to the fixed law of gravity, from so simple a beginning endless forms most beautiful and most wonderful have been, and are being, evolved (Darwin 1859, pp. 459–460).

[46] Cosmides, Tooby, and Barkow (1992, p. 5, who cite Tooby and Cosmides 1990a and 1990b).

[47] In his book on *How the Mind Works,* Steven Pinker elaborates on the constraints that human's evolutionary history imposes: [Natural and sexual] selection operates over thousands of generations. For 99% of human existence, people lived as foragers in small nomadic bands. Our brains are adapted to that long-vanished way of life, not to the brand-new agricultural and industrial civilizations. They are not wired to cope with anonymous crowds, schooling, written language, government, police, courts, armies, modern medicine, formal social institutions, high technology, and other newcomers to the human experience. Since the modern mind is adapted to the Stone Age, not the computer age, there is no need to strain for adaptive explanations for everything we do. Our ancestral environment lacked the institutions that now entice us to nonadaptive choices, such as religious orders, adoption agencies, and pharmaceutical companies, so until very recently there was never a selection pressure to resist the enticements. Had the Pleistocene savanna contained trees bearing birth-control pills, we might have evolved to find them as terrifying as venomous spiders (Pinker 1997, p. 42). For other easily assessable discussions of natural and sexual selection on the evolutions the attributes of organisms, see Dawkins (1989), Rubin (2002), Buss (2004).

[48] Levitan (2006).

[49] Rose (1998).

[50] Paul Rubin argues that one unheralded reason economists must teach, at a nontrivial class time cost, their students how people can achieve mutual gains from trade if the trading partners produce and trade according to the "law of comparative advantage" is that students' brains are hardwired, from the experience of our ancestors during the Pleistocene epoch, to think that all of economic life is a zero-sum game, with the gains of anyone coming at the expense of someone else, which, the student infers, means that an increase in the welfare of one trading partner must be at the expense of the other trading partner (Rubin 2003, Chap. 1).

[51] Smith stressed that division of labor was a key source of productivity and human welfare growth, but that the economic gains to be had from "the division of labor is limited by the extent of the market" (1776, p. 19).

[52] Clark (2007).

[53] Gilbert notes that the brain appeared on Earth a half billion years into our past with the brain gradually enlarging for the next 430 million years to the size of the first primates' brains. Over the next 70 million years, the brain reached the size of the first protohumans: "Then something happened—no one knows quite what—and the soon-to-be-human brain experienced an unprecedented growth spurt that more than doubled its mass in a little over 2 million years, transforming the one-and-a-

quarter-pound brain of *Homo habilis* to the nearly 3-pound brain of *Homo sapiens* (Gilbert 2006, p. 10, citing Banyas 1999).

[54] For reviews of much of the psychological and neuroscience literature on modern human's mental proclivities and limitations, Gilbert (2006) and Thompson and Madigan (2005).

[55] It does not follow that people should not be expected to take no risks, and as argued in a later chapter, there are good reproductive reasons for why males should be expected to be more risk taking than females.

[56] Lay (1988); Buehler, Griffin, and Ross (1994); and Lowenstein (1996). For an intensely analytically evolutionary explanation for why the young can be "expected to discount the future more rapidly than their elders," see Rogers (1994).

[57] Hoch (1985); Griffin, Dunning, and Ross (1990); and Vallone, Griffin, Lin, and Ross (1990).

[58] Buehler, Griffin, and Ross (1994); and Lay (1988).

[59] Silk (2003), Scher and Ferrari (2000), and Shapiro and Krishman (1990).

[60] This doesn't mean that people can learn to improve their cost/benefit calculations for current and future time periods. Indeed, much education, especially economic education, seems designed to enable students to see why they should override their natural tendencies to focus on immediate and salient costs and benefits. While economists might like to think that their courses deal with *positive* issues, not *normative* ones, their courses are replete with norms for welfare and profit maximizing dictums, not the least of which are:

- All costs—out-of-pocket expenditures and opportunity costs—should be treated equally.
- Sunk costs should be ignored.
- Production should continue until the marginal cost (including all costs) equals marginal benefit.
- Future costs and benefits (and revenues) should be discounted for time and risk and adjusted for variance in outcomes and not be further subjectively weighted.
- And the list of normative economic lessons could be easily extended.

[61] Best Buy announced its plans to discontinue rebates in 2005 and OfficeMax announced its intentions to discontinue rebates in 2006, as reported by ZDNet.com, June 20, 2006 at http://news.zdnet.com/2100-1040_22-6090290.html (accessed January 13, 2008).

Chapter 12

THE QUESTION OF QUEUES

University of Chicago Professor Gary Becker is my kind of economist. Over his long and illustrious career, he has applied economic reasoning to an ever-expanding range of topics before other economists thought to do so: education, race and gender discrimination, crime, marriage and family, baseball, household production, suicide, altruism, fertility, addiction and habits—and my list is hardly complete.[1] He has been a force within the profession to redefine *economics* not so much by the topics covered (money or markets or business), but as a *way of thinking* about human behavior. He has co-authored a book with a title that captures the expansive range of his analytics, *The Economics of Life.*[2]

For his considerable creativity in extending the boundaries of the discipline in virtually all directions, he won the 1992 Nobel Prize in Economics.[3] His central methodological concern in virtually all of his writing has never been that economics explains *all* behavior, but he is obviously convinced that economists should try to see how much of the observed differences in people's behavior can be productively explained by the prices they face (whether explicit or implicit), the incomes they receive (whether in money or nonmoney forms), and the wealth they have (whether assessed in work and social skills or financial and physical assets).[4]

More directly for the purposes in this chapter, Becker has shown a knack for recognizing ordinary, day-to-day experiences we all encounter that are puzzling, especially in light of conventional economic analysis. He then often offers counter-intuitive solutions to the puzzles.

For example, Becker notes at the start of one of his journal articles how when he taught at Columbia University, he puzzled with his students in class over why in Palo Alto, California (where he has long been a senior fellow at the Hoover Institution) there was a seafood restaurant that didn't take reservations and that always had long queues for tables at dinner time.[5] A similar restaurant, with a similar menu and meal prices, across the street often had empty tables (even though the food was more or less the same quality in both restaurants, or so Becker surmised). His students were not able to offer a satisfactory solution for the difference in the wait time for the two restaurants. I am sure almost all readers have waited for a half-hour or more for a table, which should be confounding for those who believe restaurants can be expected to seek maximum profits.

I also posed a similar puzzle to my MBA students at the University of California, Irvine, many of whom had significant management experience, involving queues in a different retail sector. At the time, I needed about eighteen strips of sod to cover an area bare of grass in my backyard. I went to Home Depot because, at the time, sod was selling there for $1.69 for a 4' × 18" strip. When I arrived at 8:30 in the morning, I was told by a clerk in the garden shop that the store had sold out. The clerk admonished me, "You have to get here by 7:30 to be assured of getting sod because we almost always sell out by 8:00," an assessment she made with complete confidence. I arrived at the store the next morning at 7:25, only to learn that the daily shipment of sod had already been sold. The clerk's reaction, "Yes, some days our daily shipment is gone within minutes after the load is dropped just after 7:00." I tried other Home Depots in the area that day, and they were also sold out of sod. I went back to the first Home Depot the next morning at 7:00, only to learn that the sod shipment was late. I observed a line-up of pick-up trucks with crews ready to pounce on the sod shipment when it arrived.

Readers might rightfully wonder why Becker, other economists, and I have been puzzled by the queue at the restaurant, the home supply store, and elsewhere, as if queues are an oddity. They really aren't. They are indeed common, at grocery stores, at concerts, at airports, at fast food restaurants, at bank ATMs, at...—well, all over town. They are so common that many of us spend a nontrivial amount of time trying to avoid queues. Indeed, you might be forgiven for wondering whether queues or prices are used more frequently to ration goods and services. Because of the prevalence of queues across so many markets, no one should be surprised that economists' explanations for queues are varied and run the gamut from psychological and sociological forces at work to ethical considerations to market forces that cause many queues to be mutually beneficial for sellers and buyers.

QUEUES AS A PRICING PUZZLE

Why queues? That question reflects a pricing puzzle for me (and many other economists) on two levels. First, on a personal level, I wasted a lot of time in my attempts to buy sod, going to the several Home Depots in my area of the country and returning three times to the closest store to my home, which was 8 miles away, before I was able to buy the amount of sod I wanted. I simply could not understand Home Depot's *modus operandi.* I would have gladly paid substantially more (maybe two or three times more than the selling price per strip) for the sod because, at a much higher price, my total expenditure on sod would have been less than the total cost I incurred from the wasted time and the gasoline used in my search. I would have been happier with a higher price (assuming the higher price discouraged others from buying sod at the time I wanted it), and Home Depot would have pocketed more dollars. Home Depot was leaving money on the table and was forcing me to incur more costs than the money left on the table, or so it seemed to me at the time. But is that always the case when consumers confront a shortage?

From a professional perspective, the sod shortage was a puzzle because at the time I posed the sod puzzle to my MBA students I had recently gone over supply-and-demand analysis in class. Central to that analysis is the widely accepted and parroted deduction that the price of a good in a competitive market will move to where the "market clears," or to where the "quantity supplied exactly matches the quantity demanded"—and I had every reason to believe the market for sod in my area was reasonably competitive (given that its price seemed to be, so to speak, dirt-cheap).

Figure 1 captures the basic graphical argument that all good economics professors teach and rely on to understand the world about them, and that I had gone over with my MBA students before I posed the sod puzzle. As conventional, the demand curve is downward sloping, showing that the value of additional units falls as more and more is consumed and, hence, that the price must fall to induce a larger quantity consumed. The supply curve is upward sloping, showing that the price must rise to induce firms to produce more of the good (because their marginal cost of production rises with expanded output). In this simple model, the shortage of restaurant seats and sod strips, and the resulting queues, can be explained by the fact that the price of each was at, say, P1, below the intersection of the supply and demand curves. At that price, the quantity supplied is Q1 and the quantity demanded is Q2. The shortage (positively related to the length of the queue) is the difference, Q2–Q1.*

* For a more detailed discussion of how equilibrium is achieved in a competitive market setting, with supply and demand curves, see McKenzie and Lee (2006, chapter 6) and video modules 2.1, 2.2, 2.5, and 3.1 at http://media.merage.uci.edu/McKenzie/Modules.html.

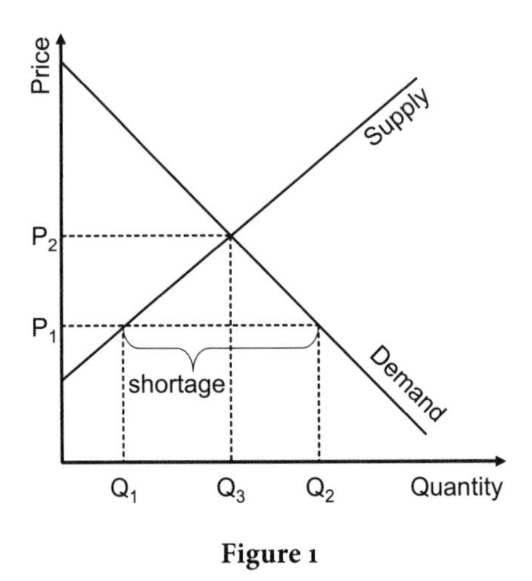

Figure 1

Why don't sellers facing such a market condition raise the price charged to P2? By raising the price from P1 to P2, they obviously gain revenues, while not incurring more costs for the Q1 units that they are willing to sell at P1. Of course, at P2, they can justify selling more units, Q3 instead of Q1. The shortage is relieved and any queue that has developed is eliminated, totally. Had Home Depot charged the market clearing price, I would not have had to waste time going to the store several times.

THE EASY SOLUTIONS FOR QUEUES

Some queues have easy explanations, not the least of which is that markets are imperfect, or don't always work with the fluidity and foresight that economists assume in their classroom discussions.

TIME FOR MARKET PRICE ADJUSTMENTS

One of the easiest explanations for queues is that market-clearing prices just don't magically *happen*. They *emerge* from the interactions of real buyers and real sellers, and the interactions between buyers and sellers take time. In Figure 1, the demand curve may have, for any number of reasons (for example, a change in consumers' evaluation of the good), shifted outward and rightward;

and/or the supply curve may have, for any number of reasons (for example, an increase in production costs), may have shifted upward and leftward. P1 could have been, in other words, the market-clearing price before the shifts in supply and demand. The result of the shift? The market shortage illustrated at P1 in Figure 1. In economists' supply-and-demand model developed on their blackboards, the price adjusts instantaneously, but such adjustments in the real world necessarily take time. During the time it takes for the price to be pushed up, a shortage—and a queue—will be apparent.

With time, as the price adjusts upward, the queue will be eliminated. If the producers in the market don't systematically raise their prices, eventually, for a wide range of goods in short supply, then those producers will be bought out by investors who see a missed opportunity. As it so happens, in the 3 years since I tried to buy sod at Home Depot, the price of sod has been raised to, at this writing, $2.69, a real (inflation-adjusted) increase in the price of sod of 45% over the 3-year period.

Of course, some queues for goods emerge because when the good is introduced, sellers have little to no knowledge of the good's demand, and the price that can be charged, until the good and its initial price are tested on the market. Markets are processes by which buyers and sellers learn from experience. Many sellers simply try a price, observe a shortage, and correct the error of their ways by gradually, if not abruptly, raising the price charged. But such a learning process takes time during which queues should be expected.

FAIRNESS IN PRICES

Researchers have found that many buyers are turned off from buying goods that they consider unfairly priced. For example, researchers have asked respondents to assess the fairness of a hardware store that raises its price for snow shovels from $15 to $20 the day after a major snowstorm. Eighty-two percent of the 107 respondents felt it was unfair for the store to take advantage of the short-run, snow-induced increase in the demand for shovels.[6] The implication is that buyers can be alienated by unfair pricing strategies, which means that stores can suffer a reduction in their long-run demands by pursuing short-run profit gains. Hence, it might be argued that queues emerge more frequently in markets than supply-and-demand curve analysis suggests because producers are reluctant to charge "unfair prices" and run the risk of upsetting the customers and losing greater long-run profits (appropriately discounted) for short-run profits.[7] Queues may also carry a subtle but valuable fairness message: "The price is so reasonable (which may, in many people's minds, equate with fairness) that lots of people want to buy it." The longer the queue, the more powerful the fairness message.

While appealing and perhaps relevant in cases in which markets are hit with sudden and temporary supply and demand shocks (as might be the case when hurricanes or tornadoes, as well as snowfalls, temporarily destroy supply chains serving communities), I'm with Becker in believing that "fairness" concerns alone can be expected to fade over the long term with gradual increases in prices.[8] If consumers adopt "references prices" by which to judge the fairness of price increases, then surely, as time passes, memories of lower prices will fade and the then higher prices can become the new reference prices. In the very short run (a day or week), the price increase in response to changing demand and supply forces may not be able to affect the quantity of the good available. Buyers may therefore not see anything about the price increase that offsets their fairness concerns. Sellers have done nothing other than force a transfer of income from buyers to themselves.

Moreover, throttling all price increases on fairness grounds will no doubt bring forth a level of economic pain in the form of shortages and lots of wasted time standing in lines for lots of things (as was true in the former Soviet Union) that could seem more unfair than higher prices. In the long-run, prices can do more than reallocate income from buyers to suppliers; they can induce an increase in supply, which can, and should, be seen by those using reference prices as having some socially redeeming value. This should be the case especially if the greater supply of a product implies a more efficient allocation of the community's (and world's) resources, which can translate into higher incomes for virtually all.

VARIABILITY IN DEMAND

The standard discussion of market-clearing prices associated with Figure 1 implicitly assumes that demand and supply are stable. That is, demand and supply don't move from time period to time period.

For many products, the opposite is descriptive of the markets: Demand is variable, sometimes highly variable, from season to season, or even from week to week, if not day to day, with much variability in many markets random and, then sometimes, hard to predict. The variability of demand can be especially troublesome for sellers in local communities, and most retail products are sold locally. This certainly is surely the case for sod at Home Depot. During any given time period, Home Depot can surely predict with greater ease its sales of sod across all of the local markets it serves than it can in any particular local market, say, Statesville, North Carolina. Demand in Statesville can be affected, temporarily, by such factors as weather and the opening of new developments, as well as the opening of other home supply stores in the area.[9] Supply of many products can also vary with the weather and supply-chain disruptions that have their source in political problems around the globe. The costs of producing final products can rise and fall with the variability in the demand for key resources.

Figure 1 suggests a straightforward answer: If the price is at P1, it should be raised to P2. The price increase implies greater profits. That's not necessarily the right diagnosis, *if there is demand variability within and across local markets.* A price hike can be ill-timed, set just as demand falls for some unexpected reason, in which case the higher price can leave sod unsold, drying out on the pallet, only to be thrown away the next day with a disposal fee tacked on by the local dump (which actually can be as great as the selling price of the sod, according to a Home Depot manager).

Put another way, underlying suggestions that the market-clearing price is the "right" (profit-maximizing) price, there is the presumption that price increases have no costs. Under variable demand conditions, however, as Home Depot raises its price above P1 toward P2, it runs a greater and greater risk that the price set will be "too high." It incurs, in short, a risk cost, which can mean that its real, economic profits (book profits minus unrecorded risk cost) can be lower at a price above P1 than at P1. And do understand that risk cost is not some imagined cost that never gets captured in a company's book. Risk cost becomes real cost when sod is left unsold and must be thrown away at a cost.

Considering the risk cost afoot in the sod business (and other businesses) raises the issue of whether adjustments in the price always to the market-clearing level is the most efficient way to allocate all goods and services. That is, allocation by queues can, in many instances, involve costs, but still queues can be more cost effective than allocation by price. If so, then buyers are getting a break in the effective prices they expect to pay across an array of products.

When I asked the Home Depot clerk why the store ran out of sod with such consistency, the clerk told me, "Sod is perishable." If sod could be stored for as long as nails, we might expect the shortages of sod to be no greater at Home Depot and other home supply stores than the shortages of nails, which are rare (attributable mainly to such considerations as occasional mistakes in ordering that are hard—that is, costly—to avoid completely).

But sod is not the only good that is perishable. When the clerk gave me her "sod is perishable" explanation for the outage, I quipped, at the time somewhat mystified, "Doughnuts are perishable, but Mag's Doughnuts doesn't run out of doughnuts by 8:30 every morning!" Obviously, I had not then thought through the difference in the market for sod at Home Depot and doughnuts at Mag's Doughnuts. Mag's workers can produce doughnuts in relatively small batches throughout the day as the stocks of the various doughnuts in their showcases dwindle. Home Depot, and their suppliers, can't do the same with sod, at least not with the same facility and at the same costs. Sod must be transported to Home Depots from miles away on sizable flatbed trucks. Replenishment of the sod stock several times a day is costly. Hence, Home Depot must estimate the demand for sod over a longer time period—a day—than Mag's must estimate the demand for doughnuts—maybe hours at most. We should expect the longer the time period for estimating the demand for a highly perishable product, the greater the risk and uncertainty in estimating the demand and the more likely mistakes in prices (all other considerations equal). To avoid the costs of having unsold sod and having to incur a disposal charge, Home Depot can be expected to adopt a safer pricing strategy, one that errs on the side of having too little sod on many days.[10]

INVENTORYING CUSTOMERS

There are good reasons stores hold inventories. The demands for their products are variable (sometimes highly so) and not very predictable during any short period of time. It's cheaper to cover the carrying costs of the inventories than to incur the costs of missing sales. Indeed, inventories, even the stocks of products stacked on stores shelves, can be seen as queues on the supply side of the market, with the items lined up waiting to be sold. Not many economists would dare conclude that these queues on stores shelves mean that the products' prices are necessarily too high.

Stores sometimes have queues of customers for the same reason they stock products they sell: Queues can be seen as a readily available inventory of customers to deal with the variability and inability to predict demand for any short period of time (days of the week or hours of the day). And it is cheaper for the stores to incur the costs of not charging exactly the market-clearing price at all times than to incur the costs of foregoing sales or of frequently (if not constantly) computing and announcing prices that cause the market to clear at all times.[11]

QUEUES AS PROFIT-MAXIMIZING RATIONING MECHANISMS

Standard supply-and-demand-curve analysis assumes implicitly (really explicitly when presented in its most technical terms) that all buyers are equal, and are the same in all important regards to sellers. Under such market conditions, the price charged cannot alter the composition of actual buyers. Buyers are all alike regardless of whether P1 or P2 is charged in Figure 1. And if all customers were just alike, price discrimination across buyers would not work.

Needless to say, most real-world markets differ radically from the standard model, as casual observation of buyers in any store and at concerts will verify. Buyers differ substantially in their enthusiasm for the product, and for other complementary products, which can explain why concert venues often charge prices so low that in order to be assured of getting tickets, concert goers will camp out for hours (if not days) in long queues that sometimes extend hundreds of yards from the box offices. As explained by economist Ken McLaughlin (and reported by Steve Landsburg), when concert tickets are priced below the (presumed) market-clearing price, the concert promoters ration out of box office queues potential concertgoers who have little enthusiasm for the performance.[12] The people who buy their tickets weeks in advance or who stand in line for hours on end and get the tickets will tend to be relatively enthusiastic (if not wildly so). Their enthusiasm can enhance the value of the concert for all goers, which can fuel the enthusiasm and demand of all concertgoers. The attendees who endure wait time in the queues will also likely be inclined to buy the performers' albums, T-shirts, posters, and other paraphernalia that are sold inside the concert halls, at inflated prices, of course. Indeed, it is not unreasonable to expect that the lower the ticket prices and the longer the queues, the greater the prices of the products sold after admission can be.

THE ECONOMIC LOGIC OF QUEUES

Grocery (and, for that matter, many other) stores are notorious for having long queues at their checkout counters at the end of most work days, say, between 5:00 and 7:00 P.M., when people drop in to buy their meals for that evening or week. Queues are common at other times of most days, also, simply because at "non-rush hours" grocery stores regularly close one or several of their checkout counters.

THE ECONOMICS OF QUEUES

Again, why queues? By that question, I really mean to ask about the economic logic realities underlying queues and their lengths. That is, why are queues as long as they are, and no longer? Clearly, mistakes in estimating store traffic at various times of the day and in planning the work schedules of checkout clerks explain some queues, and their varying lengths. Instead of raising and lowering their prices, grocery (and other) stores deal with the variation in traffic by allowing their queues to lengthen and contract. Having said that, however, both shoppers and store managers must plan for queues with some *expected* (mean) length and some *expected* (mean) waiting time. They both understand that queues can be longer than expected, but also shorter than expected during different times of the day and different days of the week. If the queues are longer than expected *consistently*, then shoppers can be expected to revise their expectations and maybe shop elsewhere, where their grocery bills plus opportunity cost of waiting in line is lower. Managers can be expected to adjust the number of checkout counters they have open to minimize the incentive their customers have for moving to other stores.

The crucial point is that queues have a rational, economic foundation, grounded in the costs and benefits of people waiting in them. In planning their trips to grocery stores, shoppers can be expected to weigh off the benefits of getting the food items they need and want against their prices *plus* the opportunity cost of the time they *expect* to stand in line. They should be willing to pay higher prices for the benefit of having to spend less *expected* time standing in line, at least beyond some point. And they should be willing, up to a point, to spend more time standing in line if they are adequately compensated by the store in terms of lower prices for the products they buy. However, as noted in Chap. 11, consumers tend to place a greater subjective weight on an out-of-pocket expenditure of a given dollar amount than on an equivalent opportunity cost also measured in dollars.[13] This means that shoppers can be expected to be willing to incur more than a dollar in opportunity cost waiting in line to save a dollar on the prices of their food items.

Similarly, store managers (and their executives and owners) can be expected to see queues at their checkout counters as an economic problem, and as a source of greater profitability, at least up to some point. Store managers also should be expected to weigh off the costs and benefits of having queues. The benefits to managers from eliminating queues are transparent: The elimination (or just shortening) of queues can make shoppers happier (because their total cost of getting what they want goes down), can increase their store traffic, and can raise the prices they can charge.[14]

The managers' management problem is that cutting the length of the queues is costly. Managers would have to set up more checkout counters, open more of them for more hours of the days, and incur a greater wage bill for the greater number of checkout clerks. The greater floor space used for checkout counters is costly because of the greater construction and land costs and because the expansion of checkout counters will force the contraction or elimination of product lines. And managers must recognize that checkout counters are fixtures that can't easily and quickly be removed, which is to say that many counters, and the floor space under them, can go unused for hours each day, giving rise to nontrivial opportunity costs where land and floor space come at premium prices.[15]

OPTIMUM QUEUES

Neither grocery store managers nor shoppers would want zero queues if such were even possible. No queues at all would mean lots of costly floor space taken up with many checkout counters, a number of which would not be used during many hours of the day, if not days of the week. No queues would likely spell high prices for shoppers. Both shoppers and managers are obviously interested in facing queues of some *optimum* (expected) length.

Store managers can be expected to add checkout counters so long as they can raise prices on the products sold by more than the rise in the cost of their additional checkout counters. And managers should be able to raise their prices because shoppers will spend less time in line. If the stores can incur $50,000 for an additional checkout counter, but can increase their expected (net) revenues through higher prices and greater sales by $60,000, then they should be willing to add the checkout counter. Otherwise, they would be leaving $10,000 in added profits on the table.

Managers, however, are constrained in how many checkout counters they can profitably add. As they add a growing number of checkout counters, they will have to contract or eliminate product lines with growing profitability. In short their costs will grow with additional checkout counters. Moreover, as they shorten their lines, managers lower the opportunity costs of their shoppers, but they also have to increase the prices on the products their shoppers buy. Beyond some point, the added price paid by shoppers will exceed the shoppers' falling opportunity cost of standing in line. And as noted above, shoppers tend to have a preference for incurring a dollar of opportunity cost over a dollar in out-of-pocket expenditures on food products.

This means that the *expected* length of the checkout queues will never likely go to zero (at least not for most run-of-the-mill neighborhood grocery stores). If the *expected* wait-time were zero, then managers would likely see an incentive to take out checkout counters, save their stored costs, impose opportunity/wait-time costs on their shoppers who could then see a reduction in their grocery bills (because of the stores' lower costs and prices).

How long should the queues be? The answer is, necessarily, it all depends on the actual costs and benefits as perceived by the stores and their customers. The manager needs to balance the costs and benefits for both the store and shoppers. As the line is lengthened, the store saves costs and can lower prices. Shoppers incur more wait costs but can benefit from the lower prices. As the store takes out checkout counters one after the other, its cost savings from doing so are likely to fall (given that it will likely take out the counters that are least costly at the start); the price reductions can, as a consequence, be expected to fall as the queues lengthen. As the queues lengthen, shoppers will see the prices fall by smaller and smaller amounts at the same time their wait costs begin to escalate (since shoppers can be expected to give up their least costly opportunities when they start their wait, only to forego more and more costly opportunities as their wait time lengthens). The store manager can be expected to allow its (expected) wait times in the queues to grow so long as the (subjectively weighted) opportunity cost of the additional wait time for shoppers is lower than their savings on the prices of the things they buy. Store managers will work to avoid extending their queues until the wait time costs their shoppers incur is greater than their savings from lower prices.

The central point of this line of analysis here is that there is some *optimum* queue for every store, with one important determinant being the opportunity costs of stores' shoppers. We would tend to expect that stores that serve shoppers with relatively high opportunity costs will have shorter queues at their checkout counters than stores that serve shoppers with lower opportunity costs. That is to say, "down-market" grocery stores in low-wage neighborhoods—for example, Food4Less and Food Lion common in the Midwest and South—can be expected to have longer lines than "up-market" grocery stores in high-wage neighborhoods—for example, Gelson's, a high-end grocery store chain in Southern California.

For that matter, down-market Marshall's department stores, which cater predominately to low-income and price-sensitive shoppers, can be expected to have longer lines than up-market, boutique stores like J. Jill, which caters to much higher-income, and less price-sensitive shoppers. Hence, we can predict that as the wage rates in a shopping area rises (relative to the cost of floor space in stores), we should not be surprised if the lengths of queues fall, and vice versa.

PREMIUM TICKETS

Many people standing in many queues have different opportunity costs, as well as different incomes and sensitivities to price changes. High school students typically have lower opportunity costs than working adults. Hence, as a general rule, they should be willing to accept longer queues than working adults. Working adults might also be expected to want higher prices on the things they buy than high school students, because the higher prices can cause high school students (and everyone else with lower opportunity costs) to buy what they want elsewhere, or just go without. By pricing the high school students out of the market, the older adults can then get what they want at a lower *total* cost (a higher price but a lower opportunity cost). By the same token, high school students might want prices below their market-clearing levels because such prices effectively price working adults out of the market. The lower money price more than compensates high school students for their opportunity (wait-time) costs.

With differences in the willingness of buyers to stand in queues, there is a clear opportunity for firms to price discriminate among buyers with the result being that both low and high-opportunity-cost buyers can get deals better suited to their economic circumstances, and the sellers can make more profits in the process. Tailored correctly, a high price accompanied by a short wait-time for high-opportunity-cost buyers can be superior to some standardized price and expected wait-time for all. Similarly, a low price accompanied by a long wait-time for low-opportunity-cost buyers can be superior to a standard price and wait-time for all buyers. When the price increase for the high-opportunity-cost buyers exceeds the price reduction for the low-opportunity-cost buyers, the result has to be greater profits for the sellers (which is clear when the same number of buyers are served).

Interestingly, for a long time airlines have recognized that they can make money by charging different segments of their markets different prices through classes of passengers: first and coach (and sometimes business) classes. Granted, first-class passengers get larger seats and better meals, along with free drinks, than passengers in the "cattle-car" section of planes don't get. First-class passengers pay a higher price because they receive a "higher level" of service, but they are also allowed to "cut line," since they have their designated area for checking bags and are allowed to board first, in spite of being at the front of the plane.

More recently, theme parks have begun to recognize the profit potential from allowing some park guests to effectively break lines at their various attractions. They've pulled off this feat, without (apparently) alienating many park patrons, in part by allowing guests to self select into longer lines with lower prices and shorter lines with higher "premium" prices and in part by having two segregated lines, with those buying the premium tickets allowed to use a separate gate for the line, which obviates the problem of premium-ticket holders actually having to butt in front of anyone.

Universal Studios' amusement park in Hollywood charges $61 for guests 8 years of age and up (and $51 for ages three to seven) for a one-day pass. Universal charges $109 for a one-person "front of the line pass" for one day, which, to keep the queues at attractions in check, is limited in supply. It goes one step further and charges $199 per one-day ticket for its "VIP Experience," which includes front-of-the-line privileges and a guided tour of the back lots of its movie studios. Similar pricing plans have been introduced at Universal Orlando, Six Flags Over Georgia (Atlanta), Disneyland (Anaheim), and SeaWorld (Orlando).[16]

Why don't grocery stores provide similar front-of-the-line service? One obvious difficulty they confront is that they don't charge admission, but they could sell "tickets" to shoppers who want to go to the front of the line over the internet or at the store entrances. Granted, shoppers with such tickets might be reluctant to use them if they had to actually break in front of other shoppers at the checkout counters open to all shoppers, but that problem could be solved by the store setting up a special checkout counter apart from all the others (perhaps reached by going into a separate room). Grocery stores can avoid selling tickets and simply have a separate checkout counter with an electronic sign at the entrance that reads something to this effect: "Anyone going through this checkout counter will have 10% added to their total shopping bill" (with the percentage added changing for peak and off-peak times of the day). Both the ticket and sign methods of price discriminating might seem like a totally bad idea for both shoppers and the stores, unless it is realized, conceptually, that there is at least the *potential* for mutually beneficial trades. Many shoppers might want to pay money to save more of their highly valued time, and the stores might be able to generate more additional revenues than they incur in added costs from operating the separate "premium" checkout counters.[17]

The obvious problem with such "premium" pricing schemes for grocery stores is that no stores (that I know about) use them. Why do airlines and amusement parks sell means by which their customers can save wait time, but grocery stores don't? Is it that they've not thought of the idea? Maybe so, but I doubt such could be the case, given how many grocery store executives and managers are, because of competitive pressures, constantly trying to devise ways of making more money with little in the way of added costs.

A more likely explanation for the difference in the pricing strategies among airlines, amusement parks, and grocery stores comes from the observation that airlines and amusement parks on the one hand and grocery stores on the other differ dramatically in one important regard, the degree of diversity in their customer bases. Airlines' and amusement parks' customers are highly diverse along any number of social and economic dimensions, but for our purposes their customers appear to differ most prominently in terms of the opportunity costs of their time. Airlines and amusement parks, whose customer base can extend over the entire country and all economic classes, can sell front-of-the-line, premium tickets because there are any number of passengers who are willing to pay more than other passengers are willing to pay in order to save time.

Grocery stores, on the other hand, tend to be situated in neighborhoods that tend to have residents who have self-selected to live in their neighborhood because of shared values and similar household incomes—which can be construed to mean similar opportunity costs. Grocery stores simply don't consider differential pricing because their customers tend to want the same things—and want more or less the same prices and are willing to accept more or less the same wait times (which, in economic terms, simply means the additional revenue collected from the limited price discrimination can't cover the added cost of the checkout counter). This does not mean that grocery stores can't and won't price discriminate by opportunity costs. They do, but they impose different prices in *different* stores—in different neighborhoods, across which the residents' values and opportunity costs of the residents can differ markedly.

Note also that grocery stores have serious competitors for the dollars of customers who have high opportunity costs and low tolerance for long queues in the form of convenience: liquor and drug stores, and a host of restaurants that provide take-out meals, all situated within given neighborhoods. Grocery store holding companies also own different store chains that have different quality and pricing levels, and different wait times. For example, the Kroger Co. owns Ralph's grocery stores, a chain with moderately high quality products and prices and moderately short checkout queues, that tend to be located in neighborhoods with above-average incomes. Kroger also owns Food4Less grocery stores, a chain that has lower prices and longer queues. Food4Less stores tend to be located in low-income neighborhoods.[18]

Shoppers with high opportunity costs can simply choose to patronize any of the higher price outlets that have lower *expected* wait times. If grocery stores were to try to institute the equivalent of "premium shopping tickets," shoppers can go elsewhere for the same length lines. Doing that is not nearly as easy for guests of amusement parks and passengers of airlines to do, which helps explain why they can charge premium prices for shorter queues.

CONTRIVED SHORTAGES AND BUYER LOYALTY

Northwestern Law School Professors David Haddock and Fred McChesney reason that many market shortages and queues emerge because businesses don't want to risk raising their prices and losing loyal customers in the face of transient surges in demand or contractions in supply.[19] Their argument for privately contrived shortages and queues builds on points made earlier about the variability of supply and demand over time, but they develop their argument by starting with the costs buyers and sellers must incur to form long-term mutually beneficial relationships.

Before settling on products to purchase, Haddock and McChesney posit, with good reason, that buyers undertake market searches at some cost, judging the price/quality ratios of available market options. Similarly, firms incur some costs in trying to reduce the cost of the market searches of prospective buyers and assuring prospective buyers that their products are the best available in terms of their price/quality ratios, and will remain the best options into the future. Once buyers and sellers settle on each other, they develop relationships often organized around unwritten contracts with both parties assuring each other of some loyalty. Sellers commit to holding the line on prices and product qualities while the buyers commit to continuing their purchases at the stores by truncating their market searches.

Because of their investment in devising their expected-to-be ongoing relationships, both buyers and sellers have reason to want to maintain their relationships, or not break them for essentially transient considerations. Sellers don't want to give buyers a reason to renew their searches of market options because of the costs involved in replacing loyal customers, and buyers don't want to have reason to renew their costly searches of market options. Hence, if demand increases because of what is believed to be a passing fad or production costs spike, sellers may not raise their prices because they want their customers' loyalty and long-term business. Sellers might earn more short-run profits by hiking the price in response to an increase in demand, but a price increase can destroy customer price expectations and undermine their "goodwill" that, in turn, can cause their most loyal buyers to renew their market searches and, perhaps, take their business elsewhere.[20] Accordingly, as Haddock and McChesney note,

> When natural disasters like Hurricane Andrew occur, many merchants choose shortages and queues over price increases. Foreign and domestic auto companies have sometimes maintained prices below market-clearing levels, rationing their product among dealers and discouraging them from increasing price, and episode reminiscent of Henry Ford's underpricing of the original Model T. L. L. Bean once responded to an upsurge in demand by refusing to send catalogues to those who were not already on its mailing list. Newspapers typically do not vary the number of papers printed or the price charged, even on days when a particularly newsworthy event makes it likely that the issue will sell out.[21]

The law professors point out that Parisian restaurants count on their loyal local patrons to fill their tables when tourists are not around. Hence, the restaurants don't raise their prices during the high tourist season. Queues may arise during the tourist season, but the queues can be managed by the restaurants giving priority to local patrons and by taking reservations only from French-speaking (especially frequent and loyal) customers.

Haddock and McChesney draw several important deductions that can explain why sellers often ration by queues rather than by higher prices:

- First, they point out that established firms should be expected to hold the line on prices in face of transient shortages more so than transient firms. As the *New York Times* reported on the aftermath of Hurricane Andrew, "The big companies performed far differently than the price-gougers selling ice, water and lumber from the back of pickup trucks at wildly inflated prices…But unlike the carpetbagging vendors, who drove away at sunset, the big companies have a long-term stake in the South Florida market. For them the good will of local customers…is a valuable asset."[22] The "big companies" that hold the line on prices, however, invariably limits the quantity any one customer can buy in part to serve more loyal customers but also to prevent shoppers reselling at higher prices.

- Second, the inclination of firms to use price to deal with shortages depends upon customer search costs. If search costs are low, as is the case with gasoline, then prices can be expected to bob about with the changing and transitory forces of supply and demand: "Consumers obtain information about gas prices almost costlessly, as a by-product of just driving past gas stations, rather than searching out prices. Under the model here, there would be no reason for gas stations to 'hold the line' on price as demand and cost changes occurred."[23]

- Third, high and variable rates of inflation can be expected to cause businesses to use prices to eliminate shortages. This is because inflation forces buyers to continually engage in searches for price information on the levels and relative positions of various product prices. Inflation itself can, consequently, be a source of shorter queues but added inflation.[24]

- Fourth, Haddock and McChesney conclude, "Intentional shortages will more likely emerge when (a) customer demands or input costs are rising unexpectedly but the seller can predict they will move back toward long-run equilibrium levels, or (b) unexpected demand increases are believed to be permanent but will later be matched by long-term production" increases.[25] This means that even when temporary shortages exist, firms can be expected to continue to stimulate demand with advertisements. They might continue to advertise because they want to make the short-run increase in demand permanent and/or they anticipate expanding their capacities and want to achieve a sales volume that will result in the efficient utilization of the capacity coming online later.

- Fifth, firms that hold the line on prices in face of shortages and queues can be expected to manage their shortages. They can refuse orders from new customers. They can also limit the amount purchased by each customer, in part to allocate the available stock among more loyal customers but also in part to prevent some customers from buying up large quantities of the stocks only to resell what they have purchased to others at a higher prices (from the backs of their pickups). Of course, this means that the greater the ease of resale, the less willing firms will be to hold the line on their own prices, and the shorter the queues will be.

BANDWAGON EFFECTS AND QUEUES

Gary Becker has postulated that for some goods—meals at the (Palo Alto) restaurant mentioned at the start of the chapter, books, concerts, and theatrical performances—shortages occur because the profit-maximizing price is *above*, not *below*, the market-clearing price. While this is not always the case, it can be true, according to Becker, when the consumption of the good is sufficiently *social* in nature, that is, the demands of individuals is positively influenced by the number of other people enjoying the good. That is, people can jump on the "consumption bandwagon" for certain goods because other people are either on the bandwagon or are expected to make the jump. A sufficiently strong bandwagon effect can mean that the price can be positively related (within a range) to quantity of the good consumed. In Becker's words, "Suppose that the pleasure from a good is greater when many people want to consume it, perhaps because a person does not wish to be out of step with what is popular or because confidence in the quality of the food, writing, or performance is greater when a restaurant, or book, or theater is more popular." [26] He postulates, in other words, that in addition to the bandwagon effect, consumers can also buy the good because of the "snob" and "conspicuous consumption" effects, not just because of the intrinsic use value of the good by itself.[27] He notes that Stephen Hawking sold over a million copies of his book *A Brief History of Time,* "Yet I doubt," Becker probably correctly muses, "1% of those who bought the book could understand it. Its main value to the purchasers has been a display on coffee tables and as a source of pride in conversations at parties."[28]

Becker organizes his formal argument around the Palo Alto restaurant mentioned at the start of the chapter that regularly has queues in the evening for its fixed seating capacity, represented in Figure 2 by the vertical supply curve at S1. This means that the (short-run) seating is unaffected by the price, clearly an assumption Becker makes to simplify and facilitate the development of his argument. Assuming the demand, D1, is downward sloping, as is conventionally assumed, the market-clearing price is P2. The only way a queue can emerge in this model of the market is for the price to be below P2, say, at P1 (which readers will note is similar to the line of argument developed in Figure 1, with only a change in the slope of the supply curve). The shortage in this figure is the gap between the quantity demanded, Q2, and the quantity supplied, Q1.

This line of argument implicitly assumes that the demands of individual restaurant patrons (or book buyers or theater goers) is unaffected by the enjoyment of other patrons, by the number of patrons, and by the characteristics of the other patrons—which will not always be the case, especially when the goods are consumed socially. And meals at restaurants are social happenings, necessarily. Restaurant goers acquire information on the quality of the restaurant's food by taking notice of who and how many others want to eat there. Book buyers buy books because others are buying them and talking about their themes in social

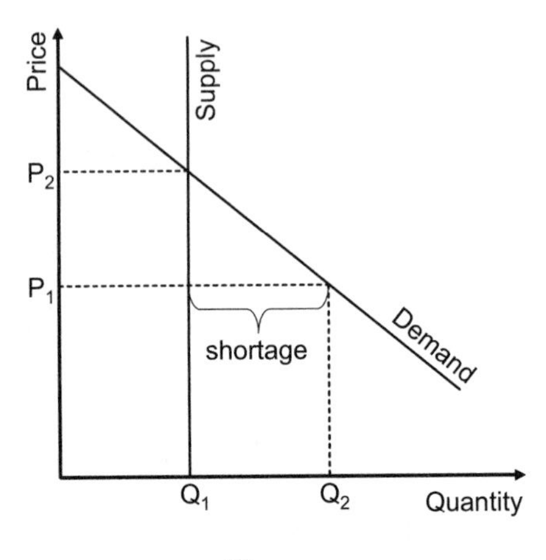

Figure 2

situations. Concert goers can feel ill at ease if the hall is sparsely filled, and can feed off the rapture that others around them feel on seeing and hearing the performance.

Hence, Becker argues that demand for eating at a restaurant can build on itself, as the number of buyers increases: First, the value all initial buyers receive from consuming the food can rise with the number of buyers. Second, as the number of buyers increases, more patrons will want to join the market. Hence, if the social interaction is sufficiently strong, Becker postulates, price can rise with quantity demanded, and the result can be that the total quantity bought can go up with price. The price increase itself can further stimulate an increase in the value of the good demand because, as Dwight Lee and I have argued, the price rise can change the composition of the restaurant patrons.[29] That is, the price increase can force out of the restaurant's markets "undesirables" (however the actual and prospective patrons define them), increasing the value of the restaurant experience for the remaining patrons and causing more people to want to go to the restaurant. Do you doubt the validity of this assumption? Consider that many people go to pricy restaurants precisely because they can have a high degree of confidence that their dining experience will not be marred by unruly and loud children and babies.

The central point Becker seeks to make is that there is no necessary reason that demand for *socially consumed* goods will be, throughout the entire range of possible prices, always downward sloping. There can be an upward sloping range, as depicted by D1 in Figure 3 (which is drawn directly from Becker's

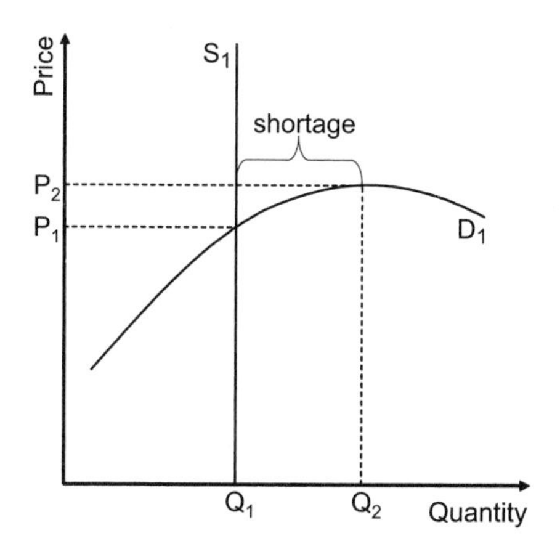

Figure 3 Source: Gary Becker 1991, p. 1113.

work[30]). The demand curve might reasonably be expected to bend downward beyond some point, Becker summarizes, when the restaurant becomes excessively crowded and noisy and the queue becomes excessively long, with the service and food quality possibly also suffering. The supply curve is S1, the same as in Figure 2.

In this graphical illustration, supply and demand intersect at the restaurant's capacity, Q1, which makes the market-clearing price P1. But P1 is hardly the revenue-maximizing price. This is because the price can still be raised from P1 toward P2 with the quantity demanded continuing to expand. P2 is the profit-maximizing price, simply because P2 necessarily yields more revenue than P1: P2 × Q1 is greater than P1 × Q1. With costs remaining the same, when quantity is held at Q1, a price of P2 increases profit above what can be had at the market-clearing price.

There are two interesting points drawn from Becker's line of analysis.

- First, a shortage, which translates into a queue, emerges at a price *above* the market-clearing price. The shortage at P2 equals the number of seats demanded, Q2, minus the number of seats available, Q1.

- Second, the restaurant owner (or seller of any such social good) has to choose the price with some care. A price above P2 can have a dramatic impact on the number of people wanting to eat at the restaurant. A price higher than P2 can cause the number of patrons who show up to collapse to zero. The "demand curve" in the figure bends down beyond Q2 because of the crowding, beyond Q2, lowering the value of eating at the restaurant for all patrons.

While I feel some reluctance to accept Becker's upward sloping demand curve, his theory does seem to describe what is so often observed, shortages and long queues at many restaurants in the face of what appear to be relatively high prices. However, his analysis is short on explaining exactly how the price that is above the market-clearing price, along with a shortage, can be achieved. Some restaurants might be fortunate in choosing an initial price of P2 with the quantity demanded magically equaling Q2 in Figure 3, which results in the maximum shortage (or length of queue) at the restaurant. But in real market settings sans magical influences, such a high initial price might preclude a pronounced bandwagon effect, because there may be few initial patrons. There is a real chicken-and-egg problem for providers of socially consumed goods subject to bandwagon effects, which reduces to an important question: How can sellers get the bandwagon rolling?

Restaurants might start the bandwagon rolling by doing as we suggested in our discussion of network effects earlier in the book (with the network effects being similar to the "social effects" Becker has in mind): initially, the firm can charge a low price that can stimulate the initial demand and create queues, with the queues signaling that this is the "in place" that, in turn, causes an increase in the value of dining at the restaurant to all patrons, resulting in more patrons being attracted to the restaurant. But such a strategy is problematic, given that a low price might attract many customers who can…well, pay only a low price, which can turn off celebrity and other A-list customers who can signal to many others that the restaurant is on its way to being the "in place." Moreover, while there can be some value in such a pricing strategy, problems can emerge when the price is later raised. The price increase can be seen as "unfair." Then again, it might signal the growing value of dining at the restaurant. We really can't say a priori how the restaurant can pull off the above-market-clearing price with a queue.

Perhaps Becker is correct to effectively "punt" on how to solve the chicken-and-egg problem. He suggests that restaurants (and other firms) who can build queues with high prices would be expected to advertise a lot, as well as pay attention to amenities that make a lot of people conclude that patrons will be coming to the restaurant because a lot of others will be there. Becker writes that such advertising expenditures can "have a multiplier effect when consumers influence each other. Advertising that raises the demands of some consumers also indirectly raises the demands of other consumers since higher consumption by those vulnerable to publicity campaigns stimulate the demands of others."[31] He also suspects that such restaurants will be subjected to faddism, or large and rapid swings in buyer tastes, a point that has a ring of truth from casual observation of the restaurant industry.

Setting aside the problems associated with Becker's bandwagon theory of queues, it seems transparent to me that some restaurants do (for a short time at least) solve their dilemma of building demand. Sometimes they do it by reputation. At one time Planet Hollywood was an "in" place to dine in Hollywood, partly because of the novelty of the movie set décor and partly because movie stars and producers founded the restaurant. When the company opened another Planet Hollywood near my home, the queues were initially long, in spite of the fact that patrons were asked to pay relatively high prices for mediocre food. But in a few short years, the restaurant was closed as fickled patrons tired of theme-based restaurants.

The initial demand for Apple's iPhones is, perhaps, a prime example of what Becker had in mind in drawing the upward sloping demand curve and the resulting shortage at a "high" price above the market-clearing quantity. As noted at the start of Chap. 1, Apple announced in January 2007 its planned introduction of the iPhone the following June. In the interim, there had been so much media buzz about the cell phone, mainly because the phone had the look and feel and some of the capabilities of Apple's wildly successful and "cool" iPod, that the queues outside of Apple and AT&T stores began to form days before the June 29 release. When the doors of the stores opened, the stock of iPhones were quickly depleted. My AT&T store reported selling out of its two to three dozen iPhones within 45 minutes, in spite of the iPhones prices—$499 and $599, depending on storage capacity—being far above other cell phones.[32]

SINGLE VERSUS MULTIPLE QUEUES

Firms should not only try to optimize on the length of their queues, they should seek the best structure for their queues. The two most prominent queue structures are the single queue, under which buyers form a single line and move from the front of the line to the next open service counter, and multiple queues, under which buyers form several lines, one for each service counter. Airlines and banks generally use the single queue structure for coach passengers (along with a single queue for first-class passengers). Grocery stores and fast-food restaurants typically use multiple queues.

Perhaps the most important economic advantage of the multiple-queue structure is that multiple queues tend to be faster than the single queue. That is, more customers can typically be checked out in a given amount of time, which also means that the *expected* (average) wait time for customers is lower, than for single queues. Multiple queues tend to have shorter *expected* waits because customers have to pay attention only to one service clerk and can be expected to move forward to be checked out with little delay. A single queue, on the other hand, requires customers to be attentive to all service counters, and many customers fail to notice open counters and move to an open counter except with some delay. Indeed, customers often have to be prodded to move to an open counter. Then, time is soaked up as the customers move what is usually a longer distance to the open checkout counter than is true for multiple queues.[33] Multiple queues offer customers more control over the time spent standing in line, given that multiple queues give customers opportunities to choose their queues based on their assessments of the transactions to be made by customers ahead of them and of the speed of the service clerks.[34]

However, multiple queues have one disadvantage: the variability in waiting times across the service queues is often greater for multiple queues than for single queues.[35] Yes, customers may know that their *expected* wait time will be shorter in multiple queues, but they also understand that from time to time they will be caught in queues with wait times far above what they expected. Granted, they might also be in queues from time to time in which their wait times are far below their expected wait times. However, customers might notice the unusually long waits, and weigh them subjectively more heavily, than the unusually short waits. The single queue might be longer but can be expected to have more movement. Hence, in spite of the longer *expected* wait times, many customers might prefer single queues because of the sense of some progress and a reduction in the anxiety felt from the variability (and unpredictability) in the wait time.[36]

Since variability of relative wait times in single and multiple-queue structures can differ substantially across retail (and other) sectors of the economy, we should not be surprised that different industries favor different structures. In general, we would expect the single-queue system will be used where the complexity of customer transactions can vary markedly, making the wait time in the individual service queues highly unpredictable. We would expect multiple-queue systems to be used where the variability of customer transactions is usually limited. Airline passengers can vary greatly in the complexity of their transactions at the service counters. They may only need to get their boarding passes and check a bag or two (all by computer), or they might have to change their tickets and find a place to stay after their flights have been cancelled. Hence, all airlines use a single-queue system.

On the other hand, the variability of the orders of fast-food customers can be limited, making multiple queues more economical. Why so many queues at grocery stores? Grocery stores often use special checkout counters for customers with fewer than ten or twenty items, reducing the variability for the rest of their checkout counters. Still, given that shoppers vary greatly in how full their carts are and how many coupons they redeem, significant variability in checkout times remains. However, grocery stores tend to use multiple queues simply because of the difficulty shoppers have in identifying open counters (checkout clerks are often hidden from view by magazine and candy racks) and in maneuvering their carts around the counters of the coiled queue.

Businesses have an obvious incentive to pick the queue structure that their customers generally prefer. The greater the preference customers have for the chosen queue structure, the longer businesses can allow their queues to become (all other considerations the same), the more costs they can save on checkout counters, and the higher the prices they can charge and the higher the profits they can earn, relative to what they would otherwise earn. Of course, competition among businesses to adopt the most preferred queue structure will ultimately check the prices charged and profits earned.

LAST-COME/FIRST-SERVED, A SOLUTION FOR QUEUE LENGTH?

Steven Landsburg, an economist I have cited before, assumes that his readers will readily concur in his claim that we all "spend too much time waiting in line. That's not some vague value judgment; it's a precise economic calculation" (with the "precise economic calculations" left unreported in his discussion).[37] He explains, drawing on the conceptual work of other economics, that the shopper who gets in line first imposes a wait cost on all others who get behind her.[38] Since shoppers at the front of the line don't consider the wait costs they impose on others (but do consider the wait costs imposed on them by others in front of them in the queue), Landsburg readily accepts the conclusions of another economist who more than two decades ago asserted that queues are not "generally socially optimal."[39] Landsburg suggests two solutions for the presumed inefficiency of the queues we all see around us.

One potential solution tendered is for shoppers in the back of the queue to pay shoppers closer to the front to change places with them. That way, the shoppers close to the front will experience a cost of holding up others in line. If shoppers keep their places in line, they will forego the opportunity cost of the buyouts. Landsburg, however, quickly dismisses this solution as being uneconomical: The required negotiations would be a "hassle," resulting in "mutually beneficial trades" going unconsummated. I agree. If such exchanges were not such a hassle, with few gains to be had from them, then we should commonly observe people making such trades, which we don't (although some older people do pay young kids to stand in line to buy tickets for concerts, thus reducing the overall wait costs, and others buy tickets from scalpers at inflated prices to avoid the wait costs).

Landsburg's next solution is one economist Rafael Hassin proposed in 1985: Switch from the common rule of queue formation of *first-come/first-served* to a rule of *last-come/first-served*. That is, late comers to queues would have rights of breaking line.

Bizarre, you think? Landsburg suggests your problem can only be that you are forgetting that the last-come/first-served rule would reduce the length of queues in general because shoppers in the back of queues would soon get out of the queues. Their withdrawal from the queues would thus lower the expected social (wait-cost) waste that first-come/first-served queues foster.

To make his point, Landsburg asks readers to imagine a line-up at a water fountain, say, in a public park (using economist Barry Nalebuff's example[40]). If late comers have the right to go to the front of the line, then many people toward the back of the line will feel the cost of having to spend more time standing in line. Some of them will, eventually, give up on getting a drink, reducing the length of the queue and the wasted time in the line. Landsburg happily muses, "That keeps the line short, which is good. In fact, it's better than good: it's ideal."[41]

Everyone else would be happy, too—right? Hardly. Not the people at the back of the line, you can bet. They might be as distressed with line breakers as they are with petty muggers. As a consequence, there need not be a movement toward queues that are "socially optimal" in terms of length. This is because the optimal queue length can be a function of more than people's objective or subjective value of their expected wait cost. People might simply view the higher expected wait cost from a first-come/first-served queue rule as being superior to the lower expected wait cost associated with the last-come/first-served rule. Why? The latter could have a higher variance (as well as longer wait-time for some people) and could be considered less fair, and further from Landsburg's presumed goal of making queues more "socially optimal."

Landsburg recognizes one big flaw in the last-come/first-served rule. People would quickly learn to game the system. They would quickly learn to drop out of the queue and return as a newcomer, with the right to cut in front of everyone else. You can imagine lots of time would be wasted as people scramble to exit and then to re-enter queues. You might also imagine that the gaming would quickly, in many queues, break down into fighting. Such prospects no doubt damn the last-come/first-served rule, and explain why the rule is rarely seen in operation.[42]

But there is a bigger, more fundamental economic problem with Landsburg's (and others') proposed last-come/first-served rule. Queues need not always and everywhere be "socially suboptimal." Notice the queue setting Landsburg chooses to elucidate his propose rule, that of a water fountain in a park, which is not part of a business. No one has any incentive at all to constrain the length of the queue at the park water fountain. That is hardly the case in our grocery store (and concert arena) examples. Stores can make money by always standing ready to move the queue length closer (if not) to some "optimal" level. They can do it, as stressed in this chapter, by varying the number of checkout counters or box offices and the prices of the many products they sell. One reason few, if any, businesses use the rule of last-come/first-served is that the rule can cause discord squared, but also because such a rule can easily shorten queues and, in so doing, make them socially suboptimal. No doubt shoppers want shorter queues, but shorter queues come with added costs for both shoppers and sellers.

CONCLUDING COMMENTS

Supply and demand are powerful forces in markets and highly useful concepts to economists as they teach their students about how competitive markets operate and the role prices play in allocating scarce resources and goods and services among buyers. But, as evident from this chapter, there is far more to market interactions of buyers and sellers than can be captured by supply and demand as price/quantity relationships, described as two curves on a graph. When shortages and queues are as prevalent as they are in real-world markets, we have to think that there are good economic reasons for them. All shortages and queues can't be chalked up to mistakes of market participants or aberrations of market forces. As has been argued in several ways in this chapter, queues can make a lot of sense. By that I mean, they can promote the economic and social goals of both buyers and sellers, which means that shortages and queues can be mutually beneficial to both sides of the market, buyers and sellers. If that were not the case—if both buyers and sellers were not made better off by them—we would expect competitive market forces activated by alert entrepreneurs to correct the problems. Shortages and queues would surely not be so common.

NOTES

[1] For a compendium of Becker's major journal contributions through the mid-1990s, see Febrero and Schwartz (1995).
[2] Becker and Becker (1997).
[3] Appropriately, Becker's Nobel Lecture, given on December 9, 1992, was on "The Economic Way of Looking at Life" (1993).
[4] Becker's analytical approach is best summarized in an article he published with his then colleague at the University of Chicago George Stigler (1977).
[5] Becker (1991).
[6] See Kahneman, Knetsch, and Thaler (1986).
[7] See McNicol (1975).
[8] Becker (1991).
[9] See De Vany and Frey (1982).
[10] See De Vany and Frey (1982).
[11] See Barro and Romer (1987).
[12] See Landsburg (1993, p. 13).
[13] See Becker, Ronen, and Sorter (1974); Weiss, Hall, and Dong (1980).
[14] Hall (1991).
[15] For longer discussions of how queues can reduce the cost that firms incur to provide other consumer benefits, see Hall (1991), Saaty (1961), and Schwartz (1975).
[16] Passy (2007).
[17] Granted, we noted earlier in the book that shoppers tend to resent having to pay a premium but look favorably on discounts (with the end result of the discount or premium being the same price paid) (Thaler 1980). If this is the case, then stores can hike their prices across the board, and give a discount to all shoppers who go through checkout counters other than the designated "premium" counter.
[18] Ralph's and Food4Less are often within short distances of one another, which means that residents in different neighborhoods can go to either. However, casual observation of the stores' shoppers reveals that shoppers have self-selected by incomes and opportunity costs.
[19] Haddock and McChesney (1994).
[20] Haddock and McChesney write,

- A demand increase may be temporary. A price rise would diminish the firm's stock of goodwill among loyal customers. Regaining the loyalty of the old clientele will be costly. A priori, the cost of regaining clientele is not necessarily less than the opportunity cost of foregone sales to transitory buyers. The firm must estimate which course is apt to be more profitable. So a firm believing that a demand increase is transitory might quite rationally restrain prices and serve only loyal buyers, thus creating excess demand and potential queues for transitory buyers (p. 567).

[21] Haddock and McChesney (1994, p. 565).

[22] As quoted by Haddock and McChesney (1994, p. 574) from Lohr (1992, pp. C1–2).

[23] Haddock and McChesney (1994, p. 568).

[24] Haddock and McChesney (1994, p. 569).

[25] Haddock and McChesney (1994, p. 568).

[26] Becker (1991, p. 1110).

[27] Leibenstein, Harvey (1950).

[28] Becker (1991, p. 1114).

[29] See Lee and McKenzie (1998).

[30] Becker (1991, p. 1113)

[31] Becker (1991, p. 1113). Becker may actually see his upward sloping demand curve as the locus of price/combinations that are realized as the normal downward sloping demand curve shifts out in response to advertising and market buzz about the product.

[32] See Friedman and White (2007) and Guglielmo (2007).

[33] See Rothkopf and Rech (1987) and Rafaeli, Barron, and Haber (2002). However, not everyone agrees that the wait time for multiple queues will tend to be shorter. See Wolff (1987, Chap. 5)

[34] See Rothkopf and Rech (1987), Averill (1973), Perlmutter and Monty (1979), and Thompson (1999).

[35] See Hall (1991) and Rafaeli, Barron, and Haber (2002).

[36] In one experimental study where the conditions of single and multiple queue structures were tightly controlled via computers, Rafaeli, Barron, and Haber found that while the single queues had longer wait times, the subjects tended to favor the single queues, because the multiple-queue system "produced violations in fairness and variations in time wasted" (2002, p. 134).

[37] Landsburg (2007, p. 125).

[38] Hassin (1985), Naor (1969), and Nalebuff (1988).

[39] Hassin (1985, p. 201).

[40] Nalebuff (1988).

[41] Landsburg (2007, p. 127).

[42] Nafebuff (1988) suggests one way the last-come/first-served rule is used in business. Cashiers often answer the phone and address questions from callers even when they face a line-up of customers. The callers, Nalebuff suggests, effectively break line. But there are obvious differences between a shopper breaking in front of other shoppers in line and callers. Callers can be put on hold or dealt with briefly. That is hardly the case when many grocery shoppers break line with a shopping cart full of groceries.

Also, stores can incur the costs of shoppers whose wait times are increased because of the calls that are taken by clerks. If they allow clerks to take calls as a matter of practice (and allow callers to break line), the stores can expect additional shoppers will use their cell phones to effectively break line. The stores will have to offset shoppers' dissatisfaction from the line-breaking by, say, providing better products at lower prices.

Chapter 13

WHY MEN EARN MORE ON AVERAGE THAN WOMEN—AND ALWAYS WILL*

The subject of this chapter—the relative pay of men and women in the workplace—might seem odd in a book on "pricing puzzles," but it really isn't out of place. To economists, labor is often treated in analyses no differently than other factors of production are treated, as a resource bought and sold under the forces of supply and demand.

Granted, in key regards, labor has marked differences from other resources, say, materials and equipment, used in producing some final good or service (a camera or restaurant meal). Workers can think, can be creative, and can talk back. Workers can also respond more or less to the workplace incentives they face.[1] The key, but not a particularly consequential, difference between the labor workers can provide, on the one hand, and some material or piece of equipment, on the other, is that, labor has a variety of special names for its prices: "pay," "earnings," "wage rate," and "salaries." Moreover, the differences in the prices paid for different types of labor raise a variety of ethical issues, especially when the price, or pay, differences are not perceived to be founded on merit, but rather on group—gender, racial, ethnic, religious—identification. The contention over pay differences seems never to be more heated than over gender-pay differences, perhaps because those differences affect such a large percentage of the population and labor force and seem to be patently unfair.

And it is a widely documented fact that males get paid *on average* significantly more than females in the USA, a persistent labor-market reality. In the USA, the

* Emory University economist Paul Rubin worked with me on an early version of a paper that has worked its way into this book as this chapter. I am indebted to Professor Rubin for allowing me to revise and extend at will our earlier joint effort.

authors of a study for the American Association of University Women found that in 2006 the median weekly earnings of full-time wage and salary women was 81% of the earnings of men, up from 62% in 1979.[2] However, a gender-pay gap cuts across all nations,[3] and industrial, occupational, and ethnic groups.[4] It has also been measured for at least two centuries.[5] Finally, female/male pay gaps have shown up in the laboratory experiments of economists.[6] Even after adjusting for any number and combination of wage-influencing variables, almost all researchers have found a significant gender-based wage gap.[7]

In this chapter, after noting the varied theoretical foundations of a substantial body of empirical (meaning statistical or econometric) research, following the work of Wayne State University Law Professor Kingsley Browne, I offer a substantially different conceptual framework for reconsidering the female/male pay gap and reinterpreting the available findings from empirical studies on the determinants of gender-pay difference.[8] I relate these differences to evolutionary psychological surveys on gender-based mating attractors and on male-against-male competition. This conceptual framework draws out the inevitable links between gender-based mating strategies and competitive labor-market behavior and outcomes. That is say, *if* females and males seek mates based on different attractors—in particular, with females and males giving relatively different weights to the actual and expected labor-market pay and on-the-job-hierarchical, corporate position of prospective mates—then competition in the mating market can be expected to lead to gender differences in the *average* pay in labor markets, everything else held constant.

More specifically, if females as a group tend to place a relatively greater weight on their prospective male mates having "good financial prospects" (as extensive research in evolutionary psychology, reviewed in this chapter, shows they do), then males should be expected to compete with other males on pay and career-path choices to signal females of their relatively greater attractiveness among males. The competition among males for female mates can be expected to drive up males' relative *average* pay.

Additionally, given the mating preferences of many females, many males can be expected to be more risk seeking than will many females be, and this increase in risk taking among males will in modern labor markets lead to higher average pay for males—because risky jobs carry "wage premiums" to compensate workers for the risk taken (because risk necessarily translates into costs workers can expect to incur).[9]

In an array of empirical studies, the (residual) gender-pay gap (left after adjusting for various factors, such as age and education, that can be expected to affect pay) has been attributed by others to gender-based discriminatory forces and restrictions in labor markets (as well as other variables). I certainly don't deny that such discriminatory forces and restrictions exist. Rather, I simply stress that the presumption that the entire pay residual reflects raw labor-market discrimination, conventionally conceived, is likely wrong and that some portion of the pay-gap residual has its foundation in discriminatory forces apart from labor markets, but in mating markets that drive a wedge between the *average* market pay of males and females and can give rise to a pay-gap residual (which may be explained by market forces that, by their nature, can't be measured very well for econometric studies).

To the extent that mating market discriminatory forces are "hard wired" into females' and males' brains (the consequence of evolutionary processes that possibly date to at least the Pleistocene epoch, as many scholars to be cited have argued), a gender-pay gap of some magnitude should be expected to persist.[10] That is to say, don't hold your breath on the average wages of males and females to ever fully align, no matter how many equal-pay-for-comparable-work laws are legislated. Such laws might narrow the wage gap somewhat (because they address discriminatory forces at work in labor markets), but should not be expected to eliminate the gap altogether (because they do not address discriminatory forces at work in mating markets). To eliminate the gap altogether would require a rewiring of male and female brains on sexual/reproductive attractors.

In this chapter, no attempt will be made to deny the merit of conventional explanations—including blatant discrimination—for the widespread persistence of some part of the female/male wage gaps. At the same time, I argue that such explanations very likely do not explain the full extent of the wage gap.

This chapter straddles and integrates theoretical contributions from five established but disparate academic subdisciplines: behavioral biology, evolutionary psychology, labor economics, experimental economics, and empirical studies. The contribution in this chapter comes mainly from showing how widely accepted conceptual points in mate selection theory can be linked to market theory to predict the observed persistence of a female/male wage gap on a global scale. In short, I develop in this chapter a radically different conceptual prism through which a mountain of scholarly empirical literature from several disciplines (encompassing survey studies on gender mate preferences and empirical studies on the determinants of female and male earnings) can be reinterpreted. To appreciate the importance and relevance of the evolutionary perspective for gender-based wage differences, we need to review the various conventional explanations.[11]

CHAPTER 13

CONVENTIONAL EXPLANATIONS FOR GENDER-PAY DIFFERENCES

Gender oppression and workplace discrimination mainly at the behest of politically and economically well-positioned males provide the most widely cited explanations for the average gender-pay gap across all disciplines and among political partisans.[12] However, among economists steeped in competitive market theory, such explanations, by themselves, are problematic. This is because profit-seeking entrepreneurs (women included) should be expected to favor the "underpaid" female workers in their drive to develop a cost and price advantage over their competitors, thus driving up the demand for and wages of female workers and driving down the demand for and wages of male workers. In the absence of market impediments or some force apart from considerations of the relative productivity of female and male workers, the gender-pay gap (that oversteps any difference in productivity of the two sexes) should largely dissipate, at least over time. Granted, some minor pay gap might linger in some markets, due to inevitable imperfections in information flows on wage differences. However, the persistence of a pronounced wage gap for as long as has been recorded can't be explained by the application of standard competitive labor-market theory. The gap shouldn't exist, or persist. Then why does it?

Economists have developed several prominent lines of argument for the persistence of the female/male wage gap. For example, gender (or any other form of) discrimination can be attributable to a group of people (employees, employers, or consumers) having a "taste for discrimination" that, as economist Gary Becker has argued, causes a person to "act as *if he* were willing to forfeit income in order to avoid certain transactions" (emphasis in the original).[13] People's tastes for discrimination can drive a wedge between group wages as surely as a tariff on imported goods can drive a wedge between prices of imported and domestic goods.[14]

Alternately, economists have argued, gender discrimination can come from the development of "dual labor markets" that are separated by class, gender, or presumed skill-level distinctions. Socioeconomic and prejudice barriers dividing the markets can prevent the movement of the sexes (or races) between the two markets, giving rise to differential wage rates that are not undermined by the kind of worker mobility that is presumed in standard competitive labor-market theory.[15]

Gender discrimination has also been attributed to market imperfections that come from the high cost of acquiring information.[16] When workers' true productivity is unknown, employers look for objective but imperfect indicators of productivity of individual workers (education or work experience, for example) based on group association. Males may receive higher pay because gender is seen as a low-cost marker of actual relative productivity of males and females.[17] In short, females may be subjected to so-called "statistical discrimination," a view of the gender-pay gap forged by my University of California-Irvine colleague Dennis Aigner.[18]

Of course, such statistical discrimination can arise because of earlier gender discrimination against females, based on their particular institutional settings, relating to restrictions encountered in acquiring education and work experience.[19] Statistical discrimination can also arise because females, given their expectation of discriminatory pay, underinvest in enhancements to their productivity (education, for example) or just in labor-market "signals" (degrees, for example) that have some prospects of indicating their true labor productivity.[20] In short, gender discrimination can be based on "adaptive behavior," which suggests that because females are guided to low-paying jobs, they adopt attitudes toward work that perpetuate their low economic status and, hence, their relatively lower pay as a group.[21]

Persistent discrimination can be attributed to limited entrepreneurial skills that restrict the entry of new employers into labor markets in which females are paid less than their equally productive male counterparts.[22] Hence, when discrimination in pay emerges, the entry of non-discriminating employers, who can increase the demand for and pay of females, is rapidly curbed by the higher production costs.

The female/male pay gap can be attributed to the relatively greater unresponsiveness of female workers to wage-rate changes, which suggests that employers who are large enough to have some control over labor demand can depress the demand for and pay of female workers more than they suppress the demand for and pay of males.[23] The pay gap can also be due to labor unions that can obtain higher-than-competitive wages. Unions must restrict membership, and their restrictions on female membership can be more severe than on males, because males dominate union policy-making decisions.[24]

If there are cultural or religious norms that hold female wages down, then there would have to be some enforcement mechanism that imposes more costs on employers for breaking with the norms than the profit to be made from hiring females over males. Government-imposed impediments to female labor-market opportunities would, of course, contribute to the persistence of a female/male wage gap primarily because government impediments carry legal enforcement, but such labor-market restrictions would be guided by the tastes for discrimination of dominate voter groups. And governments around the world have either imposed or sanctioned employment impediments on women.[25]

Any measured female/male wage gap is inevitably tainted by workers' voluntary decisions. That is to say, females earn less than males at least in part because of their choices relating to college majors (women tend to major in education, arts, and the humanities), job categories (women tend to dominate secretarial, teaching, and nursing professions), time devoted to homework, and part-versus-full-time work (women tend to dominate part-time job categories, partially because of heavy attendance to household work).[26] Moreover, many occupations with above-average pay require simple brute strength, and women, as a group, have less strength than men, although this source of differences is probably declining in modern societies with technological advances and reliance on mechanical power sources.

A DIFFERENT CONCEPTUAL FRAMEWORK

There is an explanation for the average gender-pay gap (pay residual) that is radically different than the theories reviewed, given that it is founded on mating theory within evolutionary biology and psychology. This theoretical framework posits that females, because they can bear only a very limited number of offspring, have reason to seek and choose males for mating who not only have good genes that will increase the survivability of females' own genes, but who also demonstrate a capability and willingness to provide their mates with support pre- and post-partum, if not at other times.[27] Such support, of course, can increase the survivability of the females' genes.[28] After all, thousands of years ago when babies' brains had grown to its current size, child delivery was difficult and life threatening. It was, in other words, no two-or-three-day stay in a maternity ward. Because babies' brains had grown so large, babies had to be born, in a real sense, prematurely, needing far more nurturing than the offspring of other mammals whose newborns' heads could make it through the birth canal at such a late stage in development that they could be on their feet and running within hours, if not minutes, after birth. As a consequence of the selection bias of females for males willing and able to provide support, males have long had, in order to be selected as sexual partners and mates, reason to compete among themselves to demonstrate their mating fitness to females, which can include a demonstrated ability and willingness to provide females with the support they need and seek. In many generations in the past—say, in the Pleistocene epoch, 10,000 and more years ago, when humans were hunter-gatherers and when engrained behaviors were formed and were passed on to future generations due to their success—males could demonstrate their relative superiority as mates with their hunting skills and successes.

In more modern times, male demonstrations can come in the form of pay and accumulated human capital and financial wealth as well as in the form of signals (degrees and corporate position and movements up business and social hierarchies) that indicate the prospects of providing the required support. I stress that not *all* females need to look for males who show good prospects of being able to provide support for the theory to help explain the gap in the *average* wages of males and females (and the pay residual). If only *some* (even only a sizable minority of) women seek males capable and willing to provide support, then males have an incentive to compete on income and corporate position vectors to increase their odds of being selected by females as sexual partners and long-term mates.[29]

RISKY BEHAVIOR

The second major factor leading to a wage gap is male behavior with respect to risk. Mammalian males universally have a higher variance in number of offspring than do females. In the evolutionary environment of long bygone eras, polygamy was virtually universal. The relative size of human males and females is consistent with a situation in which males averaged two to three mates.[30] This has several implications for human behavior. Here we will concentrate on those that are relevant for wage differences.[31]

Consider some risky activity. It can be a real (physical) risk, as in hunting a lion or driving a car too fast. It might also be a financial or economic risk, as in gambling or in accepting employment in occupations where annual and lifetime earnings can vary considerably across individual workers. If the result of the male's risky activity is a loss, then perhaps the individual will be unable to find a mate and will leave no genes in the gene pool. On the other hand, at least in the evolutionary environment, if the result is success, then the individual might reap considerable rewards that can be attractive to females, which can mean that the male can leave a large number of descendants. Assume for example that the gamble has a 40-60 chance of success, and that a male who accepts the gamble and succeeds has three mates. Then even though the economic payoff for the gamble is negative, the genetic payoff is positive. On average, those of our male ancestors who accepted such gambles would have had more offspring than those of our ancestors who did not accept the gamble. Those of us living today would be disproportionately descended from those males who accepted the risk, and if there is a genetic basis for risky behavior, then those of us living today would have those genes. If the number of offspring for a female is limited by her ability to carry and nurture children, then there would have been no such payoff for females, and so females today would be *relatively* less inclined to accept risks.

The major point is that this increased variance in outcomes can induce many males to take more risks than do many females. The investment difference discussed above might indicate that the mean of the male wage distribution would be greater than the mean of the female distribution. The increased risk accepted by males would mean that the variance would be higher. Even with the same mean, a higher variance would indicate that there would be more high-earning males (and of course more low-earning males) than females. But in modern times (and perhaps in all times) the lower tail of the distribution of wages can be truncated by public policies intent on helping the unfortunate. If nothing else, minimum-wage laws mean that anyone earning less than the minimum becomes unemployed and therefore is eliminated from the wage distribution. There are also many who are unable to work, and there are many more homeless males than females. In other times and places, the lower part of the distribution might simply starve; this may not occur now to the same extent (at least not in wealthy countries), but in any event males have a shorter life expectancy than females. Part of this shorter life expectancy is due to males taking physical risks. Male accident rates are always higher than female rates. This is another aspect of the male risk preference identified above.

Thus, if we begin with male and female earning distributions with the same mean but with males having a larger variance, the workings of the labor market are likely to eliminate the bottom part of the distribution. Then the male *mean* in this truncated distribution will be greater than the female *mean*. Moreover, if, as argued, the male mean is in any way larger, then this difference will be further magnified by risk taking, especially since jobs carrying risks, as noted, tend to carry a wage premium (which compensates workers for the risk costs they incur), which suggests that the male-female earnings gap measured in terms of money wages exaggerates, to some extent, the actual welfare gap between males and females, because no allowance is made for the *risk cost* incurred by males and females as *groups*.

THE LINKAGES BETWEEN MATING AND LABOR MARKETS

My argument for the persistence of the female/male pay gap is founded on a mate selection bias among females that impacts males' competition for females, which drives a wedge between the average wages of females and males. Supporting evidence (but not definitive proof) for this linkage can be found in the survey and experimental work of behavioral biologists and evolutionary psychologists, in the laboratory findings of experimental economists, and in the regression equations of econometricians.

EVIDENCE FROM BEHAVIORAL BIOLOGY AND EVOLUTIONARY PSYCHOLOGY

To test the tenets of mate selection theory, researchers in behavioral/evolutionary biology and evolutionary psychology have turned to a variety of surveys and experiments that have been mainly aimed at determining how females and males separately assess various attributes of "attractiveness" of the opposite sex for friendship, sexual relations, and marriage. The findings from this substantial literature need only be briefly summarized here, given that the literature has been reviewed in detail and at length elsewhere.[32]

Respondents in 33 countries on six continents were asked to rate 18 mate selection characteristics (including attractiveness, financial prospects, ambitiousness, chastity, and industriousness).[33] Across all of the 37 samples taken in the study, males rated their potential female mates' physical attractiveness and youth more highly than did females rate males on such dimensions. Females, on the other hand, rated their mates' potential earning capacity relatively more highly (generally twice as highly) than males in all seven surveys done in the USA and all but one of the surveys done in other parts of the world. One of the 18 attributes was "good financial prospects." In a comparison of 37 surveys, the difference between the ranks of male and female preferences for financial prospects varied. In all societies, females gave a higher ranking to a potential mate's financial prospects than did males.[34]

Other studies have found that the more attractive females are, the better able they are to marry males with "high status, occupational or social."[35] Indeed, female attractiveness tends to be a better predictor of their mates' socioeconomic status than other female characteristics, including their intelligence, education, and premarital economic and social status.[36]

Across all 33 countries surveyed, males prefer to marry, and do marry, females who are, on average, close to three years younger, which supports the view that mate preferences do impact mate selection and suggests that women select spouses based on males' ability and willingness to provide support, since male income and work status are directly related to age, or so it has been argued.[37] Such a mating-age gap also lends support to the proposition that males are selecting on the basis of female fertility.

In trying to establish and retain a relationship with the opposite sex, both males and females consider humor and niceness as being important attractors with equal frequency. However, males tended to deploy tactics involving "resource displays" (flashing money and driving expensive cars) more frequently, while females tended to use "appearance enhancements" (putting on makeup and going on diets) and threats of infidelity (flirting and showing interest in other males) more frequently.[38]

Males who marry in any given year earn about 50% more than the males of the same age who do not marry, suggesting, according to the researcher, that females select mates based on their command over resources.[39]

In a survey of 200 small non-urban societies, researchers found considerable variation over what females and males consider sexually attractive in the opposite sex. However, the researchers also found that, generally speaking, males tended to place far more weight on the physical attractiveness of females than females placed on the physical attractiveness of males, and that "the attractiveness of a man usually depends predominantly upon his skills and prowess rather than on his physical appearance."[40] When the study was redone and expanded to include 300 non-urban, non-Western societies, much the same conclusion was drawn.[41]

In a study of female and male attractions, as revealed in personal advertisements in tabloids, researchers have found that females judge males based on three factors in descending order: (1) sincerity, (2) age, and (3) financial security.[42] Again, males were far more likely to place emphasis on the "good looks" of prospective mates.

In a survey of over 1,100 personal ads in newspapers and magazines, females sought financial resources in a prospective mate eleven times as frequently as males.[43] In a study of female/male dating behavior in a Washington, D.C. dating service, the best predictor of males who were asked out for dates was higher social and economic status, whereas physical attractiveness was the only predictor of females who were asked out on dates by males.[44] Other researchers have found that while females do place some emphasis on good looks, males' high status could compensate for their lack of attractiveness.[45]

At the same time, it needs to be noted that, according to one study of the importance of "beauty in the workplace," good-looking men received an average wage premium over their not-so-good-looking male workers with comparable jobs of about 10%, according to University of Texas economist Daniel Hammermesh and his collaborators on a number of studies on the impact of "beauty" on workplace pay and success.[46] Goods looks among female workers apparently provides no wage premium (perhaps because good looks in women can be interpreted as undercutting, wrongly or rightly, their workplace smarts and productivity).

Given the wage premium good-looking males receive, females may be attracted to good-looking males partially because of the importance of their physical attributes. Females may also see good looks as a reasonably good signal of males' career success and support capability, which may be a more important source of attraction.

One research team had pictures taken of male models dressed in three status-revealing costumes: (1) high status (a physician dressed in a designer blazer and a Rolex watch), (2) medium status (a high school teacher dressed in a white shirt), and (3) low status (a waiter-in-training dressed in a Burger King uniform).[47] The researchers found that females were far more willing to engage in various levels of a relationship with the high status males (even when the males were rated as unattractive) than with the low and medium status males. In another similar study, other researchers found that high-status dress always increased females' attraction to males.[48]

Yet another set of researchers videotaped actors and actresses playing "high dominance" and "low dominance" roles in separate tapes.[49] Females gave the actors higher ratings on dating interest and sexual attractiveness in their high dominance roles than in their low dominance roles, suggesting that females are attracted to male dominance displayed. The males' ratings of the females were not affected by the roles the females played.

Females' preferences for male mates taller than they are[50] and their rating of taller males as being more attractive may be explained by the fact that male height can equate with male dominance, and with higher income, given that taller males tend to earn more and are more likely to be promoted than shorter males, with taller males receiving a wage premium averaging 10–15% more than their shorter counterparts.[51] Taller males are also more frequently sought after by females in their personal advertisements,[52] receive more responses to their personal advertisements for dates,[53] and tend to have more attractive women friends.[54]

Another research team asked male and female college students to indicate, among other factors, the "minimum percentiles" in income any prospective spouse must have to be acceptable for marriage.[55] Females indicated an average minimum acceptable percentile of 70. Males indicated an average minimum acceptable percentile of 40.

Male medical students report that their education has the effect of increasing the number of their prospective sexual and marital partners, causing the male medical students to "seek and enjoy more transitory relationships."[56] These findings and many others, especially his own international comparisons of mate attractors, caused evolutionary psychologist David Buss to conclude,

> Woman across all continents, all political systems (including socialism and communism), all racial groups, all religious groups, and all systems of mating (from intense polygamy to presumptive monogamy) place more value than men on good financial prospects. Overall, women value financial resources about 100% more than men do, or roughly twice as much.[57]

EVIDENCE FROM EXPERIMENTAL ECONOMICS

Experimental economists have only since the 1990s turned their attention to developing laboratory experiments that provide suggestive evidence on the extent and causes of the female/male pay gap observed in the workplace across many societies, ethnic groups, and occupations. The experimental evidence on gender-pay differences is, accordingly, far more limited than the mountain of econometric evidence (to be covered below). Nonetheless, the evidence generally supports this chapter's central thesis, which is that the gender-pay gap could be tied to basic biological forces driving mate selection that, in turn, affect the relative competitiveness of females and males in labor markets.

In a laboratory version of the "battle of the sexes game," one researcher asked 300 paired subjects from the USA and Sweden, all of whom knew the gender of their co-players (but nothing else about them), to divide $100 (between themselves and their co-players, all without the pairs being able to communicate with each other).[58] In the trials that were the focus of the investigation, the subjects were limited to two possible divisions of the $100:

(1) the "hawkish strategy": $60 for the subject making the division and $40 for the co-player and

(2) the "dovish strategy": $40 for the subject and $60 for the co-player.[59] The subjects would be rewarded only if their co-players' division made the sum for each player equal to $100.

The study's author also found that the females had an average "experimental earnings" that was 78% of the males' average[60] and that in the USA and Sweden, both females and males tended to take the division favoring themselves ($60/40) with far greater frequency when they knew they were playing against females.[61]

Other researchers had 60 subjects (recruited university students), evenly divided between females and males, work their way through computer-based mazes with different levels of difficulty within a restricted (fifteen-minute) time period.[62] In noncompetitive trials—that is, when the subjects were paid an appearance fee plus two shekels for each maze they solved without regard to how many mazes other subjects solved—there was no statistically significant difference in the count of mazes solved by males and females (although in the noncompetitive piece-rate trials females solved 87% as many mazes as did the males).[63] However, when the subjects were put in mixed-gender "tournaments" (in which all subjects were paid an appearance fee, but in which only the subject with the most mazes solved would be paid twelve shekels per maze solved), a *significant* pay gap emerged, with females solving on average 72% as many mazes as did males. However, as the researchers note, "[T]he increase in the gender gap in performance between the noncompetitive and the competitive treatment is driven by an increase of the performance of men and basically no change in the performance of women."[64] Additionally, the authors found that the average performance of males in mixed gender tournaments was, statistically speaking, much the same as it was in male-only tournaments, whereas females' average performance was elevated significantly in the shift from mixed-gender tournaments to female-only tournaments.[65] This means that the gender-pay gap narrowed in the shift from mixed-gender trials to single-sex trials, a finding that caused the authors to speculate that there may be some truth to the often heard view that females perform relatively better in female-only educational environments.[66]

ECONOMETRICS EVIDENCE

Economists generally agree that comparisons of the *mean* earnings of female and male workers is not very instructive, although such mean comparisons are frequently cited by policy partisans and are widely reported in the press.[67] This is because mean earnings differentials of females and males can be explained by a host of factors, including differences between female and male workers in terms of their ages, the number of hours worked and at what times of the day, education, job preferences and category, work-related risks assumed, work experience, and the continuity of work experiences.[68] When economists have used regression analysis (sophisticated statistical analysis that allows for a separation of the effects of various variables on male and female earnings), they have found that some of the female/male wage gap is indeed attributable to such factors in varying degrees—with the variance related to the exact nature of the dependent variable (the exact measure of earnings, for example) used, the group of workers studied, the exact combination of independent work-related variables (age, education, and hours worked, for example) used in the regression analysis, and the years covered by the study.[69] However, almost all econometric studies on the wage gap have found that after adjusting for different combinations of independent variables, males still earn significantly more than females, with the male wage premium (after adjusting for other variables) ranging from 7% to 61% of the female wage.[70] Consider the following sample of findings from two econometric studies (one old and one relatively recent) that mirror findings of many other studies (several of which are briefly reviewed in a long endnote to this chapter[71]):

Using mean census data for females and males 25 and older with incomes for 1959, economists James Gwartney and Richard Stroup found the ratio of female mean earnings to be 33% of males.[72] When adjusted for several variables, females' earnings were 39% that of males. Practically the same results were obtained for 1969 data. When they restricted their data to full-time, full-year workers in 1959, females earned on average 56% of males before adjustments and 58% after adjustments.

The General Accounting Office (GAO) evaluated the female/male pay gap by using the Panel Study of Income Dynamics, a nationally represented longitudinal data set, which included a variety of work-related, family, and demographic variables.[73] This means the GAO was able to track the work histories of males and females ages 25 to 65 during the 1983–2000 period. The GAO found that on average when compared with males, females in the study sample had 75% as many years of work experience, worked 78% as many hours per year, were 76% as likely to work full time, and had three times the number of weeks out of the labor force per year. After controlling for a variety of variables (including industry, occupation, race, marital status, and job tenure), the GAO found that on average females earned annually 80% of what males earned.

The male wage premium, which is that portion of the female/male wage gap that is not explained by the independent variables in the econometric equations, has often been attributed to rank discrimination against females, or the mistaken prejudice on the part of employers that females are not as productive as males. Such prejudices, the argument goes, have tempered the relative demand for female workers and have led to social, cultural, and legal restrictions on the ability of females to earn as much as their male counterparts.[74] The persistence of a male wage premium has also been explained by the relatively greater "psychic benefits" that employers (male and female) receive from hiring predominantly (or only) males or that customers (male and female) receive from buying predominantly from males.[75]

Others have countered that the male wage premium might not be as great as the econometrics studies suggest. Then again, it might be greater than the studies indicate. This is the case because econometric studies do not include (because they cannot measure adequately) important determinants of absolute and relative worker wages, for example, the "quality" of work done, work intensity, the entertainment value of some work, job-related "risks" assumed, and "dedication" to jobs of different worker groups.[76]

EXPLAINING THE NARROWING PAY GAP

However measured, the female/male pay gap has narrowed significantly over the past four decades. As noted earlier, female workers on average earned 59% of what males earned in 1962. In 2002, female workers earned 77% of what male workers did. As noted at the start of the chapter, in another study for the American Association of University Women conducted in 2006 the pay gap between the median weekly earnings of full-time male and female workers narrowed by 19% points between 1979 and 2006.[77] A part of the explanation can be attributed to the stagnation in the real average wages of males during the last three decades of the twentieth century. During that period, females' average real wages continued to rise.[78] However, economists have found that the wage gap has narrowed for several other reasons, including the following:

While there is survey evidence that occupational crowding resulting from occupational segregation is linked to differences in female/male work preferences,[79] the occupational segregation of the sexes began to decline in the 1970s.[80]

Self employment among females began rising substantially in the mid-1970s, hiking females' measured relative incomes (which are not adjusted for the additional risk taking).[81]

The absolute and relative rise during the last quarter of the twentieth century in the economic rewards to workers with higher skills and greater education disproportionately benefited females as females acquired relatively more skills and education. Female work experience also improved.[82]

Females have shifted toward higher paying occupational categories,[83] perhaps reflecting a decrease in the gender differences in college majors[84] and the relative rise in females' math SAT scores.[85]

The GAO found in regression analysis that the number of children variable was associated with a 2.1% *increase* in average annual earnings of males and a 2.5

decrease in the average annual earnings of females.[86] This suggests that the drop in the birth rate from 15.6 to 14.4 per 1,000 in the population during the 1983–2000 study period could have narrowed, albeit slightly, the female/male pay gap by increasing the labor force participation in full-time jobs of females and the annual earnings of females and lowering the average earnings of males.

A labor economist found that a 1-year delay in the first childbirth from age 24 to 25 can increase a woman's career earnings by 10%, partly because the childbirth delay can result in an increase in the number of hours worked over a career by 5%.[87] Hence, an as-yet-undetermined portion of the narrowing of the male/female pay gap can be attributed to the ongoing rise in the average age at which women have their first births by 3 years, from an average age of first births at 21.4 years in 1970 to almost 25 years in 2000, according to researchers at the National Center for Health Statistics.[88] The delay in births may be related to the spreading use of contraceptives and abortions as methods of managing births and to the escalating costs of child rearing (attributable in part to the relative rise in women's wages), but it may also be related to the fact that young married couples now place relatively less importance on children for a satisfying marriage.[89] This is to say, some portion of the narrowing male/female pay gap may be explained by the emergence and spread of birth-control technologies and to the rise in the relative costs of children vis-à-vis other things couples can do with their money (buy boats or take vacations).

The evidence of the impact of equal-opportunity/equal-pay laws on the female/male pay gap is mixed. One labor-market researcher found that during the 1967–1974 period the Equal Employment Opportunity Legislation (or the Civil Rights Act of 1964) increased the real earnings of female workers by nearly 5% (1976) and lowered the female/male pay gap by nearly 10% and 14% in the 1967–1974 and the 1968–1975 periods, respectively.[90] However, other researchers found that in the 1960s and 1970s the impact of equal opportunity laws was quantitatively small and statistically insignificant.[91]

All in all, the economic rationale for "statistical discrimination" against females may have diminished as females improved their job skills, increased their work experiences, delayed their first childbirths, and increased their commitment to their jobs. Of course, with a reduction in statistical (or any other form of) discrimination against female workers, they could have had greater incentives to acquire more education, shift their college majors, change their occupational goals, and dedicate themselves to their careers.[92]

In citing the econometric literature on the female/male wage gap, I do not mean to settle the debate over exactly what is the mean male wage premium due to pure discrimination or other considerations. Frankly, that debate will not likely ever be settled because, as acknowledged by the GAO, there is no consensus among researchers on the magnitude of the premium, and because economists are unable to impose laboratory-type controls on their investigations of real world labor markets.[93] Rather, my purpose here is to point out that the preponderance of the econometric evidence leads to a substantive conclusion for both economic and biological sciences: *Females tend to earn significantly less than their male counterparts in the workplace.* This tendency has been observed on a worldwide scale and has been persistent in spite of major political and legal efforts to eliminate discrimination.

THE FEMALE/MALE WAGE GAP: HARD WIRED OR CULTURAL?

An overriding the thesis of this chapter is straightforward: A pay gap should be expected between female and male workers because females look to males for support of themselves and their children and because males have a higher variance of earnings than do females (which can reflect greater risk taking on the parts of males who can receive a wage premium for the risks they take). A gap in the socioeconomic status of females and males should also be expected, given that males' socioeconomic status can be a strong signal of their ability to provide support. The pay and status gaps should be expected because males' pay and status can make them more attractive to females, thus engendering a level and intensity of competition among males for pay and status that females do not need to seek (at least not with the same intensity). The studies reviewed suggest that males' ability to offer support and socioeconomic status to females influences females' assessments of males' relative attractiveness.

However, the evidence does not support the inference that none of the residual female/male pay difference, as found in econometric studies, can be attributable to sexual discrimination founded on social and cultural norms, ignorance, and chauvinism. The evidence and conceptual arguments made in this chapter suggests that the female/male pay gap has two constituent components. One component of the gap can be attributed to males' competition among themselves for the attention of females, an innate (conscious or subconscious) drive built into their psychological makeup. The other component can be attributed to discriminatory labor-market restrictions or prejudices against females. Exactly what portion of the actual residual pay gap is attributable to each component is not known and, perhaps, cannot be known.

CHAPTER 13

A SUMMARY ASSESSMENT

The evidence reviewed is, admittedly, suggestive but ambiguous. Understandably, some researchers have argued that females look to males' pay and socioeconomic status because of extant restrictions on females' incomes and socioeconomic status, a line of argument that suggests that females' should be expected to be less concerned with their mates' socioeconomic status as they gain income and status.[94] Interestingly, the evidence, albeit limited, from separate studies of female college students, feminist leaders, and female medical students suggests that many females become more concerned with the pay and socioeconomic status of their prospective mates as they, the females, gain status and income due to the abatement of labor-market discriminatory practices.[95] However, such evidence does not deny the basic point being made, that restrictions on females' income and socioeconomic status can be a reason many females seek males who can enhance the females' living standard and socioeconomic status over and above what it would otherwise be.

The abatement, if not elimination, of labor-market restrictions on the pay and status of females, regardless of how they are imposed, could still dampen females' demand for males who can be supportive. The pay gap could thus be narrowed in two ways with the abatement of labor-market restrictions:

First, females' relative wages could rise, making females inclined to seek males with higher pay and socioeconomic status but still with less of a pay and status premium.

Second, with females' interest in male pay and socioeconomic status dampened, although still prevalent, males may no longer be driven *to the same extent* to compete among themselves in terms of relative pay and status, thus causing their *relative* (not absolute) pay to fall. That is to say, females who earn more may seek men who earn even more. However, it does not follow that the female/male pay and status gap will not narrow.

What does seem suggestive by both theory and available evidence on mate selection regarding what females seek in male mates is that the female/male pay and status gaps can be narrowed by policy changes but will not ever be completely eliminated. If a complete closure of the gaps could be imagined, for purposes of argument, many males would still be driven to earn more than other males in order to improve their chances of attracting females. Given males' heavy emphasis on the physical attractiveness of females and not on females' income earning capability, females' competitiveness should be diverted toward enhancing their relative physical appearance. Hence, female/male pay and status gaps can open up in three key ways:

First, male competition for demonstrated evidence of their relative ability to support females can drive up male pay and socioeconomic status above the pay and socioeconomic status of females.

Second, given male attention to female appearance, females can relax to the extent to which they compete for pay and socioeconomic status because females can enhance their living standard by tapping the support (inflated by male competitiveness) males are willing to provide to females and their children. In short, male competitiveness for female attention can be treated by females as a "wealth effect" that suggests that females do not have to earn the same income or achieve the same socioeconomic status to achieve any given wealth level to which females and males may equally aspire.

Third, it is possible (although hardly certain) that females may not be inclined to allocate the same time and resources to the development of their skills to raise their pay and socioeconomic status. To the extent that females divert their competitiveness to catch the attention of males by doing what is necessary to be more attractive physically, then females must divert (marginally, if not significantly) time away from elevating their pay and acquiring socioeconomic status. Females, however, may see the diversion of their resources into appearance competition as productive, given that, as the evidence suggests, more physically attractive females can attract males with higher pay and socioeconomic status. That is, all other considerations equal, females and males may be driven at some fundamental level to seek more or less the same life-cycle living standard. Females and males, as distinct groups, simply tend to use different means to accomplish the same ends.[96]

A complete closure of the gaps would suggest that females would no longer be driven to prefer mates who are relatively more supportive of themselves and their children, which suggests that a basic tenet of mate selection theory—that attraction is "hard wired" into females' psychological makeup from their days as hunter-gatherers during the Pleistocene epoch—is misleading, or altogether wrong. If female/male attraction is truly "hard wired," then females' drive for male support should be left largely (if not totally) unaffected by transient or short-term shifts in social and cultural norms or even in the abandonment of restrictions on female pay and the array of employment opportunities. In other words, there should be some minimum pay gap toward which the prevailing wage gap can be driven, but the gap cannot be completely closed—at least not in the short or intermediate term (say, a few generations).

All of this research suggests that if mechanisms (jobs, education, or insurance) for support of females, other than male mates, emerge and endure for a long stretch of time, then, presumably, females would not have to look to males for their support and the support of their children. Males would then not be so inclined to compete based on their pay and socioeconomic status (because such behavior might not be as attractive to females), which suggests that the female/male pay gap should be expected to narrow (or perhaps close, depending on how long the support mechanism lasts).

If divorce is facilitated, females might become, eventually, less inclined to judge males based on their pay or socioeconomic status and more inclined to judge males based on their ability to make binding marriage commitments (thus increasing the expected value of more limited support levels). With divorce made easier, males can be expected to divert resources from increasing their pay and higher socioeconomic status to developing a reputation for credible commitments, the net result of which can be a reduction in the female/male wage gap.

The basic thesis of this chapter is that male competitiveness for attracting females is a biological force with heretofore-unrecognized market consequences. It would seem to follow that with a relative rise in the ratio of males to females in the population (a consequence, for example, of a cultural bias that results in, say, female fetuses being aborted more frequently than male fetuses, as has occurred in China over the past several decades[97]) males would have to compete more aggressively for the attention of females.[98] The result could be a rise in the pay gap for two reasons already noted in another context.

First, the competition among males could drive their absolute wages up.

Second, females could relax in seeking higher wages, given the wealth effect of the heightened competition among males to earn higher pay and provide support.

CONCLUDING COMMENTS

In concluding sections of chapters, it is very tempting, and common, for authors to assert considerable confidence in the theoretical and empirical components of their studies. I cannot do that here. However, I can say that the broad range of research studies reviewed is strongly suggestive of two conclusions, one weak and one stronger.

The weaker of the two conclusions is that the literature reviewed suggests that there could very well be a biological foundation to the gender-pay gap that is not widely appreciated. Enthusiasm for that conclusion must necessarily be tempered by the fact that the gender-pay gap can be driving differential mating preferences between the sexes.

The stronger conclusion that can be drawn from this literature review is that commonly made claims that *all* of the average gender-pay gap (or pay-gap residual) has its foundations in narrowly defined labor-market discrimination is clearly suspect, given the variety of findings from econometric, experimental, and mating-market studies.

Accordingly, I repeat a point made at the start of this chapter, those wanting full equality in pay for comparable work should not hold their breaths, a point I make in spite of my fondest wish that a more politically correct assessment of future trends in the relative wages of males and females could be pressed.

NOTES

[1] Dwight Lee and I have taken up many of these issues in a book on *Managing Through Incentives* (McKenzie and Lee 1998).

[2] See Dey and Hill (2007, p. 6)

[3] See Blau and Kahn (1997); Anker and Hein (1986); Psacharopoulos and Tzannatos (1992); DeNavas-Walt, Cleveland, and Webster (2003); and Rose and Hartmann (2004).

[4] Caiazza, Shaw, and Werschkul (2003); and Bureau of Labor Statistics (2003).

[5] Goldin (1989).

[6] Holm (2004); and Gneezy, Niederle, and Rustichini (2003).

[7] For reviews of the economic literature on labor-market wage differences, see Alexis (1974), Marshall (1975), Cain (1986), and Gunderson (1989).

[8] See Browne (1995, 1998, and 2002). My own work on linking labor-market outcomes to mate selection, done in separate papers with Paul Rubin (which has been converted to this chapter) and with Steven Frank (Frank and McKenzie 2006) was done before I learned of the Kingsley Browne pioneering work.

[9] Browne (1995 and 1998).

[10] As noted in Chap. 11, Jerome Barkow, Leda Cosmides, and John Tooby explain why certain modern behaviors can be expected to be "hard wired":

What we think of as all human history—from, say, the rise of the Shang, Minoan, Egyptian, Indian, and Sumerian civilizations—and everything we take for granted as normal parts of life—agriculture, pastoralism, governments, police, sanitation, medical care, education, armies, transportation, and so on—are all the novel products of the last few thousand years. In contrast to this, our ancestors spent the last two million years as Pleistocene hunter-gatherers, and, of course, several hundred thousand years before that as one kind of forager or another. These relative spans are important because they establish which set of environments and conditions defined the adaptive problems the mind was shaped to cope with: Pleistocene conditions, rather than modern conditions. This conclusion stems from the fact that the evolution of complex design is a slow process when contrasted with historical time. Complex, functionally integrated designs like the vertebrate eye are built up slowly, change by change, subject to the constraint that each new design feature must solve a problem that affects reproduction better than the previous design. The few thousand years since the scattered appearance of agriculture is only a small stretch in evolutionary terms, less than 1% of the two million years our ancestors spent as Pleistocene hunter-gathers. For this reason, it is unlikely that new complex designs—ones requiring the coordinated assembly of many novel, functionally integrated features—could evolve in so few generations...Moreover, the available evidence strongly supports this view of a single, universal panhuman design, stemming from our long-enduring existence as hunter-gatherers (1992, p. 5).

[11] For a more extensive review of the evolutionary biology/psychology literature relating to pay differences, see Browne (2002).

[12] Cain (1986), Blau (1984), and Reskin and Hartmann (1986).

[13] Becker (1971, p. 14). While Becker stays with the standard assumption in microeconomics that individuals as consumers and workers operate independently, Krueger (1963) and Alexis (1974) assume that group preferences for a "taste for discrimination" are somehow determined jointly, or collectively, the net effect of which is for the powerful group to suppress the wages of the disfavored group.

[14] Becker (1971), Krueger (1963). A tariff on an imported good can be to drive up the price consumers have to pay for the imported good, but the price of the imported good will not rise by as much as the tariff, which means that the after-tariff price the importer receives for the product decreases. On the other hand, the higher price consumers pay for the imported good can cause the demand to shift to the domestic good, thus raising the price domestic producers can charge, which is one important reason domestic producers favor tariffs on their imported competitors.

[15] Piore (1970) and Bergman (1971). For an array of studies on barriers to the advancement of women at work, see Tinker (1990).

[16] Phelps (1972), Arrow (1972 and 1973), McCall (1972), Thurow (1975), and Aigner and Cain (1977).

[17] Aigner and Cain (1977).

[18] From this perspective, the prospects of females' withdrawal from the labor force for childbirth can cause females to be paid less than males. Use of the prospects of childbirth as an observed indicator of commitment to full-time work and, hence, productivity, can cause females who plan to have children to be overpaid while females who do not plan to have children to be underpaid (Thurow 1975, p. 178).

[19] Marshall (1974).

[20] Spence (1973).

[21] Piore (1970), Arrow (1973).

[22] Becker (1971, pp. 44–45).

[23] Madden (1973).

[24] Gould (1977) and Hill (1977).

[25] Becker (1971, pp. 81–83).

[26] O'Neill (1983a, pp. 19 and 22; 1983b and 1984); and Gunderson (1989).

[27] Of course, another strategy is for females to seek good genes in their mates and deceive other males into providing support for the children. Male jealousy has evolved to control this behavior. This can lead to "mate guarding" and isolation of women. This extreme form of jealousy by no means occurs in modern western societies.

[28] Dawkins (1976).

[29] Frank and McKenzie (2005) have provided a simple mathematical model of how differences in mating attractors imply a gender-pay gap, and vice versa, in a world of constrained resources.

[30] Wilson (1978, p. 20).

[31] For a more discussion of the implications of polygyny for political behavior, see a book by Paul Rubin on *Darwinian Politics* (2002).

[32] Buss (1985, 1992, and 2003); Buss and Barnes (1986); and Buss, et al. (1990).

[33] Buss (1989 and 1990).

[34] The male–female difference in ranks for financial prospects varied between 1 and 7, with an average of 3.7 (Buss 1989, Buss et al. 1990). Another psychological study provided data on changes in mate preferences over time in the USA (Buss 2003). In 1939, males ranked financial prospects of mates 17 and females ranked this attribute 13 among 18 different characteristics. In 1996, the ranks were 13 for males and 11 for females. Both sexes have given increased weight to financial prospects, but the difference between the sexes has narrowed. This pattern suggests that mating preferences have adjusted to changing financial opportunities for males and females (Buss 2003).

[35] Elder (1969), Taylor and Glenn (1976), Udry and Eckland (1984).

[36] Elder (1969).

[37] Buss (1989).

[38] Buss (1988a).

[39] Trivers (1985).

[40] Ford and Beach (1951, p. 94).

[41] Gregersen (1982).

[42] Harrison and Saeed (1977).

[43] Weidenbaum [see p. 24 in Buss].

[44] Green, Buchanan, and Heuer (1984).
[45] Townsend and Levy (1990a).
[46] See Hammermesh and Biddle (1994) and Pfann, et al. (2000).
[47] Townsend and Levy (1990b).
[48] Hill, Nocks, and Gardner (1987).
[49] Sadalla, Kenrick, and Vershure (1987).
[50] Beigel (1954), Gillis and Avis (1980)
[51] Gillis (1982).
[52] Cameron, Oskamp, and Sparks (1978).
[53] Lyn and Shurgot (1984).
[54] Feingold (1982).
[55] Kendrick, Sadalla, Groth, and Trost (1990).
[56] Townsend (1987, p. 440).
[57] Buss (2003, p. 25).
[58] Holm (2004).
[59] In trials not reported here, Holm gave the subjects a third option, $50 for both the subject and his/her co-player.
[60] Holm (2004, p. 18).
[61] When playing against other females, Swedish females took the hawkish strategy with a frequency that was 89% greater than when playing against males (66.7% versus 35.3%). When playing against females, American females took the hawkish strategy with a frequency that was 131% greater than when playing against males. The frequency with which the hawkish strategy was taken by males in both Sweden and the USA was less affected by the gender of their opponents (Holm 2004; p. 8, table 1).
[62] Gneezy, Niederle, and Rustichini (2003).
[63] Gneezy, Niederle, and Rustichini (2003, p. 1057).
[64] Gneezy, Niederle, and Rustichini (2003, p. 1057).
[65] Gneezy, Niederle, and Rustichini (2003, p. 1061).
[66] Gneezy, Niederle, and Rustichini (2003, p. 1072).
[67] See Rose and Hartman (2004). For a sample of the widespread press coverage of the Rose and Hartman study in mid-2004, see Bernstein (2004) and Madrick (2004). For press coverage of the female/male pay gap by occupation developed by the U.S. Census Bureau (Weinberg 2004), see Nyhan (2004).
[68] In their review of only twenty of the numerous female/male wage-gap econometric studies, Treiman and Hartmann (1981, pp. 20–37) found 35 identifiable categories of independent variables used to explain the wage gap. See also Cain (1986, pp. 750–752).

[69] See Cain (1986, pp. 750–752); Triemann and Hartmann (1981, pp. 20–37); Blau and Kahn (1997). Treiman and Hartmann (1981) found that from the array of econometric studies, several general conclusions can be drawn, among which are the following: (1) increasing the number of variables included in the regression equations reduces the adjusted female/male pay gap; (2) the inclusion of variables for occupation and establishment reduces the pay gap; (3) the inclusion of household responsibilities reduces the pay gap; (4) the pay gap is smaller in the public sector than in the private sector; and (5) including a variable for labor-market experience, and continuity of experience, tends to reduce the pay gap.

[70] See Treimann and Hartmann (1981, pp. 20–37); Cain (1986, pp. 750–752). The econometric studies cited use some combination of the following explanatory variables: education, age, race, mental ability, formal training, actual labor-market experience, proxy for labor-market experience, marital status, health, hours of work, length of service with current employer, size of city of residence, region of residence, various measures of parental background, quality of school absenteeism record, number of children, plans to stop work for reasons other than training, urban/rural, turnover, occupation (census one and three-digit), occupational prestige, Duncan scale of a socioeconomic index, industry, union membership, type of employer (government/private, sex segregated/integrated, size of work force), supervisory status, percentage female in work force, median income of male incumbents, local labor-market conditions, length of trip to work, veteran status, and migration status (Cain 1986, p. 752). Only one study (Malkiel and Malkiel 1973) was able to get the adjusted pay gap down to 1%, and then only by looking at salaries in a single company and by switching from adjusting the female mean salaries (the usual method employed by econometricians) to adjusting the male mean salaries. (As indicated above, when the female mean salary was adjusted, the pay gap was 14%.)

[71] Consider these additional findings of a male/female wage gap that persists after statistical adjustments have been made for different collections of worker traits that might affect their wages rates:

Featherman and Hauser (1976) found in their assessment of the annual earnings of married workers in 1973, that females earned 38% of males unadjusted and 48% adjusted (with a variable for occupation included in the regression analysis, which is important to note because job choice can itself be the product of gender discrimination).

- Sawhill (1973), using annual earnings from the Current Population Survey (CPS) data for workers 14 and older, found that in 1966 females earned 46% of males unadjusted and 56% of males adjusted.

- Suter and Miller (1973) also used CPS data for 1966, but restricted their analysis to workers 30 to 44 years old. They found that females earned 39% of males unadjusted and 62% adjusted (including a variable for occupation).
- Using annual earnings data from the General Social Survey, Roos (1981) found that for 1974 through 1977, females 25 to 64 earned 46% of males unadjusted and 63% adjusted (including a variable for occupation).
- Fuchs (1971), using a sample of census hourly wage-rate data for 1959, determined that females earned 60% of males unadjusted and 66% adjusted.
- Using annual earnings data from the National Longitudinal Survey for 1966, Treiman and Terrell (1975) found that white females earned 42% of their male counterparts unadjusted and 67% adjusted.
- Focusing on the annual salaries of professional full-time workers in one company for 1966 and 1969–1971, Malkiel and Malkiel (1973) found that females earned 66% of males unadjusted and 77% adjusted (including a variable for occupation).
- In a survey of college faculty's annual salaries, Astin and Bayer (1972) found that female faculty members earned 78% of their male counterparts unadjusted and 87% adjusted (including a variable for occupation).
- In another survey of college faculty's annual salaries, Johnson and Stafford (1974) found that while female faculty members in six disciplines (anthropology, biology, economics, mathematics, physics, and sociology) start their careers at salaries close to their male colleagues, they earn only 85% of their male counterparts after 15 years (with the authors attributing the salary differential largely to females' career interruptions and to their having jobs at institutions emphasizing teaching, not research).

[72] Gwartney and Stroup (1973).
[73] General Accounting Office (2003).
[74] Reskin and Hartmann (1986); Blau, Ferber, and Winkler (1998).
[75] Becker (1971.)
[76] Fuchs (1988, Chap. 3); Gunderson (1989); Heckman (1980); General Accounting Office (2003).
[77] Dey and Hill (2007, p. 6).
[78] Blau (1998, pp. 129–132).
[79] Daymount and Andrisani (1984, p. 412).
[80] Beller (1985); Jacobsen (1994); and Blau, Simpson, and Anderson (1997).
[81] Devine (1994a).
[82] Goldin (1989); Blau and Kahn (1997).
[83] O'Neill and Polachek (1993), Wellington (1993).
[84] Brown and Corcoran (1997).

[85] Blau (1998, p. 138).

[86] General Accounting Office (2003, p. 32 and table 2).

[87] Miller (2005).

[88] Mathew and Hamilton (2002).

[89] The Pew Research Center reported that in 1990, 65 percent of surveyed couples said having children was "very important" to a good marriage. By 2007, only 41 percent said having children was "very important" to a good marriage. Having children ranked eighth in young couples assessed attributes of a good marriage, behind chore-sharing, good housing, adequate income, happy sexual relationship, and faithfulness (Taylor, Funk, and Clark, 2007).

[90] Beller (1979 and 1980).

[91] Leonard (1984) and Oaxaca (1977).

[92] Blau (1998, pp. 138–140). It should be noted that the narrowing of the overall female/male wage gap does not imply an across-the-board narrowing of the wage gap for all age cohorts. Weinberger and Kuhn (2004) have found that the wage gap for older female workers with college degrees did not narrow during the 1989–1999 period. The decline in the *overall* pay gap among college-educated women appears to have been driven by the fact that "younger cohorts of women face a smaller disadvantage, relative to men, than did older cohorts of women at the same age" (Weinberger and Kuhn 2004, p. 2).

[93] General Accounting Office (2003).

[94] Coombs and Kenkel (1966); Hill, Rubin, and Peplau (1979); Dion (1981); Rosenblatt (1974); Murstein (1980).

[95] Wiederman and Allgeier (1992); Freedman (1979); Townsend (1987 and 1989).

[96] What actually happens to relative female/male pay depends upon the relative amount of time that each sex diverts from earning an income and achieving socioeconomic status toward looking "attractive." Females may divert more time and resources toward improving their physical appearance and sexuality. However, males may divert more time seeking to give the (mistaken) impression of relatively higher pay and socioeconomic status by, say, buying expensive suits and cars and by honing their skills at office politics that might improve their short-term chances of promotion.

[97] Hudson and den Boer (2004).

[98] Hudson and den Boer (2004) and Rubin (2002) indicate that this demographic situation will also have political ramifications.

~

BIBLIOGRAPHY

Abrahamson, Eric and David H. Freedman. 2007. *A Perfect Mess: How Crammed Closets, Cluttered Offices, and On-The-Fly Planning Make the World a Better Place.* New York: Little, Brown, and Company.

Adams, Cecil. 1992. Why do prices end in 99? *The Straight Dope*, February 21, accessed on February 20, 2006 at http: //www.straightdope.com/classics/a3_166.html

Aguiar, Mark and Erik Hurst. 2004. Consumption vs. Expenditures. New York: National Bureau for Economic Research, research paper 10307, accessed on March 15, 2007 at http: //www.nber.org/papers/W10307.pdf

Aguirregabiria, Victor. 1999. The dynamics of markups and inventories in retailing firms. *Review of Economic Studies* 66: 275–308.

Aigner, Dennis J. and Glen G. Cain. 1977. Statistical theories of discrimination in the labor market. *Industrial and Labor Relations Review* 30: 175–187, accessed on June 25, 2004 at http: //pcift.chadwyck.com/pcift/htxview?template=basic.htx&content=ft_frame.htx&action=ShowImage&body=gif&KW=&SFT=N&Page=2&AID=3121-1977-030-02-000004&JID=&from=toc&CONTROL=ON&OTINDEX

Akerlof, George A. 1970. The market for "lemons": Quality uncertainty and the market mechanism. *Quarterly Journal of Economics* 84 (August, 3): 488–500.

———. 1991. Procrastination and obedience. *American Economic Review* 81 (2): 1–19.

Alexis, Marcus. 1974. The political economy of labor market discrimination: Synthesis and exploration. In A. Horowitz and G. von Furstenberg, *Patterns of Discrimination*. Lexington, Mass.: D.C. Heat/Lexington Books.

Allen, Marcus and William H. Dare. 2004. The effects of charm listing prices on house transaction prices. *Real Estate Economics* 32 (4): 695–713.

Alpert, Mark I., John N. McGrath, and Judy I. Alpert. 1984. Magic Prices: An extension. *Proceedings and Abstracts*, 13th Annual Meeting of the Western Regional Conference. Atlanta: American Institute for Decision Science.

Anderson, Eric T. and Duncan I. Simester. 2003. Effects of $9 price endings on retail sales: Evidence from field experiments. *Quantitative Marketing and Economics* 1 (1): 93–110.

Anker, Richard and Catherine Hein. 1986. Inequalities in Third World employment: Statistical evidence. In Anker, R. and Hein, C. (eds) *Sex Inequalities in Urban Employment in the Third World*. London: Macmillan Press, pp 63–115.

Apogee Research, Inc. 2003. Analysis of options for child safety seat use in air transportation. Report prepared for the Federal Aviation Administration. July.

Arrow, Kenneth J. 1972. Models of job discrimination. In A. H. Pascal (ed) *Racial Discrimination in Economic Life*. Lexington, Mass.: Lexington Books, pp 83–102.

———. 1973. The theory of discrimination. In O. A. Ashenfelter and A. Rees (eds) *Discrimination in Labor Markets*. Princeton, N.J.: Princeton University Press, pp 3–33.

Arthur, W. Brian. 1989. Competing technologies, increasing returns, and lock-in by historical events. *Economic Journal* 99 (394, March): 116–131.

———. 1990. Positive feedbacks in the economy. *Scientific American* 262 (February): 92–99.

———. 1996. Increasing returns and the new world of business, *Harvard Business Review* July-August: 100–109.

Associated Press. 2007. Rush to ethanol could produce glut. *Los Angeles Times*, June 14, C2.

Astin, Helen S. and Alan E. Bayer. 1972. Sex discrimination in academe. *Educational Record* 53: 101–118.

Averill, James R. 1973. Personal control over aversive stimuli and its relationship to stress. *Psychological Bulletin* 80 (4): 286–303.

Banerjee, Abhijit and Lawrence H. Summers. 1987. On frequent flyer programs and other loyalty-inducing economic arrangements. Cambridge, Mass.: Harvard Institute for Economic Research, Harvard University, discussion paper 1337, September.

Banyas, C. A. 1999. Revolution and phylogenetic history of the frontal lobes. In B. L. Miller and J. L. Cummings (eds) *The Human Frontal Lobes*. New York: Guilford Press, pp 83–106.

Barro, Robert J. and Paul M. Romer. 1987. Ski-lift pricing, with application to labor and other markets. *American Economic Review* 77 (Dec., 5): 875–890.

Bastiat, Frédéric. 1845. *Economic Sophisms*, Irvington-on-Hudson, NY: The Foundation for Economic Education, Inc., trans. and ed. Arthur Goddard, Library of Economics and Liberty, 1996, accessed on January 4, 2004 at http: //www.econlib.org/library/Bastiat/basSoph3.html

Battaglio, Stephen and Kirk Honeycutt. 1998. Bronfman: "Event films need event ticket prices," *Hollywood Reporter* (April 1), 3.

Bawa, Kapil and Robert W. Shoemaker. 1987a. The coupon-prone consumer: Some findings based on purchase behavior across product classes. *Journal of Marketing* 51 (4): 99–110.

———. 1987b. The effects of a direct-mail coupon on brand choice behavior. *Journal of Marketing Research* 24 (November): 370–376.

Bawa, Kapil, Srini S. Srinvasan, and Rajendra K. Srivastava. 1997. Coupon attractiveness and coupon proneness: A framework for modeling coupon redemption. *Journal of Marketing Research* 34 (4, November): 517–525.

Bayot, Jennifer. 2003. Banks offer sweeteners to paying bills online. *New York Times*, April 21, sec. C.

Becker, Gary S. 1965. A theory of the allocation of time. *Economic Journal* 75 (299): 493–508.

———. 1971. *The Economics of Discrimination*. Chicago: University of Chicago Press (originally published in 1957).

———. 1991. A note on restaurant pricing and other examples of social influences on price. *Journal of Political Economy* 99 (October, 5): 1109–1116.

———. 1993. The economic way of looking at life. *Journal of Political Economy* 101 (3): 385–409.

———. 1971. *Economic Theory*. New York: Alfred A. Knopf, Inc.

——— and Guity Nashat Becker. 1997. *The Economics of Life: From Baseball to Affirmative Action to Immigration, How Real-World Issues Affect Our Everyday Life*. New York: McGraw-Hill.

——— and George J. Stigler. 1977. De gustibus non est disputandum. *American Economic Review* 67 (2): 76–90.

——— and Kevin M. Murphy. 1988. A theory of rational addiction. *Journal of Political Economy* 96 (4, August): 675–700.

———, Michael Grossman, and Kevin M. Murphy. 1994. An empirical analysis of cigarette addiction. *American Economic Review* 84 (October, 3): 396–418.

Beigel, H. G. 1954. Body height in mate selection. *Journal of Social Psychology* 39: 257–268.

Beller, Andrea H. 1979. The impact of the equal employment opportunity laws on the male-female earnings differential. In Cynthia B. Lloyd, Emily Andrews, and Curtis L. Gilroy, *Women in the Labor Market*. New York: Columbia University Press, pp 304–330.

———. 1980. The effects of economic conditions on the success of equal employment opportunity laws: An application to the sex differential in earnings. *Review of Economic Statistics* 62 (3): 370–387.

———. 1985. Changes in the sex composition of U.S. occupations, 1960–1981. *Journal of Human Resources* 20 (2): 235–250.

Bergmann, Barbara R. 1971. The effects on white incomes of discrimination in employment. *Journal of Political Economy* 79 (2): 294–313.

Bernstein, Aaron. 2004. Women's pay: Why the gap remains a chasm. *Business Week*, June 14, accessed on June 30, 2004 at http: //netscape.businessweek.com/magazine/content/04_24/b3887065.htm

Besen, Stanley M. 1986. Private copying, reproduction costs, and the supply of intellectual capital. *Information Economics and Policy* 2 (1): 2–52.

——— and Sheila Nataraj Kirby. 1989. Private copying, appropriability, and optimal copying royalties. *Journal of Law and Economics* 32 (2, October): 255–280.

Blalock, Garrick, Vrinda Kadiyali, and Daniel Simon. 2005a. The impact of 9/11 on road fatalities: The other lives lost to terrorism, accessed on March 11, 2007 from the Social Science Research Network at http: //papers.ssrn.com/sol3/papers.cfm?abstract_id=677549

Blalock, Garrick, Vrinda Kadiyali, and Daniel Simon. 2005b. The impact of post-9/11 security enhancements on demand for air travel, (with Garrick Blalock and Daniel Simon), accessed on March 11, 2007 from the Social Science Research Network at http://papers.ssrn.com/sol3/papers.cfm?abstract_id=677563

Blattberg, Robert, Thomas Buesing, Peter Peacock, and Subrata Sen. 1978. Identifying the deal prone segment. *Journal of Marketing Research* 15 (3, August): 369–377.

Blattberg, Robert C. and Scott A. Neslin. 1990. *Sales Promotion, Concepts, Methods, and Strategies*. Englewood Cliffs, N.J.: Prentice-Hall.

Blattberg, Robert C. and Kenneth J. Wisniewski. 1987. How retail price promotions work: Empirical results. Chicago: University of Chicago, working paper 43.

Blau, Francine D. 1984. Discrimination against women: Theory and evidence. In William A. Darity (ed) *Labor Economics: Modern Views*. Boston: Kluwer-Nijhoff, pp 53–89.

———. 1998. Trends in the well-being of American women, 1970–1995. *Journal of Economic Literature* 36 (1): 112–165.

——— and Lawrence M. Kahn. 1997. Wage structure and gender earnings differentials: An international comparison. *Economica* 63: S29–S62.

———, Marianne A. Ferber, and Anne E. Winkler. 1998. *The Economics of Women, Men, and Work*, 3rd edn. Englewood Cliffs, N.J.: Prentice-Hall.

———, Patricia Simpson, and Deborah Anderson. 1998. Continuing progress? Trends in occupational segregation over the 1970s and 1980s. (September 1998). NBER Working Paper No. W6716 (September), accessed on June 24, 2004 at http: //ssrn.com/abstract=226370

Bolger, Dan. 2002. Managing inventory theft. *Furniture World*, October 1, accessed on February 20, 2006 at http: //www.furninfo.com/absolutenm/templates/Article_Retailing.asp?articleid=2205&zoneid=73

Borenstein, Seth. 2007. Putting off until tomorrow? Procrastination on rise in U.S., making more people poor, fat, and unhappy. *Tuscaloosa News,* January 12, as found January 15 at http://www.tuscaloosanews.com/apps/pbcs.dll/article?AID=/20070112/NEWS/701120343/1007/ENTERTAINMENT2

Boyd, Harper W. and William. F. Massey. 1972. *Marketing Management: A Strategic, Decision-Making Approach.* Orlando, Fla.: Harcourt, Brace, Jovanovich.

Brenner, Gabrielle A. and Reuven Brenner. 1982. Memory and markets, or why are you paying $2.99 for a widget? *Journal of Business* 55 (1): 147–158.

Brown, Charles C. and Mary E. Corcoran. 1996. Sex-based differences in school content and the male/female wage gap. *Journal of Labor Economics* 13 (3): 431–465.

Browne, Kingsley R. 1995. Sex and temperament in modern society: A Darwinian view of the glass ceiling and the gender gap. *Arizona Law Review* 37 (winter): 971.

———. 1998. *Divided Labours: An Evolutionary View of Women at Work.* New Haven, Conn.: Yale University Press.

———. 2002. *Biology at Work: Rethinking Sexual Equality.* New Brunswick, N.J.: Rutgers University Press.

Buehler, Roger, Dale Griffin, and Michael Ross. 1994. Exploring the "planning fallacy": Why people underestimate their task completion times. *Journal of Personality and Social Psychology* 67 (September): 366–381.

Bureau of Labor Statistics, U.S. Department of Commerce. 2003. *Highlights of Women's Earnings in 2002.* Washington, D.C.: U.S. Government Printing Office.

Bureau of Transportation Statistics. 2003. Accessed on July 15, 2007 at http: //www.bts.gov/

Business Software Alliance. 2006. *Third Annual BSA and IDC Global* Software *Piracy Study,* accessed on March 21, 2007 at http: //www.bsa.org/globalstudy/upload/2005%20Piracy%20Study%20-%20Official%20Version.pdf

Buss, David M. 1986. Human mate selection. *American Scientist* 73: 47–51.

———. 1987. Sex differences in human mate selection: An evolutionary perspective. In C. Crawford, M. Smith, and D. Krebs (eds) *Sociobiology and Psychology: Ideas, Issues, and Applications.* Hillsdale, N.J.: Lawrence Erlbaum Association.

———. 1988a. The Evolution of human sexual competition: Tactics of mate attraction. *Journal of Personality and Social Psychology* 54: 616–628.

———. 1988b. From vigilance to violence: Tactics of male retention in American undergraduates. *Ethology and Sociobiology* 9: 1–49.

———. 1989. Sex differences in human mate preferences: Evolutionary hypotheses tested in 37 cultures. *Behavioral and Brain Sciences* 12: 1–14.

———, et al. 1990. International preferences in selecting mates: A study of 37 societies. *Journal of Cross-Cultural Psychology* 21: 5–47.

———. 2003. *The Evolution of Desire: Strategies of Human Mating.* New York: Basic Books, New York.

———. 2004. *Evolutionary Psychology: The New Science of the Mind.* Boston: Pearson.

——— and Michael F. Barnes. 1986. Preferences in human mate selection. *Journal of Personality and Social Psychology* 50: 559–570.

Cabolis, Christos, Sofronis Clerides, Ioannis Ioannou, and Daniel Sendt. 2005. A textbook example of international price discrimination. New Haven, Conn.: working paper no. 05-26, Yale International Center for Finance, Yale University, October, accessed on the Social Science Research Network on February 19, 2007 at http://papers.ssrn.com/sol3/papers.cfm?abstract_id=807705

Caiazza, Amy, April Shaw, and Misha Werschkul. 2003. Women's economic status in the states: Wide disparities by race, ethnicity, and region. Washington, D.C.: Institute for Women's Policy Research, accessed on August 22, 2007 at http://www.iwpr.org/pdf/R260.pdf

Cain, Glen G. 1986. The Economic analysis of labor market discrimination: A survey. In Orley Ashenfelter and Richard Layard (eds) *Handbook of Labor Economics.* London: Elsevier Science Publishers, pp 693–785.

Cameron, C., P. S. Oskamp, and W. Sparks. 1978. Courtship American style: Newspaper advertisements. *Family Coordinator* 26: 27–30.

Caves, Richard E. 2000. *Creative Industries: Contracts between Art and Commerce.* Cambridge, Mass.: Harvard University Press.

Chen, Yuxin, Sridhar Moorthy, and Z. John Zhang. 2005. Research note: Price discrimination after the purchase: Rebates as state-dependent discounts. *Management Science* 51 (7): 1131–1140.

Clark, Gregory. 2007. *A Farewell to Alms: A Brief Economic History of the World.* Princeton, N.J.: Princeton University Press.

Clerides, Sofronis. 2002. Book value: Intertemporal pricing and quality discrimination in the U.S. market for books. *International Journal of Industrial Organization* 20 (10): 1385–1408.

CMS. 2006. *Advantage Update*, no. 3, accessed on February 14, 2007 at http://www.retailwire.com/Downloads/AU_3-06.pdf

Coleman, Walter S. 1990. *ATA Petition: Infant/Child Restraints* (Rules Docket, AGC-204), filed with the Federal Aviation Administration February 22, 4.

Conant, Michael. 1960. *Antitrust in the Motion Picture Industry* (Berkeley, Calif.: University of California Press).

Corts, Kenneth S. 1998. Third-degree price discrimination in oligopoly: All-out competition and strategic commitment. *Rand Journal of Economics* 29 (2): 306–323.

Cosmides, Leda, John Tooby, and Jerome H. Barkow. 1992. Introduction: Evolutionary psychology and conceptual integration. In Jerome H. Barkow, Leda Cosmides, and John Tooby (eds) *The Adapted Mind: Evolutionary Psychology and the Generation of Culture*. New York: Oxford University Press, pp 3–18.

Cowen, Tyler. 2007. *Discover Your Inner Economist: Use Incentives to Fall in Love, Survive Your Next Meeting, and Motivate Your Dentist.* New York: Dutton Adult.

Crandall, Robert W. 1975. The postwar performance of the motion-picture industry. *Antitrust Bulletin*, 2 (Spring): 49–88.

——— and Clifford Winston. 2003. "Does Antitrust policy improve consumer welfare? Assessing the evidence," *Journal of Economic Perspectives*, vol. 17 (no. 4, Fall). 3–26.

Dale, Stacy Berg, and Alan B. Krueger. 2002. Estimating the Payoff to Attending a More Selective College: An Application of Selection on Observables and Unobservables. *Quarterly Journal of Economics* 107 (4, November): 1491–1527.

Daniel, Fran. 2007. Pushing savings: With coupon use down, new approaches tried. *Winston-Salem Journal*, September 23, D1.

Darwin, Charles. 1859. *The Origins of the Specie.* New York: Gramercy/Random House, republished in 1979.

Dawkins, Richard. 1976. *The Selfish Gene.* Oxford: Oxford University Press.

———. 1989. *The Selfish Gene.* New York: Oxford University Press (originally published 1976).

Daymount, Thomas and Paul Andrisani. 1984. Job preferences, college major, and the gender gap in earnings. *Journal of Human Resources* 19: 408–428.

DeNavas-Walt, Carmen, Robert W. Cleveland, and Bruce H. Webster, Jr. 2003. *Census Bureau, Current Population Reports (P60–221): Income in the United States: 2002.* Washington, D.C.: U.S. Government Printing Office.

Devine, Theresa J. 1994. Characteristics of self-employed women in the United States. *Monthly Labor Review* 117 (3, March): 20–34.

Dhar, Sanjay K., Donald Morrison, and Jagmohan S. Raju. 1996. The effect of package coupons and brand choice: An epilogue on profits. *Management Science* 60 (1): 192–203.

——— and Stephen J. Hoch. 1996. Price discrimination using in-store merchandising. *Journal of Marketing* 60 (January): 17–30.

Daily Mail (London). 2000. Asda Axis the 99p Price Ploy. May 22, 15.

Dawes, Robyn M. 1964. Social selection based on multidimensional criteria. *Journal of Abnormal and Social Psychology* 68 (January): 104–109.

De Vany, Arthur S. 2004. *Hollywood Economics: How Extreme Uncertainty Shapes the Film Industry*. London: Routledge.

——— and Ross Eckert. 1991. Motion picture antitrust: The Paramount cases revisited. *Research in Law and Economics* 14 (November): 51–112.

——— and Gail Frey. 1982. Backlogs and the value of excess capacity in the steel industry. *American Economic Review* 72 (June, 3): 441–451.

——— and W. David Walls. 2007. Estimating the effects of movie piracy on box-office revenues. *Review of Industrial Organization* (forthcoming; complete cite when published).

Dodds, William B. and Kent B. Monroe. 1985. The effect of brand and price information subjective product evaluations. *Advances in Consumer Research* 12: 85–90.

Douglass, Elizabeth. 2007. State's drivers are guzzling less, *Los Angeles Times*, April 13, A1.

The Economist. 2004. Room, board and broadband, February 14, 59.

Education Life. 2007. "Data," *New York Times*, April 22, 10–11.

Elder, Glen H. 1969. Appearance and education in marriage mobility. *American Sociological Review* 34 (4): 519–533.

Etter, Lauren and Llan Brat. 2007. Ethanol boom runs out of gas. Wall Street Journal, October 1, A1.

Fasoldt, Al. 2004. Saving money on the high cost of printing. *Technofile*, as accessed on March 27, 2007 at http: //aroundcny.com/Technofile/texts/tec040304.html

Featherman, David L. and Robert M. Hauser. 1976. Sexual inequalities and socioeconomic achievement in the United States: 1962–1973. *American Sociological Review* 41 (3): 462–483.

Febrero, Ramon and Pedro S. Schwartz (eds). 1995. *The Essence of Becker*. Palo Alto, Calif.: Hoover Institution Press.

Federal Aviation Administration. 2005. Federal Aviation announces decisions on child safety seats, August 25.

Feingold, A. 1982. Do taller men have prettier girlfriends? *Psychological Reports* 50: 810.

Fels, Rendigs and Robert G. Uhler. 1975. *Casebook of Economic Problems and Policies: Practice in Thinking*, 1975–76 edn. St. Paul, Minn.: West Publishing Co.

Fisher, Franklin M. 1998. Direct Testimony, U.S. vs. Microsoft Corporation, Civil Action No. 98-1233 (TPJ), filed October 14, as accessed in 1998 from http: //www.usdoj.gov/atr/cases/f2000/2057.pdf

Fitzgerald, Kate. 1996. Survey: Consumers prefer sampling over coupons. *Advertising Age* January 29, 9.

Ford, Clellan Stearns and Frank A. Beach. 1951. *Patterns of Sexual Behavior*. New York: Harper and Row.

Frank, Robert H. 2007. *The Economics Naturalist: In Search of Explanations for Everyday Enigmas*. New York: Basic Books.

Frank, Steven A. and Richard B. McKenzie. 2005. The male-female pay gap driven by coupling between labor markets and mating markets. *Journal of Bioeconomics* 8: 269–274.

Friedman, David. 1990. *Price Theory: An Intermediate Text*, 2nd edn. Cincinnati, Oh.: South-Western Publishing Co.

Friedman, Josh and Ronald D. White. 2007. Some iPhones glitches reported. *Los Angeles Times*, July 2, C2, also accessed July 2, 2007 from http: //www.latimes.com/business/la-fi-iphone2jul02,1,4774908.story?ctrack=1&cset=true

Friedman, Milton. 1957. *A Theory of the Consumption Function*. Princeton, N.J.: Princeton University Press.

Fuchs, Varian R. 1971. Differences in hourly earnings between men and women. *Monthly Labor Review* 94: 9–15.

———. 1988. *Women's Quest for Economic Equality*. Cambridge, Mass.: Harvard University Press.

Einav, Liran. 2004. Gross seasonality and underlying seasonality: Evidence from the U.S. motion picture industry. Stanford, Calif.: Stanford University, working paper.

Filson, Darren, David Switzer, and Portia Besocke. 2004. At the movies: The economics of exhibition contracts. *Economic Inquiry* 43 (2): 354–369.

Fox, William F. 2002. History and economic impact. Knoxville, Tenn. Economics Department, University of Tennessee, March 13, accessed on February 20, 2006 at http: //cber.bus.utk.edu/staff/mnmecon338/foxipt.pdf

Friedman, Lawrence. 1967. Psychological pricing in the food industry. In Almarin Phillips and Oliver E. Williamson (eds) *Prices: Issues in Theory, Practice, and Public Policy*. Philadelphia: University of Pennsylvania Press.

Gabor, Andre and Clive Granger. 1964. Price sensitivity of the consumer. *Journal of Advertising Research* 4 (December): 40–44.

Garvey Corporation. 1964. *The Price Is Right*. St. Louis.

General Accounting Office. 2003. *Women's Earnings: Work Patterns Partially Explain Difference Between Men's and Women's Earnings*. GAO-04-35. Washington, D.C.: U.S. Government Accounting Office (October), accessed June 24, 2004 at http: //www.house.gov/maloney/issues/womenscaucus/2003EarningsReport.pdf

Georgoff, David M. 1971. *Odd-Even Retail Price Endings*. Ann Arbor, Mich.: Michigan State University Press.

Gerstner, Eitan and James D. Hess. 1991. Who benefits from large rebates: Manufacturer, retailer or consumer? Evidence from scanner data. *Economic Letters* 36 (1): 5–8.

Gilbert, Daniel. 2006. *Stumbling on Happiness*. New York: Alfred A. Knopf and Random House, Inc.

Gillis, John S. 1982. *Too Tall, Too Small.* Champaign, Ill.: Institute for Personality and Ability Testing.

——— and Walter E. Avis. 1980. The male-taller norm in mate selection. *Personality and Social Psychology Bulletin* 6: 396–401.

Gilmour, J. 1985. One cent less doesn't make sense. *Australian Business.* 20 (March): 34.

Ginsberg, Eli. 1936. Customary prices. *American Economic Review* 26 (2): 296.

Givon, Moshe, Vijay Mahajan, and Eitan Muller. 1995. Software piracy: Estimation of lost sales and the impact of software diffusion. *Journal of Marketing* 59: 29–37.

Glendall, Philip, Judith Holdershaw, and Ron Garland. 1997. The effect of odd pricing on demand. *European Journal of Marketing* 31 (11/12): 799–813.

Gendall, Philip, Michael F. Fox, and Priscilla Wilton. 1998. Estimating the effect of odd pricing. *Journal of Product & Brand Management* 7 (5): 421–432.

Gneezy, Uri, Muruel Niederle, and Aldo Rustichini. 2003. Performance in competitive environments: Gender differences. *Quarterly Journal of Economics* 118 (3): 1049–1074.

Goldberg Dey, Judy and Catherine Hill. 2007. *Behind the Pay Gap.* Washington, D.C.: American Association of University Women Educational Foundation, April.

Goldin, Claudia. 1989. Life-cycle labor force participation of married women: Historical evidence and implications. *Journal of Labor Economics* 7: 20–47.

Gould, William B. 1977. *Black Workers in White Unions: Job Discrimination in the United States.* Ithaca, N.Y.: Cornell University Press.

Grabmeier, Jeff. 2006. Procrastinators get poorer grades in college, study finds. As found on the Ohio State Research web site, http: //researchnews.osu.edu/archive/procrast.htm (on January 20, 2007).

Gregersen, Edgar. 1982. *Sexual Practices: The Story of Human Sexuality.* New York: Franklin Watts.

Griffin, Dale W., David Dunning, and Lee Ross. 1990. The role of construal processes in overconfident predictions about self and others. *Journal of Personality and Social Psychology* 59 (December): 1128–1139.

Grow, Brian. 2005. The great rebate runaround. *Business Week* (November 23), as found on January 10, 2006 at http: //www.businessweek.com/bwdaily/dnflash/nov2005/nf20051123_4158_db016.htm

Gueguen, Nicolas and Celine Jacob. 2005. Nine-ending price and consumer behavior: An evaluation in a new context. *Journal of Applied Sciences* 5 (2): 383–384.

Guglielmo, Connie. 2007. Sales of iPhones beating forecasts (Bloomberg News report). *Orange County (Calif.) Register.* July 4, 2007, marketplace 1.

Gunderson, Morley. 1989. Male–female wage differentials and policy responses. *Journal of Economic Literature* 27 (1): 46–72.

Haddock, David D. and Fred S. McChesney. 1994. Why do firms contrive shortages? The economics of intentional mispricing. *Economic Inquiry* 32 (October, 4): 562–581.

Hafner, Katie and Brad Stone. 2007. iPhone owners crying foul over price cut. *New York Times*, September 7, C1.

Hall, Randolph W. 1991. *Queuing Methods for Service and Manufacturing*. Englewood Cliffs, N.J.: Prentice-Hall.

Hamermesh, Daniel S. and Jeff E. Biddle. 1994. Beauty and the labor market. *American Economic Review* 84 (December, 5): 1174–94.

Hamer, W. G. 2007. The cost of water and water markets in Southern California, USA. Transactions of the Wessex Institute, June 11, as accessed on June 11, 2007 from http: //library.witpress.com/pages/PaperInfo.asp?PaperID=17561

Harper, Donald V. 1966. *Price Policy and Procedure*. New York: Harcourt, Brace and World.

Harris, Richard J. and Mark A. Joyce. 1980. "What's fair? It depends on how you phrase the question," *Journal of Personality and Social Psychology* 38: 165–179.

Harris, William T. 1996. Captive audiences and the price of popcorn. *Pennsylvania Economic Review* 5 (4): 39–46.

Harrison, Albert A. and Laila Saaed. 1977. Let's make a deal: An analysis of revelations and stipulations in lonely hearts advertisements. *Journal of Personality and Social Psychology* 35: 257–264.

Hassin, Refael. 1985. On the optimality of first come last served queues. *Econometrica* 53 (1): 201–202.

Hawkins, Edward R. 1954. Price policy and theory. *Journal of Marketing* 18 (January): 233–240.

Hayek, Friedrich A. 1945. The use of knowledge in society. *American Economic Review* 35 (4, September): 519–530, accessed March 9, 2007 at http: //www.econlib.org/Library/Essays/hykKnw1.html

Headley, Robert K. 1999. Motion Picture Exhibition in Washington, D.C., accessed August 21, 2007 at http: //www.lib.umd.edu/RARE/Exhibits/Headley/book.html

Hecker, JayEtta Z. 2003. *Flood Insurance: Challenges Facing the National Flood Insurance Program*, Washington, D.C.: U.S. General Accounting Office, testimony before the Subcommittee on Housing and Community Opportunity, Committee on Financial Services, House of Representatives, April 1, accessed on March 16, 2007 at http: //www.gao.gov/new.items/d03606t.pdf

Heckman, James J. 1980. Sample selection bias as a specification error. In J. P. Smith (ed) *Female Labor Supply: Theory and Estimation*. Princeton, N.J.: Princeton University Press, pp 206–248.

Heeler, Roger and Adam Nguyen. n.d. Price endings in Asia, accessed on February 23, 2006 from http: //130.195.95.71: 8081/WWW/ANZMAC2001/anzmac/AUTHORS/pdfs/Heeler.pdf

Heilman, Carrie, Kyryl Lakishyk, Kent Nakamoto, and Sonja Radas. 2005. The effect of in-store free samples on short- and long-term purchasing behavior. Charlottesville, Va.: McIntire School of Commerce, University of Virginia, working paper, June, as accessed on March 27, 2007 at http: //www.commerce.virginia.edu/faculty_research/Research/Papers/FreeSamples_Posted_June24_2005.pdf

Hill, Herbert. 1977. *Black Labor and the American Legal System*. Washington, D.C.: Bureau of National Affairs.

Hill, Charles T., Zick Rubin, and L. A. Peplau. 1979. Breakups before marriage: The end of 103 affairs. In L. Levinger and O. C. Moles (eds) *Divorce and Separation*. New York: Basic Books.

———, E. S. Nocks, and L. Gardner. 1987. Physical attractiveness: Manipulation by physique and status displays. *Ethology and Sociobiology* 8: 143–154.

Hock, Stephen J. 1985. Counterfactual reasoning and accuracy in predicting personal events. *Journal of Experimental Psychology: Learning, Memory, and Cognition* 11 (4): 719–731.

Hoch, Stephen J., Xavier Dreze, and Mary Purk. 1995. *Exploring Relationship Marketing*. Cleveland, Ohio: American Greetings Research Council.

Hogl, S. 1988. The effects of simulated price changes on consumers in a retail environment—Price thresholds and price theory. *Proceedings*. Lisbon: Esomar Congress.

Holdershaw, J. L. 1995. *The Validity of Odd Pricing*. Palmerston North, New Zealand: Massey University, unpublished MBS thesis.

Holm, H. J. n.d. Sex differences or paranoia? Gender differences in experimental discriminatory behavior. Lund, Sweden: Department of Economics, University of Lund, working paper, accessed on July 8, 2004 at http: //www.nek.lu.se/publications/workpap/Papers/WP00_1.pdf

Hudson,Valerie M. and Andrea M. den Boer. 2004. *Bare Branches: The Security Implications of Asia's Surplus Male Population*. Cambridge: MIT Press.

Jackson, Thomas Penfield. 1999. U.S. District Court, Findings of Fact, U.S. Government v. Microsoft Corporation civil action 98-1232, November 5.

Jacobsen, Joyce P. 1994. Sex segregation at work: Trends and predictions. *Social Science Journal* 31 (2): 153–169.

Johnson, George E. and Frank P. Stafford. 1974. Lifetime earnings in a professional labor market. *Journal of Political Economy* 82 (May/June): 549–570.

Johnson, William R. 1984. The economics of copying. *Journal of Political Economy* 93 (1): 158–174.

Jolson, Marvin, Joshua Weiner, and Richard Rosecky. 1987. Correlates of rebate proneness. *Journal of Advertising Research* 27 (1): 33–43.

Kahneman, Daniel. 1994. New challenges to the rationality assumption. *Journal of Institutional and Theoretical Economics* 150 (1): 18–36.

Kahneman, Daniel, Jack L. Knetsch, and Richard Thaler. 1986. Fairness as a constraint for profit seeking: Entitlements in the market. *American Economic Review*. 76 (4): 728–741.

Kahneman, Daniel and Amos Tversky. 1979. Prospect theory: An analysis of decisions under risk. *Econometrica* 47: 313–327.

———. 2000a. Choices, values, and frames, as reprinted in *Choices, Values, and Frames*, Daniel Kahneman and Amos Tversky (eds). Cambridge, U.K.: Cambridge University Press 2000, pp 1–16.

———. 2000b. Prospect Theory: An analysis of Decision under risk, as reprinted in *Choices, Values, and Frames*, Daniel Kahneman and Amos Tversky (eds). Cambridge, U.K.: Cambridge University Press, 2000, pp 1–16.

——— (eds). 2000. *Choices, Values, and Frames*. Princeton, N.J.: Princeton University Press.

Kahneman, Daniel, Richard H. Thaler, and Jack L. Knetsch. 1990. Experimental tests of the endowment and the Coase theorem. *Journal of Political Economy* 98 (6): 1325–1348.

Kashyap, Anil K. 1995. Sticky prices: New evidence from retail catalogs. *Quarterly Journal of Economics* 110 (1): 245–274.

Kendrick, D. T., E. K. Sadalla, G. Groth, and M. R. Trost. 1990. Evolution, traits, and the stages of human courtship: Qualifying the parental investment model. *Journal of Personality* 50: 97–116.

Kim, Hyeong Min. 2006. Consumers' responses to price presentation formats in rebate advertisements. *Journal of Retailing* 82 (4): 309–317.

King, Thomas B. 1992. Coming soon: Cut-rate films on Tuesday. *Wall Street Journal*, February 13, B1.

Kinzie, Susan. 2006. Swelling textbook costs have college students saying "pass." *Washington Post*, January 23, A1, accessed on August 23, 2007 at http://www.washingtonpost.com/wp-dyn/content/article/2006/01/22/AR2006012201290.html

Klein, Joel I., et al. 1998. Complaint, United States of America v. Microsoft Corporation, civil action no. (none given), May 20, 1998, as accessed in 1998 from http: // www.usdoj.gov/atr/cases3/micros/1763.htm

Knight, Dana. 2006. Season of Gift Cards. *Indianapolis Star.com*, December 27, accessed on February 21, 2007 at http: //www.indystar.com/apps/pbcs.dll/article?AID=/20061227/BUSINESS/612270386

Kolstad, James L. 1989. Require safety seats for little travelers, *USA Today*, May 24, 12A.

Kreul, Lee M. 1982. Magic numbers: Psychological aspects of pricing. *Cornell Hotel and Restaurant Administration Quarterly* 23 (1): 70–75.

Krishna, Aradhna and Robert W. Shoemaker. 1992. Estimating the effects of higher coupon face values on the timing of redemptions, the mix of coupon redeemers, and purchase quantity. *Psychology and Marketing* 9 (November/December): 453–467.

Krishnan, Trichy V. and Ram C. Rao. 1995. Double couponing and retail pricing in a compound product category. *Journal of Marketing Research* 32 (4, November): 419–432.

Krueger, Anne O. 1963. The economics of discrimination. *Journal of Political Economy* 71 (5): 481–486.

Lal, Rajiv and Carmen Matutes. 1994. Retail pricing and advertising strategies. *Journal of Business* 67 (3, July): 345–370.

Lammers, Bruce H. 1991. The effect of free samples on immediate consumer purchase. *Journal of Consumer Marketing* 8 (2): 31–37.

Landsburg, Steven E. 1993. *The Armchair Economist: Economics and Everyday Life*. New York: Free Press.

———. 2007. *More Sex Is Safer Sex: The Unconventional Wisdom of Economics*. New York: Free Press.

Lambert, Zarrel V. 1975. Perceived prices as related to odd and even price endings. *Journal of Retailing* 51 (3): 13–22.

Lawton, Christopher. 2007. What's behind cheaper printer cartridges: H-P, Kodak, others slash prices but also volume of ink; A higher cost per page. *Los Angeles Times*, May 9, D1.

Lay, Clarry H. 1988. The relation of procrastination and optimism to judgments of time to complete an essay and anticipation of setbacks. *Journal of Social Behavior and Personality* 3 (September): 201–214.

Lee, Dongwon, Haipeng Chen, Daniel Levy, Robert J. Kauffman, and Mark Bergen. 2006. Making sense of ignoring sense: Price points and price rigidity under rational inattention. Working paper. Minneapolis, Minn.: Carlson School of Management, University of Minnesota, accessed on April 20, 2006 at http: //www.misrc.umn.edu/workshops/2006/spring/dongwon.pdf.n

Lee, Dwight R. 1969. Utility analysis and repetitive gambling. *American Economist* 13 (Fall, 2): 87–91.

——— and Richard B. McKenzie. 1998. How the client effect moderates price competition. *Southern Economic Journal* 64 (3): 741–752.

Leibenstein, Harvey. 1950. Bandwagon, snob, and Veblen effects in the theory of consumers demand. *Quarterly Journal of Economics* 64 (May, 2): 183–207.

Leonard, Jonathan S. 1984. Antidiscrimination or reverse graphics, Title VII, and affirmative action on productivity. *Journal of Human Resources* 19 (2): 145–174.

Levedahl, William J. 1986. The pricing of cents-off coupons: Multipart pricing or price discrimination? *Quarterly Journal of Business and Economics* 25 (Fall): 56–68.

Leeds, Jeff. 2001. Record industry says Napster hurt music sales: Data show a 38.8% drop-off in CD singles last year. *Los Angeles Times*, February 24, C1.

Levitan, Daniel J. 2006. *This Is Your Brain on Music*. New York: Dutton Books.

Lewin Tamar. 2003a. When books break the bank. *New York Times*, September 16, accessed on February 19, 2007 at http: //www.nytimes.com/2003/09/16/nyregion/16BOOK.html?ex=1379131200&en=47ae57d313426aa0&ei=5007&partner=USERLAND

———. 2003b. Students find $100 textbooks cost $50, purchased overseas. *New York Times*, October 21, accessed on February 19, 2007 at http: //www.nytimes.com/2003/10/21/education/21BOOK.html?ex=1382068800&en=bb0a793986b18c60&ei=5007&partner=USERLAND

Liberman, Nira, Michael D. Sagristano, and Taacov Trope. 2002. The effect of temporal distance on Level of Mental Construal. *Journal of Experimental Social Psychology* 38 (6): 523–534.

Lindstedt, Sharon. 1999. Tops supermarkets in Western New York entice shoppers with free food samples. *Buffalo News*, May 24, B3, as accessed July 26, 2006 at http: //www.commerce.virginia.edu/faculty_research/Research/Papers/FreeSamples_July26_2006.pdf

Lifson, Thomas. 2004. The college tuition scam. *American Thinker*, accessed on February 15, 2007 at http: //www.americanthinker.com/2004/05/the_college_tuition_scam.html

Lightfoot, Jim. 1990. The need for child safety restraints aboard aircraft. *Congressional Record*, February 27.

Little, John D. C. and Joan Ginese. 1987. Price endings: Does 9 really make a difference? Cambridge, Mass.: Sloan School of Management, MIT, working paper.

Litwak, Mark. 1986. *Reel Power: The Struggle for Influence and Success in the New Hollywood*. New York: William Morrow.

Locay, Luis and Alvaro Rodriguez. 1992. Price discrimination in competitive markets. *Journal of Political Economy* 100 (October, 5): 954–965.

Loewenstein, George F. 1987. Anticipation and the evaluation of delayed consumption. *Economic Journal* 97 (September): 666–684.

———. 1996. Out of control: Visceral influences on behavior. *Organizational Behavior and Human Decision Processes* 65 (3): 272–292.

——— and Erik Angner. 2003. Predicting and indulging changing preferences. In George F. Loewenstein, D. Read, and R.F Baumeister (eds) *Time and Decision*. New York: Russell Sage Foundation pp 351–391.

———, Ted O'Donoghue, and Matthew Rabin. 2003. Projection bias in predicting future utility. *Quarterly Journal of Economics* 118 (4): 1209–1248.

Lohr, Steve. 1992. Lessons from a hurricane: It pays not to gouge. *New York Times*, September 22, C1-2.

Lynn, Michael and B. A. Shurgot. 1984. Responses to lonely hearts advertisements: Effects of reported physical attractiveness, physique, and colorization. *Personality and Social Psychology Bulletin* 10: 349–357.

Madden, Janice Fanning. 1973. *The Economics of Sex Discrimination*. Lexington, Mass.: D.C. Heath and Co.

Madrick, Jeff. 2004. Economic scene: The earning power of women has really increased, right? Take a closer look. *New York Times*, June 10, C2, accessed on June 30, 2004 at http: //query.nytimes.com/gst/abstract.html?res=FB0B11F73C540C738DDDAF089 4DC404482

Malkiel, Burton G. and Judith A. Malkiel. 1973. Male-female pay differentials in professional employment. *American Economic Review* 63. 693–705.

Maltz, Elliot and Vincent Chiappetta. 2002. Maximizing value in the digital world. *Sloan Management Review* 43 (Spring): 77–84.

Marshall, Alfred. 1920. *Principles of Economics.* New York: Macmillan and Co., Ltd., accessed from Library of Economics and Liberty on March 2, 2007 at http: //www.econlib.org/library/Marshall/marP1.html

Marshall, Ray. 1974. Economics of racial discrimination. *Journal of Economic Literature* 12: 849–871.

Mason, J. Barry, and Morris L. Mayer. 1990. *Modern Retailing: Theory and Practice*, 5th edn. Homewood, Ill.: BPL/Irwin.

Mathew, T. J. and Brady E. Hamilton. Mean age of mother, 1970–2000. *National Vital Statistics Reports.* 51 (1). Hyattsville, Md.: National Center for Health Statistics, December 11, as accessed on July 3, 2007 at http: //www.cdc.gov/nchs/data/nvsr/nvsr51/nvsr51_01.pdf

McKenzie, Richard B. and Gordon Tullock. 1975. *The New World of Economics: Explorations into the Human Experience*. Homewood, Ill.: Richard D. Irvine.

——— and John T. Warner. 1987. *The Impact of Airline Deregulation on Highway Safety.* St. Louis: Center for the Study of American Business, Washington University, December.

——— and William F. Shughart. 1988. Has deregulation of airlines caused more air fatalities? *Regulation*, no. 3/4: 42–47.

——— and Dwight R. Lee. 1998. *Managing Through Incentives: How to Develop More Collaborative, Productive, and Profitable Organization*. New York: Oxford University Press.

——— and Dwight R. Lee. 2006. *Microeconomics for MBAs: The Economic Way of Thinking for Managers*. Cambridge, U.K.: Cambridge University Press.

——— and Dwight R. Lee. 2006. *Microeconomics for MBAs: The Economic Way of Thinking for Managers.* Cambridge, U.K.: Cambridge University Press downloadable video modules at \\www\home\mckenzie\public_html\ModulesaftePublication101206.html

——— and Dwight R. Lee. 2007. *In Defense of Monopoly: How Market Power Fosters Creative Production.* Ann Arbor, Mich.: University of Michigan Press.

McLaughlin, Katy. 2002. Cranky customers: Claiming that holiday rebate. *Wall Street Journal*, December 3, D2.

McNicol, David L. 1975. The two price systems in the copper industry. *Bell Journal of Economics* 6 (Spring, 1): 50–73.

Miller, Amalia R. 2005. The effects of motherhood timing on career path. Charlottesville, Va.: Department of Economics, University of Virginia, working paper, July, as accessed on July 3, 2007 at http: //www.virginia.edu/economics/papers/miller/fertilitytiming-miller.pdf

Monroe, Kent B. 1973. Buyers' subjective perceptions of price. *Journal of Marketing Research* 10 (1): 70–80.

———. 1979. *Pricing: Making Profitable Decisions*. New York: McGraw-Hill.

Moorthy, Sridhar and Dilip Soman. 2003. On the marketing of rebates: Having your cake and eating it too. Toronto: Rotman School of Management, University of Toronto, working paper.

Morris, George H. 1979. *Book of Strange Facts and Useless Information*. New York: Doubleday.

Mowen, John and Maryanne Mowen. 1991. Time and outcome valuation: Implications for marketing decision making. *Journal of Marketing* 55 (October): 54–62.

Murph, Darren. 2007. Thieves swiping HOV exemption stickers from hybrids, *Engadget*, April 22, accessed June 19, 2007 at http: //www.engadget.com/tag/thieves/

Nagle, Thomas T. 1987. *The Strategy and Tactics of Pricing*. Englewood Cliffs, N.J.: Prentice-Hall.

Nalebuff, Barry. 1988. Puzzles: Cider in your ears, continuing dilemma, the last shall be first, and more. *Journal of Economic Perspective* 2 (2, Spring): 149–156.

Naor, P. 1969. The regulation of queue size by levying tolls. *Econometrica* 37 (1, January): 15–24.

Narasimhan, Chakravarthi. 1984. A price discrimination theory of coupons. *Marketing Science* 3 (2): 128–146.

National Safety Council 1989. *Accident Facts: 1989*. Chicago: National Safety Council.

Nelson, Phillip. 1970. Information and consumer behavior. *Journal of Political Economy* 78 (2, March/April): 311–329.

Neslin, Scott A. and Farral G. Clarke. 1987. Relating the brand use profile of coupon redeemers to brand and coupon characteristics. *Journal of Advertising Research* 27 (February/March): 23–32.

Nevo, Aviv and Catherine Wolfram. 2002. Why do manufacturers issue coupons? An empirical analysis of breakfast cereals. *Rand Journal of Economics* 33 (2, Summer): 319–339.

New York Times. 2007. Sales of ink and laptops push H.P. past forecast. *New York Times*, August 17, 2007, C3.

Norr, Henry. 2000. The how and why of rebates. *San Francisco Chronicle*, December 18, T21.

Novos, Ian E. and Michael Waldman. 1984. The effect of increased copyright protection. *Journal of Political Economy* 92 (April): 236–246.

Nyhan, Paul. 2004. Pay gap still divides sexes: Study finds women earn 74 cents to men's dollar. *Seattle Post-Intelligencer*. June 4, accessed on June 30, 2004 at http: // seattlepi.nwsource.com/business/176322_gender04.html?searchpagefrom=1&searchdiff=26

Oaxaca, Ronald L. 1977. The persistence of the male-female earnings differentials. In F. Thomas Juster (ed) *The Distribution of Economic Wellbeing*. Cambridge, Mass.: Ballinger Co., pp 303–354.

Odell, Patricia. 2006. Senator Schumer asks FTC to standardized rebates. *PROMOMagazine.com*, January 4, accessed July 3, 2007 at http: //promomagazine.com/legal/standardized_rebates_010406/index.html

O'Neill, June. 1983a. The determinants and wage effects of occupational segregation. Washington, D.C.: Urban Institute, working paper.

———. 1983b. *The Trend in Sex Differentials in Wages,* a paper presented at a conference on Trends in Women's Work, Education, and Family Building. Washington, D.C.: Urban Institute.

———. 1984. Earnings differentials: Empirical evidence and causes. In G. Schmid and R. Weitzel (eds) *Sex Discrimination and Equal Opportunity*. London: Gower Publishing Co., pp 69–91.

——— and S. Polachek. 1993. Why the gender gap in wages narrowed in the 1980s. *Journal of Labor Economics* 11 (1): 205–228.

Orbach, Barak Y. and Liran Einav. 2005. Uniform prices for differentiated goods: The case of the movie-theater industry. Olin Discussion Paper No. 337. Cambridge, Mass.: John M. Olin Center for Law, Economics & Business, Harvard Law School, October), accessed on February 14, 2006 at http: //www.stanford.edu/~leinav/papers.htm

Oyer, Paul. 1998. Fiscal year ends and nonlinear incentive contracts: The effect of business seasonality. *Quarterly Journal of Economics* 113 (1, February): 149–189.

Palmon, Oded, Barton A. Smith, and Ben Z. Sopranzetti. 2004. Clustering in real estate prices: Determinants and consequences. *Journal of Real Estate Research* 26 (2): 115–147.

Parmar, Arundhati. 2003. Promoting promotions. *Marketing News*, July 7, 3.

Passell, Peter. 1997. The new economics of higher education. *New York Times*, April 22, accessed on February 15, 2007 at http: //www.econ.ucsb.edu/~tedb/eep/news/college.html

Passy, Charles. 2007. What you get with VIP status at theme parks. *Wall Street Journal*. May 17, D1.

Perlmutter, Lawrence C. and Richard A. Monty. 1979. *Choice and Perceived Control*. Hillsdale, N.J.: Lawrence Erlbaum.

Pfann, Gerard A., Jeff E. Biddle, Daniel S. Hamermesh, and Ciska M. Bosman. 2000. Business success and businesses "beauty capital." *Economic Letters* 67 (May, 2): 201–207, May.

Phelps, Edmund S. 1972. The statistical theory of racism and sexism. *American Economic Review* 62: 659–661.

Pigou, A. C. 1962. *The Economics of Welfare*, 4th edn. London: Macmillan.

Pindyck, Robert S. and Daniel L. Rubinfeld. 2004. *Microeconomics*, 6th edn. Englewood Cliffs, N.J.: Prentice Hall.

Pinker, Stephen. 1997. *How the Mind Works*. New York: W. W. Norton & Company.

Piore, Michael J. 1970. Jobs and training. In Samuel. H. Beer (ed) *The State and the Poor*. Cambridge, Mass.: Winthrop Press.

Pogue, David. 2007. Paying more for a printer, but less for ink. *New York Times*. May 17, C1.

Pontoniere, Paola. 2006. Deforestation—The dark side of Europe's thirst for green fuel. *Pacific News Service*, February 28, accessed May 7, 2007 from http: //news.pacificnews.org/news/view_article.html?article_id=37e104044bb11d19d566ac8f3621c63f

Poole, Robert. 2007. A new solution to enforcing HOV occupancy. *Surface Transportation Innovations*, no. 40, February, as found on July 23 at http: //www.reason.org/surfacetransportation40.shtml

Prag, Jay and James Casavant. 1994. An empirical study of the determinants of revenues and marketing expenditures in the motion picture industry. *Journal of Cultural Economics* 18: 217–235.

Psacharopolos, G. and Tzannatos, Z. 1992. *Case Studies on Women's Employment and Pay in Latin America*. Washington: D.C.: World Bank.

Quigley, Charles J., Jr. and Elaine M. Notarantonio. 1992. An exploratory investigation of perceptions of odd and even pricing. In Victorian L. Crittendon (ed) *Developments in Marketing Science*. Chestnut Hill, Mass.: Academy of Marketing Science, pp 306–309.

Rabin, Matthew. 1998. Psychology and economics. *Journal of Economic Literature* 36 (March): 11–46.

Rabin, Matthew, and Richard H. Thaler. 2001. Anomalies: Risk aversion. *Journal of Economic Perspectives.* 15 (1): 219–232.

Rafaeli, Anat, Greg Barron, and Karen Haber. 2002. The effects of queue structure on attitudes. *Journal of Service Research* 5 (November, 2): 125–139.

Ravid, S. Abraham. 1999. Information, blockbuster and stars: A study of the film industry. *Journal of Business* 72: 463–492.

Radcliffe, Jim. 2007. Hybrid cheaters spotted in carpool lanes. *Orange County Register*, February 19, accessed on March 12, 2007 at http: //www.ocregister.com/ocregister/news/columns/article_1583777.php

Read, Leonard E. 1958. I, pencil: My family tree as told by Leonard E. Read. *The Freeman* (December), accessed on March 9, 2007 at http: //www.econlib.org/library/Essays/rdPncl1.html

Reibstein, David J. and Phillips A. Traver. 1982. Factors affecting coupon redemption rates. *Journal of Marketing* 46 (Fall): 102–113.

Reskin, Barbara F. and Heidi I. Hartmann. 1986. *Women at Work, Men's Work: Sex Segregation on the Job*. Washington, D.C.: National Academy Press.

Reuters. 2007. GM unveils redesigned Chevy Malibu. CNN Money.com, accessed on January 19, 2007 at http: //money.cnn.com/2007/01/02/autos/malibu.reut/index.htm?postversion=2007010311

Ricadela, Aaron and Steve Koeing. 1998. Rebates' pull is divided by hard and soft lines. *Computer Retail Week,* September, accessed on January 13, 2007 at http: //calbears.findarticles.com/p/articles/mi_hb061/is_199809/ai_hibm1P129104220

Reuters. 2007b. Ethanol boom may fuel shortage of tequila: Mexican farmers burning agave fields and replanting them with corn. MSNBC, May 29, accessed on May 30, 2007 at http: //www.msnbc.msn.com/id/18926019/

Riofrio, Melissa. 2004. The cheapskate's guide to printing. *PC World*, March 3, accessed on March 28, 2007 at http: //www.pcworld.com/article/id,114728-page,1/article.html

Rivera, Ray. 2007. Biodiesel makers see opportunity as New York seeks greener future. *New York Times*. May 28, A10.

Robinson, Joan. 1965. *The Economics of Imperfect Competition*. London: Macmillan.

Rogers, Alan R. 1994. Evolution of Time Preference by Natural Selection. *American Economic Review* 84 (3, November): 460–481.

Roos, Patricia A. 1981. Sex stratification in the workplace: Male-female differences in economic returns to occupations. *Social Science Research* 10: 195–224.

Rose, Michael. 1998. *Darwin's Spectre*. Princeton, N.J.: Princeton University Press.

Rose, Nancy L. 1987. Financial influences on airline safety. Cambridge, Mass., Sloan School of Management, Massachusetts Institute of Technology, working paper no. 1890-87.

Rose, S. J. and H. I. Hartmann. 2004. *Still a Man's Labor Market: The Long-Term Earnings Gap.* Washington, D.C.: Institute for Women's Policy Research.

Rothkopf, Michael H. and Paul Rech. 1987. Perspectives on queues: Combining queues is not always beneficial. *Operations Research* 35 (Nov./Dec, 6): 906–909.

Rotstein, Gary. 2006. Hey procrastinators, today is your day. *Pittsburgh Post-Gazette*, April 17, as found on January 16, 2007 at http: //www.post-gazette.com/pg/06107/682743-294.stm

Rubin, Paul H. 2002. *Darwinian Politics: The Evolutionary Origins of Freedom.* New Brunswick, N.J.: Rutgers University Press.

Rudolph, Harold J. 1954. Pricing for today's market. *Printer's Ink* 28 (May): 22–24.

Runge, C. Ford and Benjamin Senauer. 2007. How biofuels could starve the poor. *Foreign Affairs* (May/June), accessed May 7, 2007 from http: //www.foreignaffairs.org/20070501faessay86305/c-ford-runge-benjamin-senauer/how-biofuels-could-starve-the-poor.html

Saaty, Thomas L. 1961. *Elements of Queuing Theory.* New York: McGraw-Hill.

Sadalla, Edward K., Douglas T. Kenrick, and Beth Vershure. 1987. Dominance and heterosexual attraction. *Journal of Personality and Social Psychology* 52: 730–738.

Salter, Sean P., Ken H. Johnson, and W. Paul Spurlin. n.d. Off-Dollar Pricing, Residential Property and Marketing Time, accessed on February 23, 2006 at http: //www.fma.org/Chicago/Papers/OffDollarv2.0.pdf

Salon, Steven and Joseph Stiglitz. 1977. Bargains and ripoffs: A model of monopolistically competitive price dispersion. *Review of Economic Studies* 44 (3, October): 373–388 493–510.

Sanders, Shane, Dennis L. Weisman, and Dong Li. 2005. Child safety seats on commercial airliners: A demonstration of the cross-elasticities. *Journal of Economic Education*, forthcoming in 2008, as accessed on July 15, 2007 at http: //www-personal.ksu.edu/~sdsander/JEE.pdf

Sawhill, Isabel. 1973. The economics of discrimination against women: Some new findings. *Journal of Human Resources* 8: 383–396.

Scher, Steven J. and Joseph R. Ferrari. 2000. The recall of completed and noncompleted tasks through daily logs to measure procrastination. *Journal of Social Behavior and Personality* 15 (special issue): 225–265.

Schindler, Robert M. 1984. Consumer recognition of increases in odd and even prices. In Thomas C. Kinnear (ed) *Advances in Consumer Research.* Provo, Utah: Association for Consumer Research 11: pp 459–462.

———. 2006. The 99 price ceiling as a signal of a low-price appeal. *Journal of Retailing* 82 (1): 71–77.

——— and Alan R. Wiman. 1989. The effect of odd pricing on price recall. *Journal of Business Research* 19 (3): 165–177.

——— and Thomas Kibarian. 1993. Increased consumer sales response through use of 99-ending prices. *Journal of Retailing* 72 (2): 187–199.

——— and Patrick N. Kirby. 1997. Patterns of rightmost digits used in advertised prices: Implications for nine-ending effects. *Journal of Customer Research* 24 (3): 187–199.

——— and Lori S. Warren. 1988. Effect of odd pricing on choice of items on a menu. In Michael J. Houston (ed) *Advances in Consumer Research*, vol. 15. Provo, Utah: Association for Consumer Research, pp 348–353.

Schwartz, Barry. 1975. *Queuing and Waiting: Studies in the Social Organization of Access and Delay*. Chicago: University of Chicago Press.

Shapiro, Stewart and H. Shanker Krishman. 1999. Consumer memory for intentions: A prospective memory perspective. *Journal of Experimental Psychology: Applied* 5 (June): 169–189.

Shoemaker, Robert W. and Vikas Tibrewala. 1985. Relating coupon redemption rates to past purchasing of the brand. *Journal of Advertising Research* 25 (October/November): 40–47.

Shelley, Marjorie K. 1994. Gain/loss asymmetry in risky intertemporal choice. *Organizational Behavior and Human Decision Processes* 59 (1): 124–159.

Shim, Richard. 2002. Phillips settles rebate complaint. *New York Times*, August 7, B1.

Shugan, Steven M. 1980. The cost of thinking. *Journal of Consumer Research*. 7 (3): 99–111.

Silk, Timothy G. 2003. Why do we buy but fail to redeem? Gainesville, Fla: dissertation, School of Business, University of Florida (September), as found on January 13, 2007 at http: //catalyst.gsm.uci.edu/tools/dl_public.cat?year=2003&file_id=126&type=cal&name=Rebates-Silkjobtalkpaper.pdf

———. 2006. Getting started is half the battle: The influence of deadlines and effort on consumer self-regulation to redeem rewards. Vancouver, B. C.: Sauder School of Business, working paper.

——— and Chris Janiszewski. 2006. Managing mail-in rebate promotions: An empirical analysis of purchase and redemption. Vancouver, B.C.: Sauder School of Business, University of British Columbia, working paper, as found on February 7, 2007 at http: //www.cba.ufl.edu/mkt/docs/janiszewski/Rebate.pdf

——— and Cornelia Pechmann. 2007. How might rebates affect consumer welfare? A behavioral stage framework and analysis. Vancouver, B.C.: Sauder School of Business, University of British Columbia.

Sims, Christopher A. 1998. Stickiness. *Carnegie-Rochester Conference Series on Public Policy* 49 (1): 317–356.

———. 2003. Implications of rational inattention. *Journal of Monetary Economics* 50 (3): 665–690.

Skouras, Thanos, George J. Avalonitis, and Kostis A. Indounas. 2005. Economics and pricing: How and why they differ. *Journal of Product and Brand Management* 14 (6): 362–374.

Smith, Adam. 1904. *An Inquiry into the Nature and Causes of the Wealth of Nations.* London: Methuen and Co., Ltd. 1904. Edwin Cannan (ed), accessed from Library of Economics and Liberty on March 2, 2007 at http: //www.econlib.org/library/Smith/smWN1.html

Smith, Andrew F. 2001. *Popped Culture: A Social History of Popcorn in America.* Washington, D.C.: Smithsonian Institution Press.

Snyder, Richard G. 1988. *The Status of Infant/Child Restraint Protection in Crash Impacts*, a report presented at the FAA/Flight Safety Foundation International Aircraft Occupant Safety Conference and Workshop, October 31–November 3.

Sobel, Joel. 1984. The timing of sales. *Review of Economic Studies* 51 (3), (July): 353–368.

Soman, Dilip. 1998. The illusion of delayed incentives: Evaluating future effort-money transactions. *Journal of Marketing* 35 (November): 427–437.

Spence, A. M. 1973. Job market signaling. *Quarterly Journal of Economics* 87: 355–374.

Spencer, Jane. 2002. Rejected! Rebates get harder to collect. *Wall Street Journal*, June 11, D1.

Steel, Piers. 2007. The nature of procrastination. *Psychological Bulletin* 133 (1) (January): 65–94.

Steinberg, Sandon A. and Richard F. Yalch. 1978. When eating begets buying: The effects of food samples on obese and non-obese shoppers. *Journal of Consumer Research* 4 (March): 243–246.

Stewart, Colin. 2007. Inside technology: Not-so-brave new world. *Orange County Register*. July 4, Marketplace 1.

Stigler, George J. 1961. The economics of information. *Journal of Political Economy* 69 (3) (June): 213–225.

Stiving, Mark and Russell S. Winer. 1997. An empirical analysis of price endings with scanner data. *Journal of Customer Research* 24 (1): 57–67.

Sturdivant, Frederick D. 1970. *Managerial Analysis in Marketing*. Glenview, Ill.: Scott, Foresman and Company.

Suter, L. E. and H. P. Miller. 1973. Income differences between men and career women. *American Journal of Sociology* 78: 962–974.

Sweeney, George. 1984. Marketing, price discrimination, and welfare: Comment. *Southern Economic Journal* 50 (3) (January): 892–899.

Taleb, Nassim Nicholas. 2007. *The Black Swan: The Impact of the Highly Improbable.* New York: Random House.

Tat, Peter K. and Monle Lee. 2001. Motivational aspects of rebate redemption. *Journal of Marketing Theory and Practice* 1 (2): 52–63.

———, William A. Cunningham, and Emin Babakus. 1988. Consumer perception of rebates. *Journal of Advertising Research* 28 (4): 45–50.

Taylor, Jerry and Peter VanDoran. 2002. Did Enron pillage California? *Briefing Paper,* no. 72. Washington, D.C.: Cato Institute, August 22, as accessed March 16, 2007 at http: //www.cato.org/pubs/briefs/bp72.pdf

Taylor, Patricia A. and Norval D. Glenn. 1976. The utility of education and attractiveness for female status attainment through marriage. *American Sociological Review* 41: 484–498.

Taylor, Paul, Cary Funk, and April Clark. 2007. *Generation Gap in Values, Behaviors: As Marriage and Parenthood Drift Apart, Public Is Concerned about Social Impact.* Washington, D.C.: Pew Research Center, July 1, accessed on July 4, 2007 at http: // pewresearch.org/assets/social/pdf/Marriage.pdf

Telser, Lester G. 1980. A theory of self-enforcing agreements. *Journal of Business* 53 (January): 27–44.

Thaler, Richard H. 1980. Toward a positive theory of consumer choice. *Journal of Economic Behavior and Organization* 1: 39–60.

———. 1981. Some empirical evidence on dynamic inconsistency. *Economic Letters,* Elsevier 8 (3): 201–207.

———. 1991. *Winner's Curse: Paradoxes and Anomalies of Economic Life.* New York: Free Press.

Thompson, Richard F. and Stephen A. Madigan. 2005. *Memory: The Key to Consciousness.* Washington, D.C.: Joseph Henry Press.

Thompson, Suzanne C. 1981. Will it hurt less if I can control it? A complex answer to a simple question. *Psychological Bulletin* 90 (1): 89–101.

Tinker, Irene (ed). 1990. *Persistent Inequalities: Women and World Development.* New York: Oxford University Press.

Thurow, Lester C. 1975. *Generating Inequality.* New York: Basic Books.

Tom, Sabrina M., Craig R. Fox, Christopher Trepel, and Russell A. Poldrack. n.d. Losses loom larger than gains in the brain: Neural loss aversion predicts behavioral loss aversion, as found on January 28, 2007 at http: //w4.stern.nyu.edu/emplibrary/TomEtAl-LossAversionOnBrain (submitted).pdf

Tooby, John and Leda Cosmides. 1990a. On the universality of human nature and the uniqueness of the individual: The role of genetics and adaptation. *Journal of Personality* 58 (1): 17–67.

———. 1990b. The past explains the present: Emotional adaptations and the structure of ancestral environments. *Ethology and Sociobiology* 11: 375–424.

Townsend, Adam, Eric Carpenter, and Vik Jolly. 2007. Officials: Save water. *Orange County Register*, March 29, A1, accessed on March 29, 2007 at http: //www.ocregister.com/ocregister/homepage/abox/article_1633377.php

Townsend, John Marshall. 1987. Sex differences in sexuality among medical students: Effects of increasing socioeconomic status. *Archives of Sexual Behavior* 16: 425–441.

———. 1989. Mate selection criteria: A pilot study. *Ethology and Sociobiology* 10: 241–253.

——— and Gary D. Levy. 1990a. Effects of potential partners physical attractiveness and socioeconomic status on sexuality and partner selection. *Archives of Sexual Behavior* 19: 149–164.

———. 1990b. Effects of potential partners costumes and physical attractiveness on sexuality and partner selection. *Journal of Psychology* 124: 371–389.

Treiman, Donald J. and Kermit Terrell. 1975. The process of status achievement in the United States and Great Britain. *American Journal of Sociology* 81 (November): 563–583.

Treiman, Donald J. and Heidi I. Hartmann. 1981. *Women, Work and Wages: Equal Pay for Jobs of Equal Value.* Washington, D.C.: National Academy Press.

Trivers, Robert L. 1985. *Social Evolution.* Menlo Park, Calif.: Benjamin-Cummings.

Tullock, Gordon. 1975. Transitional gains trap. *Bell Journal of Economics* 6 (2) (Autumn): 671–678.

Tversky, Amos. 1972. Elimination by aspect: A theory of choice. *Psychological Review* 79 (4): 281–299.

——— and Eldar Shafir. 1992. Choice under conflict: The dynamics of deferred decisions. *Psychological Science* 3 (6): 358–361.

Twedt, Dik Warren. 1965. Does the "9 Fixation" in retail pricing really promote sales? *Journal of Marketing* 29 (3): 54–55.

Tyson, Jeff. 2000. How movie distribution works, as can be seen at Lycos Zone: HowStuffWorks, accessed on March 8, 2004 at http: //howstuffworks.lycoszone.com/movie-distribution4.htm

Udry, J. Richard. and Bruce K. Eckland. 1984. Benefits of being attractive: Differential payoffs for men and women. *Psychological Reports* 54: 47–56.

United States v. Paramount Pictures, Inc., 334 U.S. 131 (1948).

U.S. Department of Justice. 1999. Plaintiff's Joint Proposed Findings of Fact, United States of America v. Microsoft Corporation, Civil Action No. 98-1232 (TPJ), August 10, accessed in 1999 from http: //www.usdoj.gov/atr/cases/f2600/2613.htm

U.S. Department of Transportation, Federal Aviation Administration. 1989. *FAA Statistical Handbook of Aviation: 1988*. Washington: U.S. Government Printing Office.

U.S. Department of Transportation, Federal Aviation Administration. 1990. Miscellaneous Operation Amendments, *Federal Register*, March 1, 7414–7421.

U.S. Congress, House. 1990. A Bill to Amend the Federal Aviation Act of 1958 to Require Child Safety Restraint Systems Approved by the Secretary of Transportation on Commercial Aircraft, H.R. 4025, 101st Cong., 2d Sess.

U.S. Department of Transportation. 1990. *An Impact Analysis of Requiring Child Safety Seats in Air Transportation*, prepared for the Office of Aviation Policy and Plans, Federal Aviation Administration, Washington, Office of Rulemaking, Federal Aviation Administration, draft, June 4, iv.

USA Today. 1990. Don't require seats for tots who fly (editorial) May 24, 12A.

Vallone, Robert P., Dale W. Griffin, Sabrina Lin, and Lee Ross. 1990. The overconfident prediction of future action outcomes by self and others. *Journal of Personality and Social Psychology* 58 (April): 582–592.

Varian, Hal R. 1980. A model of sales. *American Economic Review* 70 (3) (September): 651–659.

Vedder, Richard. 2004. *Going Broke by Degree: Why College Costs Too Much*. Washington, D.C.: American Enterprise Institute.

———. 2006. Scholarships, reduced consumer welfare and information. Center for College Affordability and Productivity, September 22, accessed on February 15, 2007 at http: //collegeaffordability.blogspot.com/2006/09/scholarships-reduced-consumer-welfare.html

Vilcassim, Naufel J. and Dick R. Wittink. 1987. Supporting higher shelf price through coupon distributions. *Journal of Consumer Marketing* 4 (Spring): 29–39.

Vogel, Harold L. 2004. *Entertainment Industry Economics: A Guide for Financial Analysis*, 6th edn. Cambridge, U.K.: Cambridge University Press.

Ward, Ronald W. and James E. Davis. 1978. A pooled cross-section time series model of coupon promotions. *American Journal of Agricultural Economics* 60 (August): 393–401.

Warren-Boulton, Frederick R. 1998. Direct Testimony, State of New York *ex rel.* Attorney General Dennis C. Vacco et al. vs. Microsoft Corporation, Civil Action No. 98-1233 (TPJ), accessed in 1989 from http: //www.usdoj.gov/atr/cases/f2000/2079.htm

Waxman, Sharon. 2007. "Pirates" haul so far estimated at $401 million. *New York Times*, May 29, B1.

Weare, Christopher. 2003. *The California Electricity Crisis: Causes and Policy Actions* San Francisco: Public Policy Institute of California.

Weinberg, D. H. 2004. *Evidence from Census 2000 about Earnings by Detailed Occupation for Men and Women.* CENSR-15. Washington, D.C: U.S. Census Bureau (May), accessed on June 25, 2004 at http: //www.census.gov/prod/2004pubs/censr-15.pdf

Weinberger, Catherine and Peter Joseph Kuhn. 2004. *The Narrowing of the U.S. Gender Earnings Gap, 1969–1999.* Santa Barbara, Calif.: University of California, Santa Barbara, May 17, accessed on June 24, 2004 at http: //www.nber.org/~sewp/events/2004.05.28/Weinberger-Kuhn.GenderGap.003.pdf

Wellington, Alison J. 1993. Changes in the male/female wage gap, 1976–1985. *Journal of Human Resources* 28 (2): 383–411.

Whalen, Bernard R. 1980. Strategic mix of odd, even prices can lead to increased retail profits. *Journal of Marketing* 29 (3): 54–55.

White, B. A. 1983. Comparison shopping and the economics of manufacturers' coupons. Buffalo, N.Y.: Department of Economics, State University of New York, Buffalo, working paper 517.

Wiederman, M. W. and E. R. Allgeier. 1992. Gender differences in mate selection criteria: Sociobiological or socioeconomic explanation? *Ethology and Sociobiology* 13: 115–124.

Wilson, Edward O. 1978. *On Human Nature.* Cambridge: Harvard University Press.

Windle, Robert and Martin Dresner. 1991. Mandatory child safety seats in air transport: Do they save lives? *Journal of Transportation Research Forum* 31 (2): 309–316.

Wingfield, Nick. 2007. Apple price cut on new iPhone shakes investors. *Wall Street Journal,* September 6, A1.

Wolff, Ronald W. 1988. *Stochastic Modeling and the Theory of Queues.* Englewood Cliffs, N.J.: Prentice-Hall.

Woodyard, Chris. 2007. Carpool stickers add value to used Priuses in California. *USA Today,* March 25, as accessed on June 19, 2007 at http: //www.usatoday.com/money/autos/2007-03-25-hybrid-carpool-stickers_N.htm

Wright, Charles. 1995. Bill's world. *Australian Financial Review* November 24, 1.

Zachmann, William F. 1992. Blue believer: Why I still think OS/2 is a winner," *PC Magazine* February 25, 1992, 107.

Subject Index